Surveying Fundamentals

Surveying
Fundamentals

JACK C. McCORMAC

Department of Civil Engineering
Clemson University, South Carolina

Prentice-Hall, Inc., Englewood Cliffs, New Jersey 07632

Library of Congress Cataloging in Publication Data

McCormac, Jack C. (date)
  Surveying fundamentals.

  Includes bibliographic references and index.
  1. Surveying. I. Title.
TA545.M33   526.9   82-3765
ISBN O-13-878843-X   AACR2

Printed in the United States of America

10 9 8 7 6 5 4 3 2

Editorial/production supervision by Karen Skrable
Cover and interior design by Mark A. Binn
Manufacturing buyers: Joyce Levatino and Anthony Caruso

ISBN 0-13-878843-X

Prentice-Hall International, Inc., *London*
Prentice-Hall of Australia Pty. Limited, *Sydney*
Prentice-Hall Canada Inc., *Toronto*
Prentice-Hall of India Private Limited, *New Delhi*
Prentice-Hall of Japan, Inc., *Tokyo*
Prentice-Hall of Southeast Asia Pte. Ltd., *Singapore*
Whitehall Books Limited, *Wellington, New Zealand*

Contents

Distance Measurement 32

Taping 44

Distance Corrections 55

Mistakes, Errors, and Miscellaneous Taping Operations 70

Electronic Distance Measurement 80

Introduction to Leveling 100

Differential Leveling 117

Leveling Continued 143

Angles and Directions 161

Measuring Angles and Directions 181

Miscellaneous Angle Discussion   205

Traverse Adjustment and Area Computation 217

Miscellaneous Traverse Computations 247

Topographic Surveying 262

Land Surveying or Property Surveying   284

Surveying Astronomy   311

Volumes 337

Construction Surveying 356

Contents   xiii

 Horizontal and Vertical Curves 377

 Photogrammetry—by Dr. Donald B. Stafford 401

Control Surveys   429

State–Wide Plane Coordinates   451

Surveying—The Profession 475

Appendix 483

Index 513

Preface

The purpose of this book is to serve as an introduction to surveying and to present the elementary principles in such a manner as to encourage the reader toward further study of the subject. It is written for a one or two term course for students in civil engineering, civil engineering technology, forestry, building and construction science, architecture, agriculture, geography, and other related areas. Hopefully, it will also be useful to practicing surveyors.

The measurement of distances, elevations, and directions are considered in Chapters 1–13 while Chapters 14–25 are devoted to the applications of those measurements to the determination of land areas, preparation of topographic maps, practice of land surveying, public land surveys, computation of earthwork volumes, surveying astronomy, construction surveys, photogrammetry, control surveys, and professional registration and ethics.

The author has spent an appreciable part of this book describing the use of traditional surveying equipment—tape, level, and transit—despite the fact that they have been replaced for much work by electronic distance measuring instruments, theodolites, and other modern devices. The satisfactory use of these older pieces of equipment requires a detailed understanding of surveying and a great deal of practice while the more modern equipment requires less understanding and less practice. The author feels very strongly that the surveyor who understands and can use the basic equipment as well as the newer equipment can do far better work than a person who is only trained with modern equipment.

Numerous numerical examples and homework problems are included in the text. If the reader will solve several of each of the included sets of home problems he or she should be able to firmly fix in mind the theory involved. Answers to odd-numbered problems are provided throughout the text. A high level of math is not needed to solve these problems nor to understand the surveying theory presented.

Most students seem to enjoy surveying. Perhaps this is due to the "hands on" nature of the subject. The author hopes that through this textbook he will be able to expand the appeal of surveying to the interested reader.

JACK C. MCCORMAC

Acknowledgments

The author gratefully acknowledges the aid received from several sources. I am indebted to Ben Benson, W. P. Byrd, Robert C. Darling, Frank J. Hatfield, Michael Orlando, F. T. Quiett, and Jerome L. Spurr, who have by their suggestions and criticisms directly contributed to the preparation of this manuscript; to my own surveying professor, the late Colonel John Anderson, who patiently instructed me in his surveying class; to my associates, who have helped in my study of surveying, including J. P. Rostron, Donald B. Stafford (who wrote Chapter 22 on Photogrammetry), I. A. Trively, and the late J. M. Ford, Jr. Finally thanks are due to Mrs. Mannetta Shusterman who helped type the manuscript.

Surveying Fundamentals

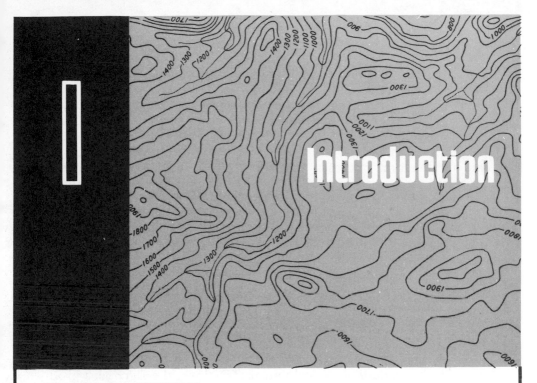

1-1 FAMOUS SURVEYORS

Many famous men in our history have engaged in surveying at some period in their lives. Particularly notable among these are several presidents—Washington, Jefferson, and Lincoln. Although the practice of surveying will not provide a sure road to the White House, many members of the profession like to think that the characteristics of the surveyor (honesty, perseverance, self-reliance, and so on) contributed to the development of these great men. Today surveying is an honored and widely respected profession. A knowledge of its principles and ethics is useful to a person whatever his future endeavors will be.

1-2 EARLY HISTORY OF SURVEYING

It is impossible to determine when surveying was first used, but in its simplest form it is surely as old as recorded civilization. As long as there has been property ownership there have been means of measuring the property or distinguishing one person's land from another. Even in the Old Testament there are frequent references to property ownership, property corners, and property transfer. For instance, Proverbs 22:28: "Remove not the ancient landmark, which thy fathers have set." The Babylonians surely practiced some type of surveying as early as 2500 B.C. because archaeologists have found Babylonian maps on tablets of that estimated age.

1

The early development of surveying cannot be separated from the development of astronomy, astrology, or mathematics because these disciplines were so closely interrelated. In fact, the term *geometry* is derived from Greek words meaning earth measurements. The Greek historian Herodotus ("the father of history") says that surveying was used in Egypt as early as 1400 B.C. when that country was divided into plots for taxation purposes. Apparently geometry or surveying was particularly necessary in the Nile valley in order to establish and control landmarks. When the yearly floods of the Nile swept away many of the landmarks, surveyors were appointed to replace them. These surveyors were called "rope-stretchers" because they used ropes (with markers on them at certain intervals) for their measurements.

During this same period surveyors were certainly needed for assistance in the design and construction of irrigation systems, huge pyramids, public buildings, and so on. Their work was apparently quite satisfactory. For instance the dimensions of the Great Pyramid of Gizeh are in error about 8 in. over a 750-ft base.[1] It is thought that the rope-stretchers layed off the sides of the pyramid bases with their ropes and checked squareness by measuring diagonals. In order to obtain the almost level foundations of these great structures, the Egyptians probably either poured water into long narrow clay troughs (an excellent method) or used triangular frames with plumb bobs or other weights suspended from their apexes as shown in Fig. 1-1.[2]

Each frame apparently had a mark on its lower bar which showed where the plumb line should be when that bar was horizontal. These frames, which were probably used for leveling for many centuries, could easily be checked for proper adjustment by reversing them end for end. If the plumb lines returned to the same points the instruments were in proper adjustment and the tops of the supporting stakes (see Fig. 1-1) would be at the same elevation.

The practical-minded Romans introduced many advances in surveying by an amazing series of engineering projects constructed through-

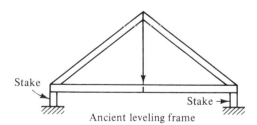

Ancient leveling frame

Figure 1-1 Ancient leveling frame.

[1] C. M. Brown and W. H. Eldridge, *Evidence and Procedures for Boundary Locations* (New York: John Wiley & Sons, Inc., 1962), p. 69.

[2] A. R. Legault, H. M. McMaster, and R. R. Marlette, *Surveying* (Englewood Cliffs, NJ: Prentice-Hall, Inc., 1956), p. 5.

out their empire.   They laid out projects such as cities, military camps, and roads by using a system of rectangular coordinates.   They surveyed the principal routes used for military operations on the European continent, in the British Isles, in northern Africa, and even in parts of Asia.

Two instruments used by the Romans were the *odometer*, or measuring wheel, and the *groma*.   The groma, from which Roman surveyors received their name of *gromatici*, was used for laying off right angles. It consisted of two crossarms fastened together at right angles in the shape of a horizontal cross with plumb lines hanging from each of the four ends.   The groma, which was pivoted eccentrically on a vertical staff, could be leveled and sights taken along its crossarms.

From Roman times until the modern era there were few advances in the art of surveying, but the last few centuries have seen the introduction of the telescope, the vernier, the theodolite, and many other excellent devices.   These developments will be mentioned in subsequent chapters.   For a detailed historical list of early instrument development, the reader is referred to "The Civil Engineer: His Origins."[3]

1-3 SURVEYING DEFINED

Surveying is the science of determining the dimensions and contour (or three-dimensional characteristics) of the earth's surface by the measurements of distances, directions, and elevations.   It also involves staking out the lines and grades needed for the construction of buildings, roads, dams, etc.   In addition to these field measurements, surveying includes the computation of areas, volumes, and other quantities, as well as the preparation of necessary maps and diagrams.   Surveying has many industrial applications: for example, setting equipment, assembling aircraft, laying out assembly lines, and so on.

1-4 PLANE SURVEYS

In large-scale mapping, adjustments are made for the curvature of the earth, and north–south lines converge to meet at the poles.   Plane surveys, however, are made on such small areas that the effect of the earth's curvature may be neglected.   In plane surveying the earth is considered to be a flat surface and north–south lines are assumed to be parallel.   Calculations for a plane surface are relatively simple, since the surveyor is able to use plane geometry and plane trigonometry.

[3] *Transactions of the American Society of Civil Engineering*, 30 (1893), 91–106.

The great majority of plane surveys are sufficiently accurate for all but the largest areas. Surveys for farms, subdivisions, buildings, highways, railroads, and in fact most man-made works are plane surveys.

It can be shown that an arc along the earth's curved surface of 11.5 mi. length is only approximately 0.05 ft longer than a plane or chord distance between its ends. Plane surveys, however, are not generally considered to be sufficiently accurate for establishing state and national boundaries. They are used in combination with surveys taking into account the earth's curvature, in subdividing the public lands of the United States.

1-5 GEODETIC SURVEYS

Geodetic surveys are those that are adjusted for the curved shape of the earth's surface. (The earth is an oblate spheroid whose radius at the equator is about 13.5 miles greater than its polar radius.) Since they allow for earth's curvature, geodetic surveys can be applied to both small and large areas. The equipment used and the methods of measurement applied are about the same as they are for plane surveys. Elevations are handled in the same manner for both plane and geodetic surveys. They are expressed in terms of vertical distances above or below a reference curved surface usually mean sea level.

Most geodetic surveys are made by government agencies, such as the former U.S. Coast and Geodetic Survey (now the National Geodetic Survey, National Ocean Survey, National Oceanic and Atmospheric Administration of the U.S. Department of Commerce). Although only a relatively small number of surveyors are employed by the National Geodetic Survey (NGS), their work is extremely important to all other surveyors. They have established a network of reference points around the U.S. that provides very precise information on locations and elevations. On this network all sorts of other surveys (plane and geodetic) of lesser precision are based.

1-6 TYPES OF SURVEYS

This section is devoted to a brief description of the various types of surveys. Most of these types of surveys employ plane rather than geodetic techniques.

Land surveys are the oldest type of survey and have been performed since earliest recorded history. They are plane surveys made for locating property lines, subdividing land into smaller parts, determining land areas and any other information involving the transfer of land from one owner to another. These surveys are also called *property surveys*, *boundary surveys*, or *cadastral surveys*.

Topographic surveys are made for locating objects and measuring the relief, roughness, or three-dimensional variations of the earth's surface. Detailed information is obtained pertaining to elevations as well as to the locations of man-made and natural features (buildings, roads, streams, and so on) and the entire information is plotted on maps (called topographic maps).

Route surveys involve the determination of the relief and the location of natural and artificial objects along a proposed route for a highway, railroad, canal, pipeline, power line, or other utility. They may further involve the location or staking out of the facility and the calculation of earthwork quantities.

City surveys are made within a given municipality for the purpose of laying out streets, planning sewer systems, preparing maps, and so on. When the term is used, it usually brings to mind topographic surveys in or near a city for the purpose of planning urban expansions or improvements.

Construction surveys are made for purposes of locating structures and providing required elevation points during their construction. They are needed to control every type of construction project.

Hydrographic surveys pertain to lakes, streams, and other bodies of water. Shore lines are charted, shapes of areas beneath water surfaces are determined, water flow of streams is estimated, and other information needed relative to navigation, flood control, development of water resources, and so on, is obtained. These surveys are usually made by a governmental agency, for example, the National Geodetic Survey, the U.S. Geological Survey, or the U.S. Army Corps of Engineers. When hydrographic and topographic surveying are combined the resulting surveys are sometimes called *cartographic surveys*.

Marine surveys are related to hydrographic surveys but they are thought to cover a broader area. They include the surveying necessary for offshore platforms, the science of navigation, the theory of tides as well as the preparation of hydrographic maps and charts.

Mine surveys are made to obtain the relative positions and elevations of underground shafts, geological formations, and so on and to determine quantities and establish lines and grades for work to be done.

Forestry and geological surveys are probably much more common than the average layman realizes. Foresters use surveying for boundary locations, timber cruising, topography, etc. Similarly, surveying has much application in the preparation of geological maps.

Photogrammetric surveys are those in which photographs (gener-

Figure 1-2 HP Model 3810A Total Station.   An instrument used for measuring distances, horizontal and vertical angles, and for computing horizontal and vertical distances.   Range is up to 1.6 km (1 mi).   Manufactured in U.S.A. by Hewlett-Packard Company.   (*Courtesy Hewlett-Packard.*)

ally aerial) are used in conjunction with limited ground surveys (used to establish or locate certain control points visible from the air).   Photogrammetry is extremely valuable because of the speed with which it can be applied, the economy, the applications to areas difficult of access, the great detail provided, etc.   Its uses are becoming more extensive each year.

Future Surveys

In the decades to come there will undoubtedly be other special types of surveying which will develop.   Surveyors might very well have to establish boundaries under the ocean, in the Arctic and Antarctic, and even on the moon and other planets.   Great skill and judgment by the surveying profession will undoubtedly be required to handle these tasks.

1-7 IMPORTANCE OF SURVEYING

As described in Section 1-2, it has been necessary since the earliest civilizations to determine property boundaries and divide sections of land into smaller pieces. Through the centuries the uses of surveying have expanded until today it is difficult to imagine any type of construction project that does not involve some type of surveying.

All types of engineers, as well as architects, foresters, and geologists, are concerned with surveying as a means of planning and laying out their projects. Surveying is needed for subdivisions, buildings, bridges, highways, railroads, canals, piers, wharves, dams, irrigation and drainage networks, and many other projects. In addition, surveying is required for the laying out of industrial equipment, setting machinery, holding tolerances in ships and airplanes, preparing forestry and geological maps, and so on.

The study of surveying is an important part of the training of a technical student even though he may never actually practice surveying. It will appreciably help him to learn to think logically, to plan, to take pride in working carefully and accurately, and to record his work in a neat and orderly fashion. He will learn a great deal about the relative importance of measurements, develop some sense of proportion as to what is important and what is not, and acquire essential habits of checking numerical calculations and measurements (a necessity for anyone working in an engineering or scientific field).

1-8 OPPORTUNITIES IN SURVEYING

There are few professions which need qualified people so much as does the surveying profession. In the U.S., the tremendous physical development of the country (subdivisions, factories, dams, power lines, cities) has created a need for surveyors at a faster rate than our schools have produced them. Construction, our largest industry, requires a constant supply of new surveyors.

There are hundreds of towns and cities throughout the United States where additional surveyors are needed. For a person with a liking for a combination of outdoor and indoor work, surveying offers attractive opportunities. There are few other fields in which a qualified person can set himself up in private practice so readily and with such excellent prospects for success.

It is necessary for a person going into private practice to meet the licensing requirements in his particular state. He[4] may be able to do his initial work under the supervision of a registered surveyor until he qualifies for a license so that he can himself practice privately. Many people are able to hold other jobs and do surveying on holidays and

---

[4] For purposes of convenience, the surveyor is generally referred to as "he" rather than "he or she" in the text. The author does not mean to say that only men are surveyors and does not mean to minimize the role of women in the surveying profession.

weekends for licensed surveyors until they themselves can become registered. Registration requirements are discussed in more detail in Chapter 25.

A large percentage of today's practicing surveyors reached their status by apprenticeships and self-study programs and perhaps have not had very much formal schooling in their field. Nevertheless, it is probable that this supply of surveyors will not be sufficient in the future.

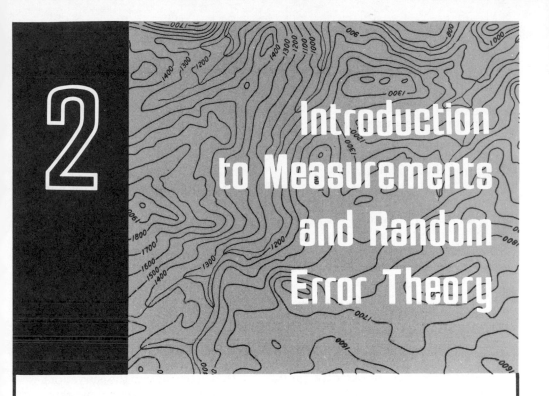

2 Introduction to Measurements and Random Error Theory

2-1 MEASUREMENT

In ordinary life, most of us are accustomed to *counting* but not as much to *measuring*. If the number of people in a room is counted the result is an exact number without a decimal, say 9 people. It would be ridiculous to say there are 9.23 people in a room. In a similar fashion a person might count the amount of money in his or her pocket. Though the result may contain a decimal as $5.65, the result is still an exact value.

   Surveying is concerned with measurements of quantities whose exact or true values may not be determined, such as distances, elevations, volumes, directions, weights, times, etc. If a person were to measure the width of his desk with a ruler divided into tenths of an inch he could estimate the width to hundredths of an inch. If he were to use a ruler graduated in hundredths of an inch he could estimate the width to thousandths of an inch; and so on. Obviously with better equipment he can estimate, an answer which is closer to the exact value but he will never be able to determine the value absolutely. Thus a fundamental principle of surveying is that no measurement is exact and the true value of the quantity being measured is never known. Measurement is the principal concern of a surveyor.

2-2 NECESSITY FOR ACCURATE SURVEYS

The surveyor must have the skill and judgment necessary to make very accurate measurements. This fact is obvious when one is thinking in terms of the construction of long bridges, tunnels, tall buildings, and missile sites or with setting of delicate machinery, but it can be just as important in land surveying.

As late as a few decades ago land prices were not extremely high except in and around the largest cities. If the surveyor gained or lost a few feet in a lot or a few acres in a farm, it was usually not considered to be a matter of great importance. The instruments commonly used for surveying before this century were not very precise compared to today's equipment and it was probably impossible for the surveyor to do the quality of work expected of today's surveyor. (What will surveyors in future centuries think of twentieth-century surveying and what wonderful equipment will they have to work with?)

It is said that early surveyors in the U.S. had to complete their work very quickly when working in Indian country. The Indians were well aware that numerous settlers followed on the heels of surveyors and the Indians apparently thought if they eliminated the surveyors, the settlers might not come to steal their land. Surveyors working under these conditions were said to rather frequently neglect to set some property corners, particularly in forbidding forests, and their rapid measurements were not always the best. In addition, many of their values were more than likely computed in the office and not measured on the ground.

Today land prices are in most areas very high and evidently the climb has only begun. In many areas of high population and in many popular resort areas, land is sold by so many dollars per square foot or so many hundreds or even thousands of dollars per front foot; therefore the surveyor must be able to do splendid work. Even in rural areas land is frequently "sky high."

2-3 ACCURACY AND PRECISION

The terms accuracy and precision are constantly used in surveying and yet their correct meanings are a little difficult to grasp. In an attempt to clarify the distinction the following definitions are presented:

Accuracy refers to the degree of perfection obtained in measurements. It denotes how close a given measurement is to the true value of the quantity.

Precision or *apparent accuracy* is the degree of refinement with which a given quantity is measured. In other words, it is the closeness of one measurement to another. If a quantity is measured several times and the values obtained are very close to each other, the precision is said to be high.

It does not necessarily follow that better precision means better accuracy. Consider the case in which a surveyor carefully measured a distance three times with a 100-ft steel tape and obtained the values 984.72 ft, 984.69 ft, and 984.73 ft. He did a very precise job and apparently a very accurate one. Should, however, the tape be found to actually be 100.30 ft long instead of 100.00 ft, the values obtained are not accurate although they are precise. (The measurements could be made accurate by making a numerical correction of 0.30 ft for each 100 ft measured.) It is possible for the surveyor to obtain both accuracy and precision if he uses care, patience, and good instruments and procedures.

In measuring distance, precision is defined as the ratio of the error of the measurement to the distance measured and it is reduced to a fraction having a numerator of unity. If a distance of 4200 ft is measured and the error is later estimated to equal 0.7 ft, the precision of the measurement is 0.7/4200 = 1/6000. This means that for every 6000 ft measured the error would be 1 ft if the work were done with this same degree of precision.

A method frequently used by surveying professors to define and distinguish between precision and accuracy is illustrated in Fig. 2-1. It is assumed that a person has been having a little target practice with his rifle. His results with the first target shown were very precise because the bullet holes were quite close to each other. They were not accurate, however, as they were located some little distance from the bull's eye.

The marksman fired accurately at Target 2 as the holes were

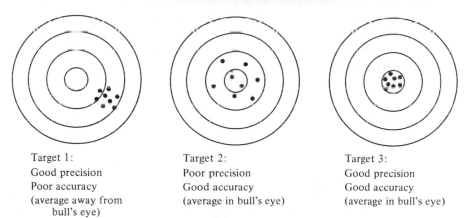

Target 1:   Target 2:   Target 3:
Good precision   Poor precision   Good precision
Poor accuracy   Good accuracy   Good accuracy
(average away from   (average in bull's eye)   (average in bull's eye)
  bull's eye)

Figure 2-1 Accuracy and precision.

placed relatively close to the bull's eye. The shots were not precise, however, as they were scattered quite a bit with respect to each other.

Finally, in Target 3, the shots were both precise and accurate as they were placed in the bull's eye close to each other. *The objective of the surveyor is to make measurements that are both precise and accurate.*

2-4 ERRORS AND MISTAKES

There is no one whose senses are sufficiently perfect to measure any quantity exactly and there are no perfect instruments with which to do the measuring. The result is that all measurements are imperfect. A major concern in surveying is the precision of the work. This subject is repeatedly mentioned as we discuss each phase of surveying.

The ever-present differences between measured quantities and the true magnitudes of those quantities are classified herein as either mistakes or errors.

A *mistake* (or blunder) is a difference from a true value caused by the inattention of the surveyor. For instance, he may read a number as 6 when it is actually 9, he may record the wrong quantities in his field notes, or he may add a column of numbers incorrectly. The important point here is that mistakes are caused by the carelessness of the surveyor and carelessness *can be eliminated* by careful checking.

Every surveyor will make occasional mistakes, but if he learns to carefully apply the checks to his work which are described in the chapters to follow, he will eliminate these mistakes. Any true professional surveyor will not be satisfied with his work until he has made sure that any blunders are detected and eliminated.

An *error* is a difference from a true value caused by the imperfection of a person's senses, by the imperfection of his equipment, or by weather effects. Errors *cannot be eliminated but they can be minimized* by careful work combined with the application of certain numerical corrections.

2-5 SOURCES OF ERRORS

There are three sources of errors: people, instruments, and nature. Accordingly, errors in measurement are generally said to be personal, instrumental, and natural. Some errors, however, do not clearly fit into just one of these categories and may be due to a combination of factors.

Personal errors occur because no surveyor has perfect senses of sight and touch. For instance, in estimating the fractional part of a scale he cannot read it perfectly and will always be either a little large or a little small.

Instrumental errors occur because instruments cannot be manufactured perfectly and the different parts of the instruments cannot be adjusted exactly with respect to each other. Moreover, with time the wear and tear of the instruments causes errors. Although the past few decades have seen the development of more precise equipment, the goal of perfection remains elusive. A reading from a scale will undoubtedly contain both a personal and an instrumental error. The observer cannot read the scale perfectly nor can the manufacturer make a perfect scale.

Natural errors are caused by temperature, wind, moisture, magnetic variations, and so on. On a summer day a 100-ft steel tape may increase in length by a few hundredths of a foot. Each time this tape is used to measure 100 ft there will be a temperature error of those few hundredths of a foot. The surveyor cannot normally remove the cause of errors such as these, but he can minimize their effects by using good judgment and making proper corrections of the results.

2-6 SYSTEMATIC AND ACCIDENTAL ERRORS

Errors are said to be systematic or accidental.

A *systematic* or *cumulative error* is one which, for constant conditions, remains the same as to sign and magnitude. For instance, if a steel tape is 0.10 ft too short, each time the tape is used the same error (because of that factor) is made. If the full tape length is used ten times, the error accumulates and totals ten times the error for one measurement.

An *accidental, compensating* or *random error* is one whose magnitude and direction are just an accident and beyond the control of the surveyor. For instance, when a person reads an angle with a surveying instrument he cannot read it perfectly. One time he will read a value which is too large and the next time he will read a value which is too small. Since these errors are just as likely to have one sign as the other, they tend to a certain degree to cancel each other or compensate for each other.

2-7 SIGNIFICANT FIGURES

This section might also be entitled "judgment" or "sense of proportion," and its comprehension is a very necessary part of the training of anyone who takes and/or uses measured quantities of any kind.

When measurements are made, the results can only be precise to the degree that the measuring instrument is precise. This means that numbers which represent measurements are all approximate values. For instance, a distance may be measured with a steel tape as being 465 ft or again more precisely as 465.3 ft or even with more care as 465.32 ft, but an exact answer can never be obtained. The value will always contain some error.

The number of significant figures which a measured quantity has is not (as is frequently thought) the number of decimal places. Instead, it is the number of certain digits plus one digit which is estimated. For instance, in reading a steel tape a point may be between 34.2 ft and 34.3 ft (the scale being marked at the one-tenth ft points) and the value is estimated as being 34.26 ft. The answer has four significant figures. Other examples of significant figures follow:

> 36.00620 has seven significant figures
>
> 10.0 has three significant figures
>
> 0.003042 has four significant figures

The answer obtained by solving any problem can never be more accurate than the information used. If this principle is not completely understood, results will be slovenly. Notice that it is not reasonable to add 23.2 cu yd of concrete to 31 cu yd and get 54.2 cu yd. One cannot properly express the total to the nearest tenth of a yard as 54.2 because one of the quantities was not computed to the nearest tenth and the correct total should be 54 cu yd.

A few general rules regarding significant figures follow:

1. Zeros between other significant figures are significant, as, for example in the following numbers each of which contains four significant figures: 23.07 and 3008.

2. For numbers less than unity, zeros immediately to the right of the decimal are not significant. They merely show the position of the decimal. The number 0.0034 has two significant figures.

3. Zeros placed at the end of decimal numbers, such as 24.3200, are significant.

4. When a number ends with one or more zeros to the left of the decimal it is necessary to specifically indicate the exact number of significant figures. The number 352,000 could have 3, 4, 5, or 6 significant figures. It could be written as $35\overline{2},000$ which has three significant figures or as $352,00\overline{0}$ which has six significant figures. It is also possible to handle the problem by using scientific notation. The number 2.500×10^3 has four significant figures while the number 2.50×10^3 has three.

5. When numbers are multiplied or divided or both, the answer should not have more significant figures than those in the factor which had the least number of significant figures. As an illustration, the following calculations should result in an answer having three significant figures which is the number of significant figures in the term 3.25.

$$\frac{3.25 \times 4.6962}{8.1002 \times 6.152} = 0.306$$

It is desirable to carry calculations to one or more extra places during the various steps; as the final step, the answer is rounded off to the correct number of significant figures.

6. For addition or subtraction, the last significant figure in the final answer should correspond to the last column full of significant figures among the numbers. An addition example follows:

$$
\begin{array}{ll}
\begin{array}{r}
33./8\backslash42 \\
361.|3| \\
81.\backslash2/4 \\
\hline
476.\,4
\end{array}
&
\begin{array}{l}
\text{Last column of} \\
\text{significant figures}
\end{array}
\end{array}
$$

2-8 INTRODUCTION TO RANDOM ERROR THEORY

The average surveyor tends to become nervous when the topic of random or accidental error theory is introduced. And yet, he or she would be well advised to become familiar with the fundamentals of the subject. Several clear, easy-to-grasp principles relating to random errors can quickly be mastered even without a detailed background in statistics and mathematics. These principles can be learned while avoiding the partial differentials and other advanced concepts needed to make absolute mathematical proofs.

As we have noted, no quantity such as distance, direction, or elevation can be measured perfectly. For every measurement there is some doubt. Even after corrections have been carefully made for systematic errors, there will still remain random errors which cannot be exactly determined or eliminated.

For this discussion it is assumed that the measured values have been corrected for blunders and systematic errors. Then the random errors described are the differences between the true values and the corrected measurements. Realistically it is impossible to eliminate all systematic errors. Despite this fact the error theory discussed here has been found to apply very well to properly conducted surveys considering the errors to be accidental or random.

For many surveys the quality of a measurement can be expressed by stating a relative error. For example, a distance may be expressed as 835.82 ft ± 0.06 ft. Such a statement indicates that the true distance measured probably falls between 835.76 ft and 835.88 ft and its most probable value is 835.82 ft. The sign or direction of the probable error is not known and thus no correction can be made. The probable error can either be plus or minus, between the limits within which the error is likely to fall. Note that this form does not specify the magnitude of the actual error, nor does it indicate the error most likely to occur.

The relative error (such as the ±0.06 ft used above) can't just be a subjective guess on the surveyor's part. It must be a logical estimate based upon his or her previous experience, upon the methods and equipment used, and upon the field conditions encountered. If no statement is given as to the relative error, the person using the data will automatically assume that the uncertainty equals ± one half of the last decimal *place*. (If a distance is recorded as 232.4 ft with no probable error given the user will assume the probable error is one half of ±0.1 or ±0.05 ft.) It is hoped that the surveyor who made the measurement understands this fact.

2-9 OCCURRENCE OF RANDOM ERRORS

In statistics, data are considered to be a sample of an infinitely large number of repetitions of a measurement for which the mean will be the true value. If a coin is flipped 100 times, the probability is that there will be 50 heads and 50 tails. Each time the coin is flipped there is an equal chance of it being heads or tails and it is apparent that the more times the coin is flipped the more likely it will be that the total number of heads will equal the total number of tails.

When a quantity such as distance is being measured random errors occur due to the imperfections of the observer and the equipment used. The observer is not able to make perfect readings. Each time a measurement is made it will be either too large or too small.

The theory of probability (to be considered in the next few sections of this chapter) can be used to study the behavior of random errors. Although probability theory is based on an infinite number of observations, it may in practice be applied with good results to situations where a fairly large number of observations are made.

For this discussion it is assumed that a distance was measured 28 times with the results shown in Table 2-1. In the first column of the table the values of the measurements are arranged for convenience in order of increasing value, while the number of times a particular mea-

TABLE 2-1

Measurement	No. or Frequency of Each Measurement	Residual or Deviation
96.90	1	−0.04
96.91	2	−0.03
96.92	3	−0.02
96.93	5	−0.01
96.94	6	0.00
96.95	5	+0.01
96.96	3	+0.02
96.97	2	+0.03
96.98	1	+0.04

Average = 96.94.

surement was obtained is given in the second column. These latter values are also referred to as the *frequencies* of the measurements.

The errors of the measurements are not known because the true value of the quantity is not known. The true value, however, is assumed to equal the arithmetic mean of the measurements. It is referred to as the *best value* or as the *most probable value*. The error of each measurement is then assumed to equal the difference between the measurement and the mean value. These are not really errors and they are referred to as *residuals* or *deviations*. For the measurements being considered here the residuals are shown in the third column of Table 2-1.

For a particular set of measurements of the same item, it is possible to take the residuals and plot them in the form of a bar graph as shown in Fig. 2-2. This figure, called a *histogram* or a *frequency dis-*

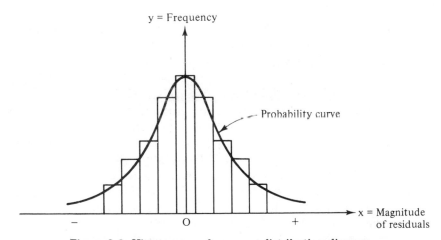

Figure 2-2 Histogram or frequency distribution diagram.

tribution diagram, has the magnitudes of the residuals plotted along the horizontal axis. Their signs, plus or minus, are shown by plotting them to the right or left of the origin (O). In addition, the number or frequency of the residuals of a particular size are shown vertically.

2-10 PROBABILITY CURVE

If an infinite number of measurements of an item could be taken and if the residuals of those values were plotted as a histogram and a curve drawn through the values the results would theoretically fall along a smooth bell-shaped curve called a *probability curve* (also called the *Gauss curve* or the *normal error distribution curve*). Such a curve shows the relationship between the size of an error and the probability of its occurrence. The curve can be represented with a mathematical equation (given at the end of this section), but the equation is seldom used in surveying measurements.

Theoretically the probability curve would be perfectly bell-shaped if an infinite number of measurements were taken and plotted. Even where a fairly large number of measurements of a particular quantity is made the histogram can be plotted and a curve drawn as shown in Fig. 2-2. Such a curve will approximate the theoretical curve and can be used practically to estimate the most probable behavior of random errors. There is little approximation involved because a curve plotted for as few as 5 or 6 determinations of a quantity is practically the same as the theoretical curve based on an infinite number of measurements.

Several important items should be noted concerning the probability curve. These include:

1. Positive and negative errors occur with the same frequency.
2. Large errors don't occur often.
3. Small errors occur more often than large ones.

It is apparent that if the measurements are taken with a different degree of precision the dimensions of the curve would be affected. The more precise the measurements, the more the errors or residuals will be concentrated near the center of the curve. For such a situation the curve will be higher and less spread out laterally. Should the measurements be taken with less precision the reverse will be true. *The effect of changes in precision, however, will only be to change the scale of the curve vertically and horizontally. The other characteristics and the usefulness of the curve will not change.*

The probability curve of Fig. 2-2 is redrawn in Fig. 2-3 and a point

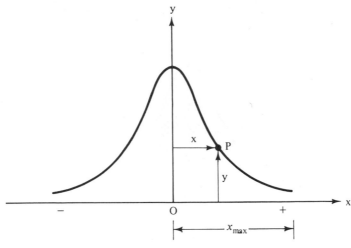

Figure 2-3

P is shown which has an ordinate equal to y and an abscissa equal to x. The y value is a number which gives the probability of an error of magnitude x.

From a theoretical standpoint a probability curve approaches the x axis at plus infinity and minus infinity. It is therefore obvious that all of the readings (or rather their residuals) are represented underneath the curve. In other words, there are no values outside of the curve and thus the sum of the values underneath the curve represents 100% of the values taken.

If the average of all the residuals for a particular set of measurements is computed and we assume that the curve is approximately symmetrical about the y axis, we see that approximately 50% of the values will be on the negative side and approximately 50% of the values will be on the positive side. (For the example presented in Fig. 2-2 the histogram is exactly symmetrical as to the number and sizes of plus and minus residuals. For many practical measurements, however, the diagram will not be perfectly symmetrical and is said to be skewed.)

Fifty percent of the area underneath the probability curve of Fig. 2-4 is shown shaded. There is a 50% chance that the error for a single measurement will fall within this area and a 50% chance that it will fall without. The value x_p shown in the figure is referred to as the *probable error* or 50% error. A particular measurement will have the same chance of having an error less than x_p as it does of being greater than x_p.

The average or 50% error can be determined by multiplying a constant 0.845 times the average numerical value of the residuals.

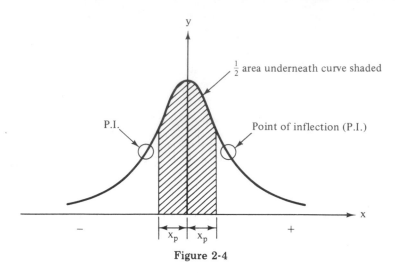

Figure 2-4

The signs ($\pm$) of the residuals are not included in the average. Letting v equals a residual, the 50% error E_{50} is as follows:

$$E_{50} = \pm 0.845 v_{average}$$

There are several ways in which errors may be denoted but the most common one is to refer them to the *standard deviation* (σ), also called the *mean square error*. If we examine the curve of Fig. 2-4, we will see that there are points of inflection (P.I.s) on each side of the curve, that is, points where the slope of the curve changes from concave to convex, or vice versa. The area underneath the probability curve between these points for a theoretical curve equals 68.3% of the total area. If a particular quantity is measured 10 times it is anticipated that 68.3% or about 7 of the 10 measurements will fall between these values and 3 will not. The residuals at the P.I.s are called the *standard deviations* or *standard errors* and can be calculated from the following expression:

$$\sigma = \pm \sqrt{\frac{\Sigma v^2}{n - 1}}$$

where Σv^2 is the sum of the squares of the residuals and n is the number of observations.

Actually the 68.3% value is not often used as such by the surveyor. How many surveyors would be willing to make a statement that a particular measurement was 632.87 $\pm$ 0.05 ft when the $\pm$0.05 ft was probably correct only 68.3% of the time? The answer is very few. He or she would be far happier if the value fell into the 90 or 95% or even higher range.

TABLE 2-2 Probabilities for
Certain Error Range

Error (±)	Probability (%)
0.50σ	38.3
0.6745σ	50.0
1.00σ	68.3
1.6449σ	90.0
1.9599σ	95.0
2.00σ	95.4
3.00σ	99.7
3.29σ	99.9

The probability of error at other positions on the curve can be determined from the following expression:

$$E_p = C_p \sigma$$

where E_p is the percentage error, C_p is a constant, and σ is the standard error.

For instance the 68.3% error occurs when $C_p = 1.00$ and the 95.4% error occurs when $C_p = 2.00$. It is impossible to establish an absolutely maximum error because this condition theoretically occurs at infinity, but many persons refer to the maximum error as being the 95.4% error (which occurs at 2.00σ) while others refer to the 99.9% error (which occurs at 3.29σ) as being the maximum. In the latter case, we see that 999 out of 1000 values would fall within this range. Several probabilities of error are summarized in Table 2-2 and represented graphically in Fig. 2-5.

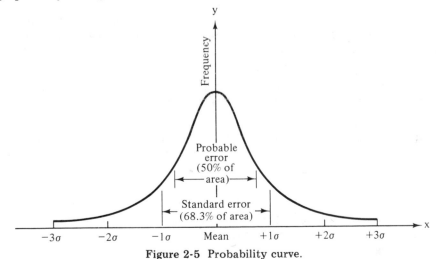

Figure 2-5 Probability curve.

Sometimes the values 2σ and 3σ (which correspond to the 95.4% and 99.7% errors respectively) are referred to as being two standard deviations or three standard deviations respectively.

The general equation of the probability curve for an infinite number of ordinates may be expressed as follows:

$$y = ke^{-h^2 x^2}$$

where y is the probability of occurrence of a random error of magnitude x; e is the base of the natural logarithms 2.718; while k and h are constants which determine the shape of the curve.

2-11 PROPAGATION OF RANDOM ERRORS

In this section we present several example problems which apply the preceding discussion of random errors to practical surveying problems. Included are calculations for single quantities as well as values for series of quantities. The accumulation of or lessening of random errors for various calculations is referred to as the *propagation of random errors*. To be able to develop equations for random error propagation for any but the simplest cases requires a substantial background in calculus. In the following analysis we restrict ourselves to straightforward algebraic equations which result from such derivations and apply them to simple numerical cases.

(a) Measurements of a Single Quantity

When a single quantity is measured several times, the probable error or 50% error of any one of the measurements can be determined from the following expression:

$$E_p = C_p \sqrt{\frac{\Sigma v^2}{n - 1}}$$

where $C_p = C_{50} = 0.6745$. For other percent errors the values of C_p vary as shown in Table 2-2. Example 2-1 illustrates the application of the equation for different percent errors.

Example 2-1

Determine the 50%, 90%, and 95% errors for the list of distance measurements shown in the first column of the following table.

Solution

Measured Values	Residual v	v^2
152.93	+0.02	0.0004
153.01	+0.10	0.0100
152.87	−0.04	0.0016
152.98	+0.07	0.0049
152.78	−0.13	0.0169
152.89	−0.02	0.0004
Avg = 152.91		$\Sigma v^2 = 0.0342$

$$E_{50} = \pm 0.6745 \sqrt{\frac{0.0342}{6-1}} = \pm 0.056 \text{ ft}$$

$$E_{90} = \pm 1.6449 \sqrt{\frac{0.0342}{6-1}} = \pm 0.136 \text{ ft}$$

$$E_{95} = \pm 1.9599 \sqrt{\frac{0.0342}{6-1}} = \pm 0.162 \text{ ft}$$

At this point we present a few remarks concerning measurements which seem to be completely out of line with the other values. When residuals are determined for a set of measurements they should be compared with the average value for those residuals. Should any of them be very large (say four or five or more times the average value) they probably should be discarded and the calculations continued with the remaining ones. A value should not be eliminated just to make things neat or symmetrical. For a value to be deleted it should be clearly set apart from the other values so that a definite mistake is indicated.

(b) A Series of Similar Measurements

Up to this point we have considered the measurement of single quantities only. Usually, however, the surveyor is involved with the measurement not just of one quantity but with a series of them. For instance he or she usually measures at least several angles or distances or elevations while working on a particular project.

For this discussion it is assumed that a set of quantities are to be measured. It is desired that each measurement be made to the same probable error so that all of them are equally reliable. Although random errors for these types of measurements tend to cancel each other somewhat, they will not completely do so for a practical number of measurements.

When a series of quantities are being measured, errors tend to accumulate in proportion to the square root of the number of measurements. This, the *law of compensation* can be written as follows:

$$E_{series} = E\sqrt{n}$$

For instance, if a series of twelve angles is measured each with a 95% error of $\pm 20''$ of arc, the anticipated total 95% error for the 12 angles would be:

$$E_{series} = \pm 20''\sqrt{12} = \pm 69.3'' = \pm 1'09''$$

As another example, suppose the 95% error in measuring a distance of one tape length (say 100 ft) is ± 0.01 ft. If a distance of 1600 ft is to be measured in 100-ft increments the 95% error for the full distance will equal:

$$E_{series} = \pm 0.01\sqrt{16} = \pm 0.04 \text{ ft}$$

Notice that for this taping measurement the 50% probable error of measuring 100 ft can be determined by using the constants given in Table 2-2:

$$E_{50} = \frac{0.6745}{1.9599}(\pm 0.01) = \pm 0.00344 \text{ ft}$$

The 50% probable error for measuring the 1600-ft length will equal

$$E_{series} = \pm 0.00344\sqrt{16} = \pm 0.0137 \text{ ft}$$

And the 95% error for measuring the 1600-ft length is

$$E_{series} = \pm 0.0137\frac{1.9599}{0.6745} = \pm 0.04 \text{ ft}$$

As one moves further into his or her study of surveying he or she may face the problem of needing to measure a distance, set of angles, etc. with a total error not exceeding a certain limit. For instance it might be necessary to measure a distance of 2000 ft with a total error of no more than ± 0.10 ft. The question would then arise: how accurately should each 100 ft be measured so that the desired limit is not exceeded? The value can be determined as follows:

$$E_{series} = \pm E\sqrt{n}$$
$$\pm 0.10 = \pm E\sqrt{20}$$
$$E = \pm 0.022 \text{ ft}; \text{ say } \pm 0.02 \text{ ft}$$

where E_{series} is ± 0.10 and $n = 20$.

(c) A Set of Like Measurements

It has been shown that when a quantity is measured several times, the probable total error will equal the probable error in a single observation times the square root of the number of observations:

$$E_{total} = E\sqrt{n}$$

Since the mean equals the sum divided by the number of observations the standard error of the mean will equal

$$E_{mean} = \frac{E\sqrt{n}}{n} = \frac{E}{\sqrt{n}}$$

If a distance is measured nine times with a probable accidental error of ±0.10 ft in each measurement, the probable error in the distance will equal the total probable error in the nine observations, $\pm0.10\sqrt{9}$, divided by the number of observations

$$\frac{\pm0.10\sqrt{9}}{9} = \frac{\pm0.10}{\sqrt{9}} = \pm0.03 \text{ ft}$$

In other words, the probable error in the mean of the nine readings equals the probable error of one observation divided by the square root of the number of observations. That is,

$$\pm\frac{0.10}{\sqrt{9}}$$

The preceding discussion clearly shows that the error of the mean varies inversely as the square root of the number of measurements. Thus in order to double the precision of a particular measured quantity, four times as many measurements should be taken. To triple the precision, nine times as many measurements should be taken.

(d) Quantities Consisting of a Series of Unrepeated Measurements

When a series of independent measurements are made with probable errors of E_1, E_2, E_3, ..., respectively, the total probable error can be computed from the following expression:

$$E_{sum} = \sqrt{E_1^2 + E_2^2 + \cdots + E_n^2}$$

Example 2-2 illustrates the application of this expression.

Example 2-2

The four approximately equal sides of a tract of land were measured. These measurements included the following probable errors: ±0.09 ft, ±0.013 ft, ±0.18 ft, and

±0.40 ft respectively. Determine the probable error for the total length or perimeter of the tract.

Solution

$$E_{sum} = \sqrt{(0.09)^2 + (0.013)^2 + (0.18)^2 + (0.40)^2}$$

$$= \pm 0.45 \text{ ft}$$

The reader should carefully note the results of the preceding calculations where the uncertainty in the total distance measured (±0.47 ft) is not very much different from the uncertainty given for the measurement of the fourth side alone (±0.40 ft). *It should thus be obvious that there is little advantage in making very careful measurements for some of a group of quantities and not for the others.*

(e) Probable Error For the Product of Two Quantities

Suppose that the sides of a rectangular piece of land have been measured each with certain estimated probable errors and it is desired to compute the area of the figure and the probable error in the resulting value. For such a situation the probable error in the area can be determined with the following expression:

$$E_{product} = \sqrt{A^2 E_b^2 + B^2 E_a^2}$$

where A and B are the measured lengths of the sides and E_a and E_b are the probable errors in those quantities.

Example 2-3 illustrates the calculation of the probable error for an area computation. This problem is one of the few cases involving random errors where a simple sketch clearly shows the theory involved in the solution. This is the idea of the Alternate Solution for this example.

Example 2-3

The measurements of the sides of a rectangle are respectively 232.60 ± 0.04 ft and 426.20 ± 0.06 ft. What is the area of the figure and the probable error?

Solution

$$\text{Area} = (232.60)(426.20) = 99134.12 \text{ sq. ft}$$

$$E_{product} = \pm\sqrt{(232.60)^2(0.06)^2 + (426.20)^2(0.04)^2}$$

$$\pm 22.03 \text{ sq. ft} \qquad (\text{say } \pm 22 \text{ sq. ft})$$

Alternate Solution   From the sketch below the area errors numbered 1 and 2 can be calculated as follows:

$$E_1 = (\pm 0.04)(426.20) = \pm 17.048 \text{ sq. ft}$$

$$E_2 = (\pm 0.06)(232.60) = \pm 13.956 \text{ sq. ft}$$

Combining the area errors 1 and 2 by the law of compensation yields

$$E_{series} = \sqrt{(17.048)^2 + (13.956)^2} = \pm 22.03 \text{ sq. ft} \qquad (\text{say } \pm 22 \text{ sq. ft})$$

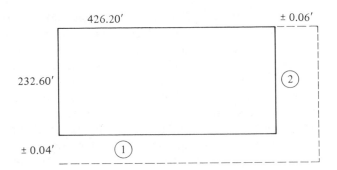

(f) Weighted Observations

In the preceding examples the assumption has been made that each of the observations were equally reliable. A common problem which the surveyor faces is that of combining the results of different measurements that were made under different conditions and circumstances, and therefore have different degrees of reliability. For such a situation it is necessary to estimate the weight (or degree of reliability) for each of the measurements before they are combined.

Relative weights may be assigned to various observations based upon (1) the judgment of the observer, (2) the number of like observations made, or (3) by assuming that the weights are inversely proportional to the square of the probable errors. If W_1 is the weight of quantity 1 and E_1 its probable error, the following expression may be written:

$$W_1 E_1^2 = W_2 E_2^2 = W_3 E_3^2 = \cdots = W_n E_n^2$$

A numerical example for this type of situation will be presented in Chapter 9 dealing with most probable elevation.

2-12 FIELD NOTES

Perhaps no other phase of surveying is as important as the proper recording of field data. No matter how much care is used in making field measurements, the effort is wasted unless a clear and legible record is kept of the work. For this reason we shall go to some length to emphasize this aspect of surveying.

The cost of keeping a surveying crew or party in the field is appreciable, sometimes a few hundred dollars per day. If the notes are wrong, incomplete, or confusing, much time and money have been wasted and the entire party may have to return to the job to repeat some or all of the work.

There are several kinds of field notebooks available, but the usual ones are $4\frac{5}{8}$ in. $\times$ $7\frac{1}{4}$ in., a size that can be easily carried in a pocket. This characteristic is quite important because the surveyor needs his hands for other work. Examples of fieldbook notes are shown throughout the text, the first one being Fig. 3-1. A general rule is that measured quantities are shown on the left-hand pages and sketches and miscellaneous notes are shown on the right-hand pages.

In keeping notes the surveyor should bear in mind that on many occasions (particularly in large organizations), persons not familiar with the locality will make use of the notes. Therefore, considerable effort should be made to record all of the information necessary for others to clearly understand the survey.

An additional consideration is that surveying notes are on many occasions used for purposes other than the one for which they were originally developed, and therefore they should be carefully preserved.

To maintain good field notes is not an easy task, but with practice the ability can be developed. The following items are absolutely necessary for the successful recording of surveying information:

1. The title of the job, date, weather, location, and names of party members should be recorded. When surveying notes are being used in the office, it may be helpful to know something of the weather conditions at the time the measurements were taken. This information will often be useful in judging the accuracy of a particular survey. Was it $110°F$ or $-10°F$? Was it raining? Were strong winds blowing?

2. Field notes should be organized in a form appropriate to the type of survey. Since other people may very well use these notes, generally standard forms are used for each of the different types of surveys. If each surveyor used his own individual forms for all surveys, there would be much confusion back in the office.

3. Measurements must be recorded in the field when taken and not trusted to memory or written on scraps of paper to be recorded at a later date.

4. Frequent sketches are used where needed for clarity. Since field books are relatively inexpensive as compared to the other costs of surveying, crowding of sketches or other data does not really save money.

5. Field measurements must not be erased when mistakes are made. A line should be drawn through the incorrect number and the corrected value written adjacent to the old value. Erasures cause suspicion that there has been some dishonest alteration of values, but a crossed out number is looked upon as an open admission of

a blunder. (Imagine a property case coming up for court litigation and a surveying notebook containing frequent erasures being presented for evidence.) It is a good idea to use red ink for making additions to the notes back in the office to clearly distinguish them from values obtained in the field.

6. Notes are printed with a sharp medium-hard (3H or 4H) pencil so that the records will be relatively permanent and will not smear. Field books are generally used in damp and dirty situations and the use of hard pencils will preserve the notes.

7. The instrument number should be recorded with each day's work. It may later be discovered that the measurements taken with that instrument contained errors which could not be accounted for in any other way. With the instrument identified the surveyor may be able to go back to the instrument and make satisfactory corrections.

8. A few other requirements include: numbering of pages, inclusion of a table of contents, drawing arrows on sketches indicating the general direction of north and the clear separation of each day's work by starting on a clean page each day. Should a particular survey extend over several days, cross references may be necessary between the various pages of that project. The numbering system generally used for surveying notes is to record the page number in the upper right corner of each right-hand page. A single page number is used for both the right- and left-hand sides.

Finally, it is essential that notes be checked before leaving the site of the survey in order to make sure that all required information has been obtained and recorded. Some surveyors keep check lists in their field books for different types of surveys. Before they leave a particular job they refer to the appropriate list for a quick check. Imagine the expense in time and money of having a survey crew make an extra trip to the site of a job some distance away in order to obtain one or two minor bits of information which had been overlooked.

2-13 OFFICE WORK AND DIGITAL COMPUTERS

Field surveying measurements provide the basis for large amounts of office work. Some of this paper work includes computations of precision, drawing of plats of land boundaries, calculations for and drawing of topographic maps, and computation of earthwork volumes.

A large percentage of these items are commonly handled today with digital computers. Once a surveyor has used a successful computer program or a programmed desk or pocket calculator for some part of his office work (such as the calculation of land areas), he will be reluctant to ever again make those calculations with his ordinary desk calculator.

It is doubtful if 5% of the adult population of the United States understands how cars actually operate, but just about 100% of those same people know what cars can do and how to make them do it. Likewise, it is doubtful if 5% of surveyors understand how electronic computers work, but all of them should know what computers can do and how they can be of service to the surveyor.

These devices can be used to perform the normal type of surveying calculations in a matter of minutes or even seconds while at the same time reducing the opportunities for mistakes to occur. The time required for looking up trigonometric functions, making interpolations, transferring numbers, etc., is eliminated because this information is programmed into the machine.

Appropriate comments are made in various places throughout the text in an attempt to advise the student of areas in which computers or programmable calculators can be particularly helpful. In addition, Sections 14-11 through 14-15 are devoted entirely to the subject.

PROBLEMS

2-1. A quantity was measured ten times with the following results: 3.625, 3.621, 3.629, 3.623, 3.624, 3.621, 3.622, 3.628, 3.620, and 3.619 ft. Determine:
  (a) the most probable value of the measured quantity
  (b) the probable error of a single measurement
  (c) the 90% error
  (d) the 95% error

  (*Ans.:* 3.623, ±0.002, ±0.005 and ±0.007)

2-2. The same questions apply as for Problem 2-1 but the following twelve quantities were measured: 162.69, 162.61, 162.65, 162.71, 162.77, 162.73, 162.71, 162.70, 162.64, 162.67, 162.80, and 162.74 ft.

2-3. The same questions apply as for Problem 2-1 but the following eleven readings were taken: 43.1, 43.6, 43.4, 43.2, 43.3, 42.9, 43.0, 43.1, 42.8, 43.5, and 43.6 minutes.   (*Ans.:* 43.2, ±0.2, ±0.5 and ±0.5)

2-4. It is assumed that the probable error in measuring 100 ft with a steel tape is ±0.02 ft. The sides of a closed figure are measured with the following results: 262.5, 348.9, 102.8, and 136.6 ft. Determine the probable error in the perimeter of the figure. Assume that the probable error in each side is proportional to the square root of the number of 100-ft spans including fractional parts.

2-5. The same question as for Problem 2-4 applies, except that the probable error per 100 ft is ±0.03 ft and the measurements are 642.8, 549.6, 302.6, 849.8 and 503.7 ft.   (*Ans.:* ±0.16 ft)

2-6. Two sides of a rectangle were measured as being 162.32 ft ±0.03 ft and 207.46 ft ±0.04 ft. Determine the area of the figure and the probable error of the area.

2-7. A particular surveying party is capable of making 100 ft measurements with a standard deviation or standard error of ±0.01 ft.

  (a) What total standard error is to be expected if a distance of 4000 ft is measured?

  (b) What total 95% error is to be expected in the 4000 ft distance?

  (*Ans.:* ±0.06 ft and ±0.13 ft)

2-8. Same questions as for Problem 2-8 except the standard error is ±0.03 ft for a 100 ft measurement.

2-9. It is desired to tape a distance of 1100 ft with a total standard error of not more than ±0.10 ft.

  (a) How accurately should each 100 ft distance be measured so that the permissible value is not exceeded?

  (b) How accurately would each 100 ft distance have to be measured so the 95% error would not exceed ±0.12 ft in a total distance of 2000 ft?

  (*Ans.:* ±0.03 ft and ±0.01 ft)

2-10. For a nine-sided closed figure the sum of the interior angles is exactly 1620°. It is specified for a survey of this figure that if these angles are measured in the field their sum should not miss 1620° by more than 1 minute. How accurately should each angle be measured?

Distance Measurement

3

3-1 INTRODUCTION

One of the most basic operations of surveying is the measurement of distance. In surveying, the distance between two points is understood to be the horizontal distance. The reason for this is that most of the surveyor's work is plotted on a drawing as some type of map. A map, of course, is plotted on a flat plane and the distances shown thereon are horizontal projections. Land areas are computed on the basis of the same horizontal measurements. This means that if a person wants to obtain the largest amount of actual land surface area for each acre of land he buys, he should buy it on the side of a very steep mountain.

Early measurements were made in terms of the dimensions of parts of the human body such as cubits, fathoms, and feet. The cubit (the unit Noah used in building his boat) was defined as the distance from the tip of a man's middle finger to the point of his elbow (about 18 in.); a fathom was the distance between the tips of a man's middle fingers when his arms were outstretched (approximately 6 ft). Other measurements were the foot (the distance from the tip of a man's big toe to the back of his heel) and the pole or rod or perch (the length of the pole used for driving oxen, later set at 16.5 ft). In England the "rood" (rod or perch) was once defined as being equal to the sum of the lengths of the left feet of 16 men, whether they were short or tall, as they came out of church one Sunday morning.[1] (Another historical distance unit of interest is the furlong. It is defined as the length of the side of a square ten acre field which is $\frac{1}{8}$th of a mile or 660 ft.) Today in the U.S. the foot is the basic unit of measurement, but most of the world uses the meter–decimal system and the American surveying profession is gradually becoming more familiar with metric units.

The meter is a unit of French origin and its application in the United States in the past has been almost entirely limited to geodetic surveys.   In 1866 the U.S. Congress legalized the use of the metric system by which the meter was defined as being equal to 3.280833 ft (or 39.37 in.) and 1 inch was equal to 2.540005 centimeters.   These values are based on the length at $0°C$ of an International Prototype Meter bar consisting of 90% platinum and 10% iridium which is kept in Sèvres, France, near Paris.   In accordance with the treaty of May 20, 1875, the National Prototype Meter 27 identical with the International Prototype Meter was distributed to various countries.   Two copies are kept by the U.S. Bureau of Standards at Gaithersburg, Maryland.

In 1960 the meter was redefined at the 11th General Conference on Weights and Measures and agreed upon by the U.S. and 35 other countries.   It was defined in terms of the wave length γ_k of the orange-red light produced by burning of the element krypton and set equal to $1,650,763.73\gamma_k$ or 3.280840 ft.   It is felt that this new definition will enable industries around the world to make more accurate measurements and also will keep them from having to check their instruments against the international prototype bar or its copies.   This new value presumably can be reproduced with great accuracy in a well-equipped laboratory.

The national geodetic control system of the U.S. is based on the old definition of the meter but this basis is being changed to the new definition and will become effective during 1983.   Until that time U.S. surveyors will continue to use the older definition.

The International Bureau of Weights and Measures has as its goal the establishment of a rational and coherent worldwide system of units. In 1960 this organization promulgated the "International System of Units," with the abbreviation SI in all languages.   SI units are currently being adopted by quite a few countries including most English-speaking nations (Britain, Australia, Canada, South Africa, and New Zealand). The purpose of the system is to standardize the units of measurements used throughout the world.   Though the U.S. Congress has not officially adopted the system it seems inevitable that they will within a few years.

[1]C. M. Brown and W. H. Eldridge, *Evidence and Procedures for Boundary Location*, (New York: John Wiley & Sons, Inc., 1962) p. 195.

To the surveyor the most commonly used SI units are the meter (m) for linear measure, the square meter (m²) for areas, the cubic meter (m³) for volumes and the radian (rad) for plane angles. In many countries the comma is used to indicate a decimal; thus to avoid confusion in the SI system, spaces rather than commas are used. For a number having four or more digits, the digits are separated into groups of threes, counting both right and left from the decimal. For instance 4,642,261 is written as 4 642 261 and 2,340.3216 is written as 2 340.321 6.

To change surveying to SI units may seem at first glance to be quite simple. In a sense this is true in that if the surveyor can measure distance with a 100-ft tape he can do just as well with a 30-meter tape using the same procedures. However, we have several hundred years of land descriptions recorded in the English system of units and stored in our various court houses and other archives. Future generations of surveyors will therefore never get away completely from the English system of units. As an illustration, today's surveyor still frequently encounters old land descriptions made in terms of so many chains (where the 66-ft chain is referred to). In some parts of the U.S. a unit of Spanish origin called the *vara* is encountered. It equals 33 in. in California and $33\frac{1}{3}$ in. in Texas. Finally it will take decades to replace the expensive instruments graduated in English units which our surveyors presently own.

There are several methods used for actually measuring distance: pacing, odometer readings, stadia, taping, and the use of various electronic devices. These methods are briefly discussed in the sections to follow.

3-2 PACING

The ability to pace distances with reasonable precision is very useful to almost anyone. The surveyor in particular can use pacing to quickly make approximate measurements or to check measurements made by more precise means. By so doing he will often be able to detect large mistakes.

A person can determine the value of his average pace by counting the paces necessary for him to walk a distance which has previously been measured more precisely (for example, with a steel tape). For most persons pacing is done most satisfactorily when taking natural steps. Others like to try to take paces of certain lengths (for example, 3 ft), but this method is tiring for long distances and usually gives results of lower precision for short or long distances. As horizontal distances are needed some adjustments should be made when pacing is

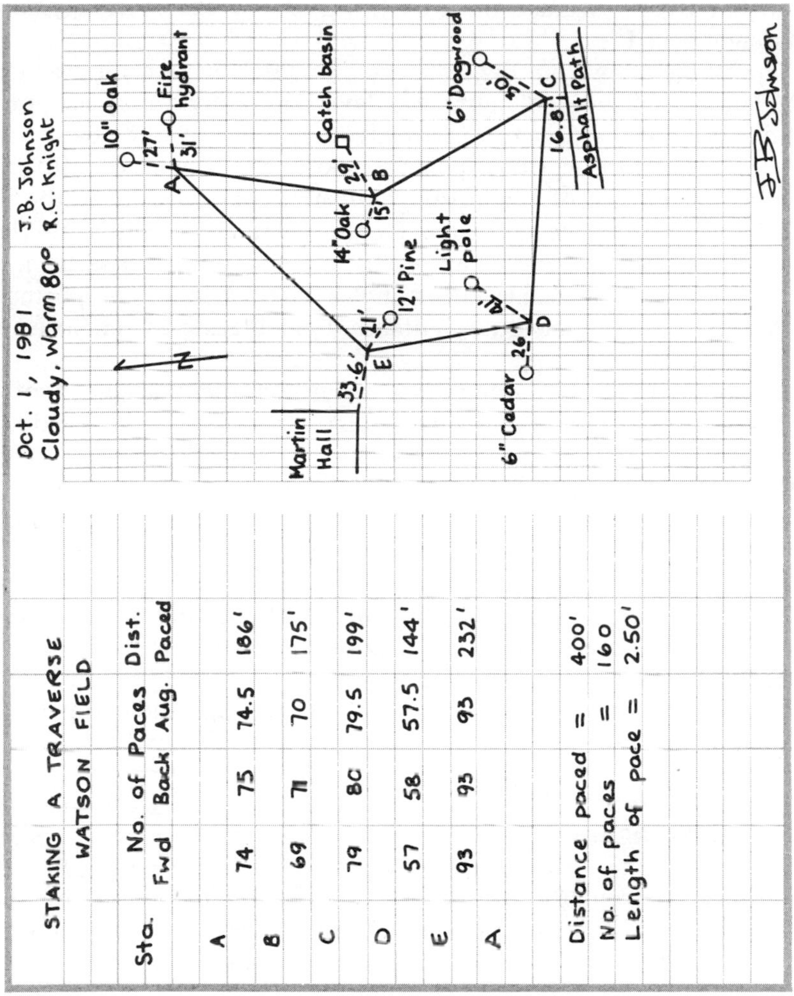

STAKING A TRAVERSE
WATSON FIELD

Oct. 1, 1981   J.B. Johnson
Cloudy, Warm 80°   R.C. Knight

| Sta. | No. of Paces | | Dist. |
	Fwd	Back	Aug. Paced
A			
	74	75	74.5	186'
B			
	69	71	70	175'
C			
	79	80	79.5	199'
D			
	57	58	57.5	144'
E			
	93	93	93	252'
A			

Distance paced   =   400'
No. of paces   =   160
Length of pace   =   2.50'

10" Oak
27'
31' Fire hydrant
A
Catch basin
29'
14" Oak
15' B
16.8' C
6' Dogwood
9'
21'
Asphalt Path
12" Pine
Light pole
33.6' 21'
E
6" Cedar 26'
D
Martin Hall
N

J.B. Johnson

Figure 3-1

done on sloping ground. As a matter of fact, the surveyor would do well to measure his pace on sloping ground as well as on level ground.

With a little practice a person can pace distances with a precision of roughly 1/50 to 1/200 depending on the ground conditions (slope, underbrush). For distances of more than a few hundred feet, a mechanical counter or *pedometer* can be conveniently used. Pedometers can be adjusted to the average pace of the user and automatically record the distance paced.

The notes shown in Fig. 3-1 are presented as a pacing example for students. A five-sided figure was layed out and each of the corners marked with a stake driven into the ground. The average pace of the surveyor was determined by pacing a known distance of 400 ft, as shown at the bottom of the left page of the field notes. Then the sides of the figure were paced and their lengths calculated. This same figure or traverse is used in later chapters as an example for measuring distances with a steel tape, measuring angles with a transit, computing the precision of the latter measurements, and calculating the enclosed area of the figure.

A note should be added here regarding the placement of the stakes. It is good practice for the student to record in his field book information on the location of the stakes. The position of each stake should be determined in relation to at least two prominent objects such as trees, walls, sidewalks, and so on, so that there will be no difficulty in relocating them a week or a month later. This information can be shown on the sketch as illustrated in Fig. 3-1.

Many surveying professors pick out in their first classes the students who have previous surveying experience and get them to work together. If this is not done the result is often that the experienced student in each party will do most of the work while the inexperienced people get little practice.

3-3 ODOMETER READINGS

Distances can be roughly measured by rolling a wheel along the line in question and counting the number of revolutions. An *odometer* is a device attached to the wheel (similar to the distance recorder used in a car) which does the counting and from the circumference of the wheel converts the number of revolutions to a distance. Such a device provides a precision of approximately 1/200 when the ground is smooth as along a highway, but results are much poorer when the surface is irregular.

The odometer may be useful for preliminary surveys, perhaps when pacing would take too long. It is occasionally used for initial route-location surveys and for quick checks on other measurements. A similar device is the *measuring wheel* which is a wheel mounted on a rod. Its user can push the wheel along the line to be measured. It is frequently used for curved lines.

3-4 TACHYMETRY

The term *tachymetry* or *tacheometry* which means swift measurements is derived from the Greek words "taklus" meaning swift and "metron" which means measurement. Actually any measurement made swiftly could be said to be tacheometric but the generally accepted practice is to list under this category only measurements made with subtense bars or by stadia. Thus the extraordinarily fast electronic distance-measuring devices are not listed in this section.

Subtense Bar

A tachymetric method which yields measurements of satisfactory precision for property surveys makes use of the subtense bar (Fig. 3-2). In Europe where the method is most commonly used a horizontal bar with sighting marks on it, usually located 2 meters apart, is mounted on a tripod. The tripod is centered over one end of the line to be measured and the bar is leveled and turned so that it is made perpendicular to the line by sighting on the other end of the line with a small attached telescope.

A theodolite (described in Section 12-12) is set up at the other end of the line and sighted on the subtense bar. The angle subtended between the marks on the bar is carefully measured (preferably with a theodolite measuring to the nearest second of arc) and the distance be-

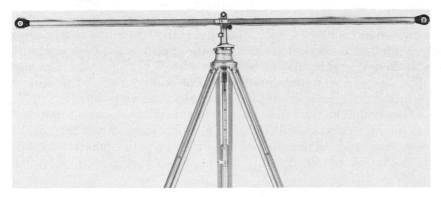

Figure 3-2 Subtense bar. (*Courtesy Kern Instruments, Inc.*)

tween the ends of the line is computed. The distance D from the theodolite to the subtense bar can be computed from the following expression:

$$D = \frac{1}{2} S \cot \frac{\alpha}{2}$$

in which S is the distance between the sighting marks on the bar and α is the subtended angle.

For reasonably short distances, say less than 500 ft, errors are small and a precision of 1/1000 to 1/5000 is ordinarily obtained. The subtense bar is particularly useful for measuring distances across rivers, canyons, busy streets, and other difficult areas. It has an additional advantage in that the subtended angle is independent of the slope of the line of sight; thus the horizontal distance is obtained directly and no slope correction has to be made.

Stadia

Although the subtense bar has been used in the United States, the stadia method is far more common. Its development is generally credited to the Scotsman James Watt in 1771.[2] The word *stadia* is the plural of the Greek word *stadium* which was the name given to a foot-race track of approximately 600 ft length.

Transit telescopes (to be discussed in Chapter 12) are equipped with three horizontal cross hairs which are mounted on the cross-hair ring. The top and bottom hairs are called the stadia hairs.

The surveyor sights through the telescope and takes readings where the stadia hairs intersect a scaled rod. The difference between the two readings is called the *rod intercept.* The hairs are so spaced that at a distance of 100 ft their intercept on a vertical rod is 1 ft; at 200 it is 2 ft; and so on. To determine a particular distance, the telescope is sighted on the rod and the difference between the top and bottom cross-hair readings is multiplied by 100. When one is working on sloping ground a vertical angle is measured and is used for computing the horizontal component of the slope distance. These measurements can also be used to determine the vertical component of the slope distance or the difference in elevation between the two points.

The stadia method has its greatest value in locating details for maps, but it is also useful for making rough surveys and for checking more precise ones. Precisions of the order of approximately 1/250 to 1/1000 can be obtained with stadia. Such precision is not usually

---

[2] A. R. Legault, H. M. McMaster, and R. R. Marlette, *Surveying* (Englewood Cliffs, NJ: Prentice-Hall, Inc., 1956), p. 39.

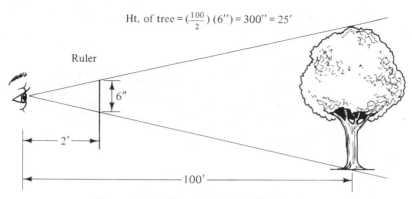

Figure 3-3 Estimating the height of a tree.

satisfactory for property surveys. The method is discussed in considerable detail in Chapter 16 where its usefulness in mapping is described.

The same principle can be used to estimate the heights of buildings, trees, or other objects as illustrated in Fig. 3-3. A ruler is held
upright at a given distance such as arms length (about 2 ft for many
people) in front of the observer's eye so that its top falls in line with
the top of the tree. Then the thumb is moved down so that it coincides
with the point where the line of sight strikes the ruler when the observer looks at the base of the tree. Finally the distance to the base of
the tree is measured by pacing or taping. By assuming the dimensions
shown in Fig. 3-3, the height of the tree shown can be estimated as
follows:

$$\text{Ht. of tree} = \left(\tfrac{100}{2}\right)(6'') = 300'' = 25'$$

Artillery students were formerly trained in a similar fashion to
estimate distances by estimating the size of objects when they were
viewed between the knuckles of their hands held at arms length.

3-5 TAPING OR CHAINING

For many centuries, surveyors measured distances with ropes, lines,
or cords which were treated with wax and calibrated in cubits or other
ancient units. These devices are obsolete today although precisely
calibrated wires are sometimes used.

In recent decades the 100-ft steel ribbon tape has become a common method for measuring distances. Such measuring is often called
"chaining," a carry-over name from the time when Gunter's chain
was introduced. The English mathematician Edmund Gunter (1581-

Figure 3-4 The surveyor's chain.

1626) invented the surveyor's chain (Fig. 3-4) in the early 1600's. His chain, which was a great improvement over the ropes and rods used up until that time, was available in several lengths including 33 ft, 66 ft, and 100 ft. The 66-ft length was the most common. (Gunter is also credited with the introduction of the words cosine and cotangent to trigonometry, the discovery of magnetic variations discussed in Chapter 11, and other outstanding scientific accomplishments.)[3]

The 66-ft chain, sometimes called the 4-pole chain, consisted of 100 heavy wire links each 7.92 in. in length. In studying old deeds and plats the surveyor will often find distances which were measured with the 66-ft chain. He might very well see a distance given as 11 ch. 20.2 lks. or 11.202 ch. The usual area measurement in the U.S. is the acre, which equals 10 square chains. This is equivalent to 66 ft by 660 ft, or 1/80th mi by 1/8th mi, or 43,560 sq. ft.

Steel tapes came into general use around the beginning of the twentieth century. They are available in lengths from a few feet to 1000 ft. For ordinary conditions, precisions of from 1/1000 to 1/5000 can be obtained, although much better work can be done by using procedures to be described later. Chapter 4 is devoted entirely to taping.

3-6 ELECTRONIC DISTANCE MEASUREMENTS

Sound waves have long been used for estimating distances. Nearly all of us have counted the number of seconds elapsing between the flash of a bolt of lightning and the arrival of the sound of the thunder and then multiplied the number of seconds by the speed of sound (about one-fifth of a mile per second). The speed of sound is 1129 ft/sec at

[3]C. M. Brown and W. H. Eldridge, *Evidence and Procedures for Boundary Location* (New York: John Wiley & Sons, Inc., 1962), p. 79.

$70°F$ and increases by a little more than 1 ft/sec for each degree F increase in temperature. For this reason alone sound waves do not serve as a practical means of precise distance measurement because temperatures along a line being measured would have to be known to almost the nearest $0.01°F$.

In the same way some distances for hydrographic surveying were formerly estimated by firing a gun and then measuring the time required for the sound to travel to another ship and be echoed back to the point of firing. Ocean depths are determined with depth finders which use echoes of sound from the ocean bottom. Sonar equipment makes use of supersonic signals echoed off the hulls of submarines to determine underwater distances.

It has been discovered during the past few decades that the use of either light waves, electromagnetic waves, infrared, or even lasers offer much more precise methods of measuring distance. Although it is true that some of these waves are affected by changes in temperature, pressure, and humidity, the effects are small and can be accurately corrected. Under normal conditions the corrections amount to no more than a few centimeters in several miles. Numerous portable electronic devices making use of these wave phenomena have been developed which permit the measurement of distance with tremendous precision.

These devices have not replaced chaining or taping, but they are widely used by private surveyors and not just by government organizations performing geodetic work. Their prices have dropped to a level where the average surveyor can consider purchasing them and they are deemed indispensable by many companies. The author thinks that they pay for themselves in one or two years at the most. For those who prefer to lease, arrangements are available with instrument companies.

Electronic distance-measuring devices have several important advantages over other methods of measurement (Fig. 3-5). They are very useful in measuring inaccessible lines or those distances which are difficult of access, for example, across lakes and rivers, busy highways, standing farm crops, canyons, and so on. For long distances (say several miles), the time required is in minutes not hours as would be required for a typical taping party. Two men, easily trained, can do the work better and faster than the conventional four-man taping crew. In addition, clients like to deal with "up-to-date, forward-looking firms" that use the latest equipment.

Electronic distance-measuring devices will probably not completely replace the steel tape, at least in the foreseeable future. The steel tape will likely be used for short distances for a long time to come because of its convenience and economy. For this reason the surveyor of today and tomorrow should become proficient with the steel tape, even though it is highly probable that he will eventually use the tape

Figure 3-5 An electronic total station instrument with which distance
and angles can be measured. (*Courtesy Hewlett-Packard.*)

only a small percentage of the time. Otherwise the chances of his making blunders and large errors in his work when he does use the tape will
be magnified.

The disadvantages of electronic distance measurements are the
cost, weight, and bulkiness of the equipment. Taping continues to be
the conventional method for measuring short distances even though the
electronic distance-measuring devices are useful there too. Chapter 7
is devoted entirely to electronic distance measurements.

3-7 SUMMARY OF MEASUREMENT METHODS

Table 3-1 presents a brief summary of the various methods for measuring distances. There is a great variation in the precision obtainable
with these different methods and the surveyor will select one which is
appropriate for the purposes of his particular survey.

TABLE 3-1

Method	Precision	Uses
Pacing	1/50 to 1/200	Reconnaissance and rough planning
Odometer	1/200	Reconnaissance and rough planning
Subtense bar	1/1000 to 1/5000	Seldom used in U.S. and then only when taping is not feasible because of terrain and when electronic distance measuring devices are not available
Stadia	1/250 to 1/1000	Very commonly used for mapping, rough surveys, and for checking more precise work
Ordinary taping	1/1000 to 1/5000	Ordinary land surveys and building construction
Precision taping	1/10,000 to 1/30,000	Excellent land surveys, precise construction work, and city surveys
Base line taping	1/100,000 to 1/1,000,000	Precise geodetic work performed by the National Geodetic Survey
Electronic distance measurement	$\pm 0.04' \pm$ 1/300,000	In the past use has been primarily for precise government geodetic work but is today commonly used for land development, land surveys, and precise construction work

Taping

4

4-1 INTRODUCTION

The general public, riding in their cars and seeing surveyors measuring distances with a steel tape, might think, "Anybody could do that. What could be simpler?" The truth is, however, that the measurement of distance with a steel tape, though simple in theory, is probably the most difficult part of good surveying. The efficient use of today's superbly manufactured surveying equipment for other surveying functions such as the measurement of angles is quickly learned. But correspondingly precise distance measurement with a steel tape requires thought, care, and experience. In theory, it is simple, but in practice it is not so easy.

4-2 EQUIPMENT REQUIRED FOR TAPING

A brief discussion of the various types of equipment normally used for taping is presented in this section. A taping party should have at least one 100-ft steel tape, two range poles, a set of 11 chaining pins, a 50-ft woven tape, two plumb bobs, and a hand level. These items are discussed briefly in the following paragraphs.

Steel tapes. Steel tapes are most commonly 100 ft long, approximately $\frac{5}{16}$ in. wide, 0.025 in. thick, and they weigh 2 or 3 lb. They are either carried on a reel or done up in 5-ft loops to form a figure 8 from which they are thrown into a convenient circle. These tapes are quite strong as long as they are kept straight but if they are tightened when they have loops or kinks in them, they will break very easily. If a tape gets wet, it should be wiped with a dry cloth and then again with an oily cloth.

Surveyor's tapes are marked at the one-foot points from 0 to 100 ft. Older tapes have the last foot at each end divided into tenths of a foot, but the newer tapes have an extra foot beyond 0 which is subdivided. Tapes with the extra foot are called *add* tapes while the ones without the extra foot are called *cut* tapes. Several variations are available, for example, tapes divided into feet, tenths, and hundredths for their entire lengths. Needless to say, the surveyor must be completely familiar with the divisions of his tape and its 0 and 100-ft marks before he does any measuring.

Tapes can be obtained in various lengths other than 100 ft. The 300- and 500-ft lengths are probably the most popular. The longer tapes, which usually consist of $\frac{1}{8}$-in. wide wire bands, are only divided at the 5-ft points to reduce costs. They are quite useful for rapid, precise measurements of long distances on level ground. The use of long tapes permits a considerable reduction in the time required for marking at tape ends, and it also virtually eliminates the accidental errors that occur while marking.

In recent years fiberglass tapes have been introduced on the market. These less expensive tapes are available in 50-ft, 100-ft, and other lengths. They are strong and flexible and they will not shrink or stretch appreciably with changes in temperature and moisture. They may also be used without hazard in the vicinity of electrical equipment.

Though the 100-ft tapes have been commonly used by surveyors for quite a few decades, it seems likely that they will, in the next few years, begin to be replaced by metric steel tapes. Probably these tapes will be available in lengths of 20, 30, 50 meters, and others. It is highly possible that the 30-m (98.4 ft) length will become the common metric tape because its length is so close to that of the present 100-ft tape. Perhaps the metric tapes will be divided into decimeters throughout their lengths with the end decimeter divided into millimeters.

For very precise taping, the *Invar* tape, which is manufactured from a nickel–steel alloy, can be used. It has the considerable advantage that its variations in length caused by temperature changes are $\frac{1}{30}$ or even less of those of the usual steel tape. Very precise taping can be accomplished with Invar tapes even on hot sunny days, but these tapes are perhaps ten times as expensive as standard tapes and they are easily bent and damaged. To avoid kinking they should be kept on reels with large diameters. They are used only for the most precise operations such as geodetic work and for checking the length of regular steel tapes. A newer tape called the *Lovar* tape has properties and costs somewhere in between those of the regular steel tapes and the Invar tapes.

Range poles. Range poles are used for sighting points and for lining up tapemen in order to keep them going in the right direction. They are usually from 6 to 10 ft in length and are painted with alternate bands of red and white to make them more easily seen. The bands are each 1 ft in length and the rods can therefore be used for rough measurements. They are manufactured from wood, fiberglass, or metal. The sectional steel tubing type is perhaps the most convenient one because it can be easily transported from one job to another.

Taping pins. Taping pins are used for marking the ends of tapes or intermediate points while taping. They are easy to lose and are generally painted with alternating red and white bands. If the paint wears off, they can be repainted any convenient bright color or they can have strips of cloth tied to them which can readily be seen. The pins are carried on a wire loop which can conveniently be carried by a tapeman, perhaps by placing the loop around his belt.

Plumb bobs. Plumb bobs for surveying are usually made of bronze and weigh from 8 to 16 oz. They have sharp replaceable points and a device at the top to which plumb bob strings may be tied.

Woven tapes. Woven tapes (Fig. 4-1) are most commonly 50 ft in length with graduation marks at 0.25 in. intervals. They can be

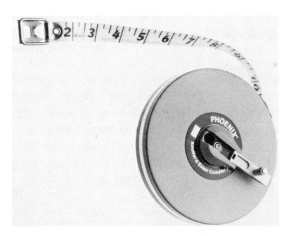

Figure 4-1 Woven tape. (*Courtesy Keuffel & Esser Co.*)

Figure 4-2 Hand level.

either nonmetallic or metallic. Nonmetallic tapes are woven with very strong synthetic yarns and are covered with a special plastic coating that is not affected by water. Metallic tapes are made with a water repellant fabric into which fine brass, bronze, or copper wires are placed in the lengthwise direction. These wires strengthen the tapes and provide considerable resistance to stretching. Nevertheless, since all woven tapes are subject to some stretching and shrinkage, they are not suitable for precise measurement. Despite this disadvantage, woven tapes are often useful and should be a part of a surveying party's standard equipment. They are commonly used for finding existing points, locating details for maps, and measuring in situations where steel tapes might easily be broken (as along highways) or when small errors in distance are not too important.

Hand levels. The hand level is widely used for taping and for the rough determination of elevations. It consists of a metal sighting tube on which is mounted a bubble tube (Fig. 4-2). If the bubble is centered while sighting through the tube, the line of sight is horizontal. Actually, the bubble tube is located on top of the instrument and its image is reflected by means of a 45° mirror or prism inside the tube so that its user can see the bubble at the same time as the terrain. The hand level is very useful to the surveyor for holding the steel tape horizontal as described in Section 4-4.

4-3 TAPING OVER LEVEL GROUND

If taping is done on fairly smooth and level ground where there is little underbrush, the tape can rest on the ground. The taping party consists of the head tapeman and the rear tapeman. The head tapeman leaves one taping pin with the rear tapeman for counting purposes and perhaps to mark the starting point. The head tapeman takes the zero end of the tape and walks down the line toward the other end.

When the 100-ft end of the tape reaches the rear tapeman, the rear tapeman calls "tape" or "chain" to stop the head tapeman. The rear tapeman holds the 100-ft mark at the starting point and aligns the head tapeman (using hand and perhaps voice signals) on the range pole which has been set behind the ending point. Ordinarily this "eyeball" alignment of the tape is satisfactory, but the use of the transit is safer and will result in better precision. Sometimes there are places along a line where the tapemen cannot see the end point and there may be positions where they cannot see the signals of the instrumentman. For such cases it is necessary to set intermediate line points before the taping can be started.

It is necessary to pull the tape firmly (see Secs. 5-4 and 5-5). This can be done by wrapping the leather thong at the end of the tape around the hand, or by holding a taping pin which has been slipped through the eye at the end of the tape, or by using a clamp. When the rear tapeman has the 100-ft mark at the starting point and has satisfactorily aligned the head tapeman, he calls "all right" or some other such signal. The head tapeman pulls the tape tightly and sticks a taping pin in the ground at right angles to the tape and sloping at 20° to 30° from the vertical. If the measurement is done on pavement, a scratch can be made at the proper point or a taping pin can be taped down to the pavement.

The rear tapeman picks up his taping pin and the head tapeman pulls the tape down the line and the process is repeated for the next 100 ft. It will be noticed that the number of hundreds of feet which have been measured at any time equals the number of taping pins which the rear tapeman has in his possession. After 1000 ft has been measured, the head tapeman will have used his 11th pin and he calls "tally" or some equivalent word so that the rear tapeman will return the taping pins and they can start on the next 1000 ft.

When the end of the line is reached, the distance from the last taping pin to the end point will normally be a fractional part of the tape. For older tapes, the first foot of the tape (from 0 to 1 ft) is usually divided into tenths, as shown in Fig. 4-3. The head tapeman holds this part of the tape over the end point while the rear tapeman

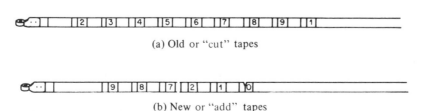

(a) Old or "cut" tapes

(b) New or "add" tapes

Figure 4-3

moves the tape backward or forward until he has a full foot mark at his taping pin.

The rear tapeman reads and calls out his foot mark, say 72 ft, and the head tapeman reads from the tape end the number of tenths and perhaps estimates to the nearest hundredth, say 0.46, and calls this out. This value is subtracted from 72 ft to give 71.54 ft and the number of hundreds of feet measured before is added. These numbers and the subtraction should be called out so the math can be checked by each partner.

For the newer steel tapes with the extra divided foot, the procedure is almost identical except that the rear tapeman would, for the example just described, hold the 71 ft mark at the taping pin in the ground. He would call out 71 and the head tapeman would read and call out plus 54 hundredths giving the same total of 71.54 ft.

A comment seems warranted here about practical significant figures as they apply to taping. If ordinary taping is being done and the total distance obtained for this line is 2771.34 ft, the 4 at the end is ridiculous and the distance should be recorded as 2771.3 or even 2771 ft because the work is just not done that precisely.

4-4 HORIZONTAL TAPING OVER SLOPING GROUND OR UNDERBRUSH

Ideally the tape should be supported for its full length on level ground or pavement. Unfortunately, such convenient conditions are often not available because the terrain being measured may be rough and covered with underbrush. For sloping, uneven ground or areas with much underbrush, taping is handled in a similar manner to taping over level ground. The tape is held horizontally but one or both tapemen must use a plumb bob, as shown in Fig. 4-4.

If taping is being done uphill, the rear tapeman will have to hold his plumb bob over the last taping pin while the head tapeman may be able to hold his end on the ground [Fig. 4-4(a)]. If they are moving

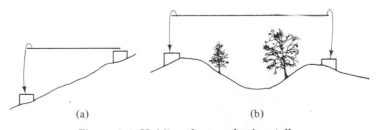

(a)   (b)

Figure 4-4 Holding the tape horizontally.

downhill, the rear tapeman may be able to hold his end on the ground while the head tapeman has to use a plumb bob. If the measurement is over uneven ground or ground where there is considerable underbrush, both tapemen may very well have to use plumb bobs as they hold their respective ends of the tape above the ground [Fig. 4-4(b)].

Considerable practice is required for a person to be able to do precise taping in rolling or hilly country. Although for many surveys the tapemen may estimate the horizontal by eye, it pays to use a hand level for this purpose. Where there are steep slopes it is difficult to estimate by eye when the tape is horizontal because the common tendency is for the downhill man to hold his end much too low causing significant error. If a precision of better than approximately 1/2500 or 1/3000 is desired in rolling country, holding the tape horizontally by estimation will not be sufficient. See Fig. 4-5.

Another problem in holding the tape above the ground is the error

Figure 4-5 Holding the tape horizontally.

caused by sagging of the tape (see Sec. 5-4). Note that both of these errors (tape not horizontal and sag) will cause the surveyor to get too much distance. In other words, either it takes him more tape lengths to cover a certain distance or he does not move forward a full 100 ft each time he uses the tape.

If the slope is less than approximately 5 ft per 100 ft (a height above the ground at which the average tapeman can comfortably hold the tape), the tapemen can measure a full 100-ft tape length at a time. If they are taping downhill, the head tapeman holds the plumb bob string at the 0-end of the tape with the plumb bob a few inches above the ground. When the rear tapeman is ready at his end, the head tapeman is lined up on the distant point, and when the tape is horizontal and pulled to the desired tension, the head tapeman lets the plumb bob fall to the ground and sets a taping pin at that point.

For slopes greater than approximately 5 ft per 100 ft, the tapeman will be able to hold horizontally only parts of the tape at a time. Assuming that they are proceeding downhill, the head tapeman pulls the tape along the line for its full length and then leaving the tape on the ground returns as far along the tape as necessary for them to hold horizontally the part of the tape from his point to the rear tapeman.

The head tapeman holds the plumb bob string over a whole foot mark and when the tape is stretched, lined, and horizontal, he lets the plumb bob fall and sets a taping pin. He holds the intermediate foot mark on the tape until the rear tapeman arrives, at which time he hands the tape to the rear tapeman with the foot mark which he has been holding. This careful procedure is followed because it is so easy for the head tapeman to forget which foot he was holding if he drops it and walks ahead. The tapemen repeat this process for as much more of the tape as they can hold horizontally until they reach the 0-end of the tape.

This process of measuring with sections of the tape is referred to as "breaking tape" or "breaking chain." If the head tapeman follows the customary procedure of leaving a taping pin at each of the positions that he occupies when "breaking tape," counting the number of hundreds of feet taped (as represented by the number of pins in the possession of the rear tapeman) would be confusing. Therefore, at each of the intermediate points the head tapeman sets a pin in the ground and then takes one pin from the rear tapeman.

4-5 TAPING ALONG SLOPES, TAPE NOT HORIZONTAL

Occasionally it may be more convenient or more efficient to tape along sloping ground with the tape held inclined along the slope. This procedure has long been common for underground mine surveys but to a

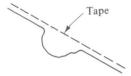

Figure 4-6

much lesser extent for surface surveys. Slope taping is quicker than horizontal taping and is considerably more precise because it eliminates plumbing with its consequent accidental errors. Taping along slopes is sometimes useful when the surveyor is working along fairly smooth slopes or when he wants to improve his precision. Nevertheless, the method is generally not used because of the problem of correcting slope distances to horizontal values. This is particularly true in rough terrain where slopes are constantly varying and the problem of determining the magnitude of the slopes is difficult.

In some cases it may be impossible to hold the entire tape (or even a small part of it) horizontally. This may occur when taping is being done across a ravine (see Fig. 4-6) or some other obstacle where one tapeman is much lower than the other one and where it is not feasible to "break tape." Here it may be practical to hold both ends of the tape on the ground.

Some means must be used for measuring the magnitudes of the slopes. A clinometer (Fig. 4-7) or a transit or a theodolite may be used for measuring the vertical angles involved. With this information the horizontal components can be computed. It is usually simpler, however, to determine elevation differences and compute slope corrections as described in Section 4-6.

Figure 4-7 Hand level and clinometer.

4-6 SLOPE CORRECTIONS

In Fig. 4-8 a tape of length s is stretched along a slope and it is desired to determine the horizontal distance h which has been measured.

It is easy for tapemen to apply an approximate correction formula for most slopes. The expression, derived below, is satisfactory for most measurements, but for slopes of greater than approximately 10 to 15%, an exact trigonometric expression or the Pythagorean theorem should be used. When a 100-ft slope distance is measured, the use of this approximate expression will cause an error of 0.0013 ft for a 10% slope and a 0.0064-ft error for a 15% slope.

It is desired to write an expression for the correction C shown in Fig. 4-8. This value, which equals $s - h$ in the figure, is written in a more practical form by using the Pythagorean theorem as follows:

$$s^2 = h^2 + v^2$$
$$v^2 = s^2 - h^2$$

from which

$$v^2 = (s - h)(s + h)$$
$$s - h = \frac{v^2}{s + h}$$

and since

$$C = s - h$$
$$C = \frac{v^2}{s + h}$$

For the typical 100-ft tape, s equals 100 ft and h varies from 100 ft by a very small value. For practical purposes, therefore, h can also be assumed to equal 100 ft when the slope correction expression is applied.

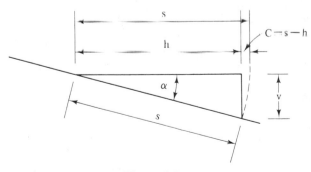

Figure 4-8

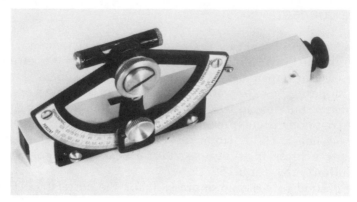

Figure 4-9 Topographic abney level. (*Courtesy The Leitz Company.*)

It is written for 100-ft tapes in the form

$$C = \frac{v^2}{200}$$

In taping it is normally convenient to measure a full tape length at a time. Therefore, in measuring along a slope it is often convenient for the head tapeman to calculate (probably in his head) the correction and set the taping pin that distance beyond the end of the tape so that they will have measured 100 ft horizontally. For a 6-ft vertical elevation difference

$$C = \frac{(6)^2}{200} = 0.18 \text{ ft}$$

For tape lengths other than 100 ft, the correction expression can be written as

$$C = \frac{v^2}{2s}$$

Sometimes for a long constant slope the tape is held on the ground and the correction is made for the entire length. Such a situation is illustrated in Example 4-1. The reader should carefully note that the correction formula was derived for a single tape length. For a distance of more than one tape length, the total correction will equal the number of tape lengths times the correction per tape length.

Example 4-1

A distance was measured on an 8% slope and found to be 2620.30 ft. What is the horizontal distance measured?

Solution   Correction per tape length $= -\dfrac{(8)^2}{(2)(100)} = -0.32$ ft

Total correction $= (26.2030)(-0.32) = -8.38$ ft

Horizontal distance $= 2620.30 - 8.38 = \mathbf{2611.92\ ft}$

5 Distance Corrections

5-1 TYPES OF CORRECTIONS

The five major areas in which the surveyor may need to apply corrections either in measuring or in laying out lines with a tape are as follows:

1. Wrong length tape or standardization error
2. Temperature variations
3. Sag
4. Slope
5. Incorrect tension

This chapter is devoted to a discussion of these corrections with the exception of slope which was presented in Chapter 4.

5-2 WRONG LENGTH TAPE OR STANDARDIZATION ERROR

Although steel tapes are manufactured to very precise lengths, with use they become kinked, worn, and imperfectly repaired after breaks. The net result is that tapes may vary by quite a few hundredths of a foot from their desired lengths. Therefore, it is wise to periodically check tape lengths against a standard. There are several ways in which this might be done. For instance, many surveying offices keep one standardized tape (perhaps an Invar type) which is used only for checking the lengths of their other tapes. Some companies take a tape which has been standardized at 100.00 ft and use it to place marks 100.00 ft apart on a concrete curb, sidewalk, or pavement. The marks are frequently used for "standardizing" their tapes. They feel that the length between the marked points will not change appreciably as temperatures vary because of the mass of the concrete and the friction of the earth.

These practices are advisable for surveyors who have extensive practices and they yield very satisfactory results for surveys where ordinary precision is desired, but they are probably not sufficient for extremely precise work. For such work, tapes can be mailed to the Bureau of Standards in Washington, D.C. For a rather large fee, they will determine the length of a tape for any specific tension and support conditions. They issue a certificate for each tape giving its length to the nearest 0.001 ft at 68°F (20°C).

Several municipal governments around the U.S., various state agencies, and a good many universities will standardize tapes, occasionally free as a service to the public. In addition, the NGS has established base lines at various locations across the country where tapes and EDM equipment can be calibrated by the surveyor.

If a tape proves to be in appreciable error from the standard, the surveyor must correct his measurements by the required amounts. He will note carefully whether such a correction is positive or negative, as explained in the following paragraphs.

The important point to grasp in making corrections is that the tape "says zero ft at one end and 100 ft at the other end" even though its correct length is 99.94 ft, 100.10 ft, or some other value. If the surveyor (ignorant of the tape's true length) uses this tape 10 times, he thinks he has measured a distance of 1000 ft, but he really has measured 10 times the actual tape length.

In measuring a given distance with a tape which is too long, the surveyor will not obtain a large enough value for his measurement and will have to make a positive correction. In other words, if the tape is too long, it will take fewer tape lengths to measure a distance than would be required for a shorter and correct length tape. For a tape that is too short, the reverse is true and a negative correction is required. It should be simple enough to remember this rule: *Tape too long, add; tape too short, subtract.*

Examples 5-1, 5-2, and 5-3 illustrate the correction of taped distances caused by wrong tape lengths. The problem of Example 5-3 is stated backward from the ones of Examples 5-1 and 5-2, and the sign of the correction is therefore reversed.

Example 5-1

A distance is measured with a 100-ft steel tape and is found to be 896.24 ft. Later the tape is standardized and is found to have an actual length of 100.04 ft. What is the correct distance measured?

Solution   The tape is too long and a + correction of 0.04 ft must be made for each tape length as follows:

$$\text{Measured value} = 896.24 \text{ ft}$$

$$\text{Total correction} = +(0.04)(8.9624) = \underline{+0.36 \text{ ft}}$$

$$\text{Corrected distance} = \mathbf{896.60 \text{ ft}}$$

Alternate Solution   Obviously, the distance measured equals the number of tape lengths times the actual length of the tape. In this case, it took 8.9624 tape lengths to cover the distance and each tape length was 100.04 ft.

$$\text{Distance measured} = (8.9624)(100.04) = 896.60 \text{ ft}$$

Example 5-2

A distance is measured with a 100-ft steel tape and is found to be 2320.30 ft. Later the tape is standardized and is found to have an actual length of 99.97 ft. What is the correct distance measured?

Solution   The tape is too short. Therefore, the correction is minus.

$$\text{Measured value} = 2320.30 \text{ ft}$$

$$\text{Total correction} = -(0.03)(23.2030) = \underline{-0.70 \text{ ft}}$$

$$\text{Corrected distance} = \mathbf{2319.60 \text{ ft}}$$

Example 5-3

It is desired to lay off a dimension of 1200.00 ft with a steel tape which has an actual length of 99.95 ft. What field measurement should be made with this tape so that the correct distance is obtained?

Solution   This problem is stated exactly opposite to the ones of Examples 5-1 and 5-2. It is obvious that if the tape is used 12 times, the distance measured (12×99.95) is less than the 1200 ft desired, and a correction of the number of tape lengths times the error per tape length must be *added*.

$$12 \text{ tape lengths} = 12 \times 100.00 = 1200.00 \text{ ft}$$

$$+12 \times 0.05 = \underline{+0.60 \text{ ft}}$$

$$\text{Field measurement} = 1200.60 \text{ ft}$$

Check   The answer can be checked by considering the problem in reverse. Here a distance has been measured as being 1200.60 ft with a tape 99.95 ft long. What actual distance was measured? The solution is as follows:

$$\text{Measured value} = 1200.60 \text{ ft}$$

$$\text{Total correction} = -(0.05)(12.006) = \underline{-0.60 \text{ ft}}$$

$$\text{Corrected distance} = \mathbf{1200.00 \text{ ft}}$$

One final example of this type of corrected measurement is shown with the field notes of Fig. 5-1 where the sides of the traverse previously

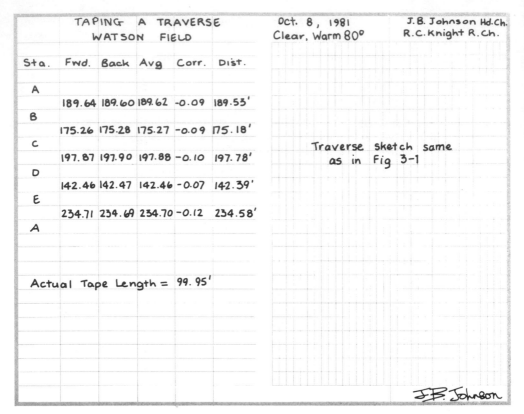

Sta.	Fwd.	Back	Avg	Corr.	Dist.
A					
	189.64	189.60	189.62	-0.09	189.53'
B					
	175.26	175.28	175.27	-0.09	175.18'
C					
	197.87	197.90	197.88	-0.10	197.78'
D					
	142.46	142.47	142.46	-0.07	142.39'
E					
	234.71	234.69	234.70	-0.12	234.58'
A					

TAPING A TRAVERSE
WATSON FIELD

Oct. 8, 1981
Clear, Warm 80°

J. B. Johnson Hd.Ch.
R.C. Knight R.Ch.

Traverse sketch same
as in Fig 3-1

Actual Tape Length = 99.95'

J.B. Johnson

Figure 5-1

paced (shown in Fig. 3-1) are measured with a tape of actual length 99.95 ft. In this case each of the sides was taped twice (forward and back), and an average value was obtained before the wrong length tape correction was applied.

5-3 TEMPERATURE VARIATIONS

Changes in tape lengths caused by temperature variations can be significant even for ordinary surveys. For precise work they are of critical importance. A temperature change of approximately 15°F will cause a change in length of approximately 0.01 ft in a 100-ft tape. If a tape is used at 20°F to lay off a distance of one mile (no temperature correction being made) and if the distance is checked the following summer with the same tape when the temperature is 100°F, there will be a difference in length of 2.75 feet caused by the temperature variation. Such an error alone would be equivalent to a precision of 2.75/5280 = 1/1920 (not so good).

Steel tapes lengthen with rising temperatures and shorten with falling ones. The coefficient of linear expansion for steel tapes is 0.0000065 per degree Fahrenheit. This means that for a 1°F rise in temperature a tape will increase in length by 0.0000065 times its length.

As described in Section 5-2, the standardized length of a tape is determined at 68°F. A tape which is 100.00 ft long at the standard temperature will at 100°F have a length of 100.00 + (32) (0.0000065) (100) = 100.02 ft. The correction of a distance measured at 100°F with this tape can be made as previously described for wrong-length tapes. The correction of a tape for temperature changes can be expressed with the formula

$$C_t = 0.0000065 \, (T - T_s)(L)$$

In this expression C_t is the change in length of the tape, T is the estimated temperature of the tape at the time of measurement, and T_s is the standardized temperature and L is the tape length.

It is clear that a steel tape used on a hot summer day in the bright sunshine will have a much higher temperature than will the surrounding air. Actually though, partly cloudy summer days will cause the most troublesome variations in length. For a few minutes the sun shines brightly and then it is covered for a while by clouds causing the tape to cool quickly, perhaps as much as 20°F or 30°F. Accurate corrections for tape temperature variations are difficult to make because the tape temperature may vary along its length with sun, shade, dampness (in grass, on ground), and so on. It has been shown that a few degrees variation may make an appreciable variation in the measurement of distance.

For the best precision it is desirable to tape on cloudy days, early in the mornings or late in the afternoons in order to minimize temperature variations. Furthermore, the use of invar tapes with their very small coefficients of expansion (0.0000001 to 0.0000002) is very helpful for precise work.

For very precise surveying, tape measurements are recorded and the proper corrections made. On cloudy, hazy days an ordinary thermometer may be used for measuring the air temperature, but on bright sunny days the temperature of the tape itself should be determined. For this purpose plastic thermometers taped to the tapes near the ends (so that their weights do not appreciably affect sag) should be used.

Regular steel tapes have not been used for geodetic work for quite a few decades because of their rather large coefficients of expansion and because of the impossibility of accurately determining their temperature during daytime operations. Until electronic distance-measuring devices were introduced almost all of the base-line measurements of the National Geodetic Survey during the twentieth century were done with Invar tapes.   Today almost all of their length measurements are made with EDMs.

If SI units are being used, the coefficient of linear expansion is 0.000 011 6 per degree Celsius ($^\circ$C).   The correction in length of a metric tape for temperature changes can be expressed by the formula

$$C_t = (0.000\ 011\ 6)(T - T_s)(L)$$

where T is the standardized temperature of the tape at manufacture (usually 20° C), T_s is the temperature of the tape at the time of measurement and L is the tape length.

It will be remembered that the expressions for temperature conversion are as follows:

$$^\circ C = \tfrac{5}{9}\ (^\circ F - 32)$$

$$^\circ F = \tfrac{9}{5}\ (^\circ C) + 32$$

5-4 SAG

When a steel tape is supported only at its ends, it will sag into a curved shape known as the *catenary*.   The obvious result is that the horizontal distance between its ends is less than when the tape is supported for its entire length.

To determine the difference in the length measured with a fully supported tape and one supported only at its end or at certain intervals, the following approximate expression may be used:

$$C_s = -\ \frac{w^2 L^3}{24 P^2} = -\ \frac{W^2 L}{24 P^2}$$

where C_s = Correction in feet;
  w = Weight of tape in pounds per foot;
  L = Unsupported length of tape in feet;
  $W = wL$ = Total weight of tape between supports;
  P = Total tension in pounds applied to the tape.

This expression, although approximate, is sufficiently accurate for surveying purposes.   It is applicable to horizontal taping or to tapes held along slopes of not more than approximately 10°.   *The effect of stretching of the metal in the tape caused by tension is neglected here, but it is considered in Section 5-5.*

To minimize sag errors, it is possible to use this formula and apply the appropriate corrections to the observed distance.   Another and more practical procedure for ordinary surveying is to increase the pull or tension on the tape in order to attempt to compensate for the effect of sag.   For very precise work, the tape is either supported at sufficient intervals to make sag effects negligible or it is standardized for the pull and manner of support to be used in the field. Examples 5-4 through 5-6 illustrate the application of the sag correction formula. The reader should particularly note in these examples the great reduction in the length correction when the tape tension is increased. The greater the tensile force applied to a tape, the longer the tape itself will become. Such stretching is neglected in the next three examples but is considered in the next section of this chapter.

Example 5-4

A steel tape weighing 2 lb is 100.000 ft long when supported continuously on a floor and pulled with a tensile force of 10 lb.

    1. If the tape is lifted from the floor and held at its ends only with the same pull, what is the distance between its ends?

    2. Repeat the problem if the pull is increased to 30 lb.

Solution   (neglecting stretching of tape due to tension)

    1. C_s = Correction in length = $-\dfrac{W^2 L}{24\, P_1^2}$

$$C_s = -\frac{(2)^2(100)}{(24)(10)^2} = -0.167 \text{ ft}$$

Corrected distance = 100.000 − 0.167 = **99.833 ft**

    2. $C_s = -\dfrac{(2)^2(100)}{(24)(30)^2} = -0.019$ ft

Corrected distance = 100.000 − 0.19 = **99.981 ft**

Example 5-5

If the tape of Example 5-4 is supported at its ends and at mid-point, as shown in Fig. 5-2, and has a 10-lb pull, what is the corrected distance between its ends? Neglect stretching of the tape.

Solution   The correction is made for each of the 50-ft spans and the weight of the tape is 1 lb for each span.

$$\text{Correction for 50-ft span} = -\frac{(1)^2(50)}{(24)(10)^2} = -0.021 \text{ ft}$$

$$\text{Total correction} = (2)(-0.021) = -0.042 \text{ ft}$$

$$\text{Corrected distance} = 100.000 - 0.042 = \textbf{99.958 ft}$$

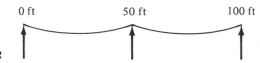

0 ft   50 ft   100 ft

Figure 5-2

This expression for the correction of tape lengths caused by sag is applicable to situations where either standard English units of distance or SI units are used, as long as the weights and tension values (W and P) are applied in consistent units. Example 5-6 illustrates the shortening of a 30-m tape due to sag when the tape is held above the ground.

Example 5-6

A steel tape weighing 0.910 kg is 30.000 m long when supported continuously across a floor and pulled with a tensile force of 5 kg. What is the sag correction when the tape is held above the ground, supported at its ends only and pulled with a tensile force of 8 kg? Neglect stretching of tape.

Solution

$$C_s = -\frac{(0.910)^2(30.000)}{(24)(8)^2} = -0.016 \text{ m}$$

Corrected distance = 30.000 - 0.016 = **29.984 m**

5-5 TENSION CORRECTIONS

Variations in tension.   A steel tape stretches when it is pulled, and if the pull is greater than that for which it was standardized, the tape will be too long. If insufficient tension is applied, the tape will be too short.   A 100-ft steel tape will change in length by approximately 0.01 ft for a 15-lb change in pull.   Since variations in pull of this magnitude are improbable, errors caused by tension variations are negligible for all except the most precise chaining.   Furthermore, these errors are accidental and tend to some degree to cancel.   For precise taping, spring balances are used so that certain prescribed tensile forces can be applied to the tapes.   With such balances it is not difficult to apply tensions within $\frac{1}{2}$ lb or closer to desired values.   As with all measuring devices it is necessary to check periodically the tension apparatus against a known standard.

Despite the minor significance of tension errors, a general understanding of them is important to the surveyor and will serve to improve the quality of his work.   The actual elongation of a tape in tension equals the tensile stress in psi over the modulus of elasticity of the steel (29,000,000 psi or 2,050,000 kg/cm^2) times the length of the tape. In the following expression the elongation of the tape in feet is represented by C_p, P_1 is the pull on the tape, A is the cross-sectional area in square inches, L is the length in feet, and E is the modulus of elasticity of the steel in psi.

$$C_p = \frac{\dfrac{P_1}{A}}{E} L = \frac{P_1 L}{AE}$$

It will be noted that the tape has been standardized at a certain pull P and, therefore, the change in length from the standardized situation is desired and the expression is written as

$$C_p = \frac{(P_1 - P)L}{AE}$$

Normal tension.   If, when the tape is suspended, it is pulled very tightly, there will be an appreciable reduction in sag and some increase in the tape length because of tension.   As a matter of fact, there is a theoretical pull for each tape at which the lengthening of the tape caused by tension equals its shortening caused by sag.   This value is referred to as the *normal tension*.   Its magnitude can be practically measured for a particular tape or it can be computed theoretically as described in the following paragraphs.

A tape may be placed on a floor or pavement, tensioned at its standardized pull, and have its ends marked on the slab.   The tape may then be held in the air above the slab supported at its ends only and pulled until its ends (as marked with plumb bobs) coincide with the marked points on the slab.   The pull necessary to make the end points

Figure 5-3 Precise distance measurements with Invar tape using spring balance for tension, taping stands or posts, and metal marking strips. (*Courtesy National Geodetic Survey.*)

coincide is the normal tension. Its value may be measured with spring balances.

A theoretical method of determining the normal tension is to equate the expression for elongation of the tape caused by tension to the expression for shortening of the tape caused by sag. The resulting expression can be solved for P_1, the normal tension:

$$P_1 = \frac{0.204\,W\sqrt{AE}}{\sqrt{P_1 - P}}$$

P_1 occurs on both sides of the equation, but its value for a particular tape may be determined by a trial-and-error method. For a normal weight 100-ft tape, this value will probably be in the range of 20 lb. As described in detail in Section 6-2, most distance measurements are too large because of the cumulative errors of sag, poor alignment, slope, and so on. As a result, overpulling the tape is a good idea for ordinary surveying because it tends to reduce some of these errors and improve the precision of the work. For such surveys, an estimated pull of approximately 30 lb is often recommended. For very precise surveying, the normal tension is applied to the tape by using accurate spring balances. See Fig. 5-3.

5-6 COMBINED TAPING CORRECTIONS

If corrections must be made for several factors at the same time (for example, wrong length tape, slope, temperature), the individual corrections per tape length may be computed separately and added together (taking into account their signs) in order to obtain a combined correction for all. Since each correction will be relatively small, it is assumed that they do not appreciably affect each other and each can be computed independently. Furthermore, the nominal tape length (100 ft) may be used for the calculation. This means that although the tape may be 99.92 ft long at 68°F and a temperature correction is to be made for a 40°F increase in temperature, the increase in tape length can be figured as (40)(0.0000065)(100) without having to use (40)(0.0000065)(99.92).

Example 5-7 illustrates the application of several corrections to a single distance measurement.

Example 5-7

A distance was measured on a uniform slope of 8% and was found to be 1665.2 ft. No field slope corrections were made. The tape temperature at the time of measurement was 18°F. What is the correct horizontal distance measured if the tape is 100.06 ft long at 68°F?

Solution   The corrections per tape length are computed, added together, and then multiplied by the number of tape lengths.

$$\text{Slope correction/tape length} - \frac{v^2}{200} = -\frac{(8)^2}{200} \qquad = \quad -0.32 \text{ ft}$$

$$\text{Temp. correction/tape length} - (50)(0.0000065)(100) = \quad -0.03 \text{ ft}$$

$$\text{Wrong length of tape/tape length} \qquad\qquad\qquad = \quad +0.06 \text{ ft}$$

$$\text{Total correction/tape length} \qquad\qquad\qquad\qquad = \quad -0.29 \text{ ft}$$

$$\text{Correction for entire distance} = (16.652)(-0.29) \quad = \quad -4.83 \text{ ft}$$

$$\text{Corrected distance} = 1665.2 - 4.83 \qquad\qquad\qquad = \textbf{1660.4 ft}$$

PROBLEMS

For each of these problems (5-1 to 5-5), distances were measured with tapes assumed to be 100.00 ft long. Later the tapes were standardized and found to have different tape lengths. Determine the correct distances measured in each case.

	Recorded Distance (ft)	Correct Tape Length (ft)	
5-1.	1692.30	100.05	(*Ans.:* 1693.15 ft)
5-2.	2893.20	99.96	
5-3.	1764.60	99.97	(*Ans.:* 1764.07 ft)
5-4.	802.73	100.04	
5-5.	4362.30	100.03	(*Ans.:* 4363.61 ft)

For each of these problems (5-6 to 5-8), distances were measured with tapes assumed to be 30.00 m long. Later the tapes were standardized and found to have different lengths. Determine the correct distances measured in each case.

	Recorded Distance (m)	Correct Tape Length (m)	
5-6.	416.32	30.02	
5-7.	719.87	30.03	(*Ans.:* 720.59 m)
5-8.	1 122.90	29.98	

5-9. The actual distance between two marks used at a university for standardizing tapes is 99.97 ft. When a certain tape was held along this line the surveyor, thinking that the distance between the marks was 100.00 ft, observed that his tape was 99.93 ft long. What is the correct length of the tape?

(*Ans.:* 99.90 ft)

5-10. Repeat Problem 5-9 for values of 100.04, 100.00 ft and 100.12 ft, respectively.

5-11. Repeat Problem 5-9 for values of 29.96 m, 30.00 m and 30.03 m, respectively.
(*Ans.:* 29.99 m)

For each of these problems (5-12 to 5-17), it is desired to lay off certain horizontal distances for building layouts. The lengths of the tapes used in Problems 5-12 through 5-15 are not 100.00 ft and not 30.00 m for Problems 5-16 and 5-17. Determine the field dimensions (or actual tape readings) which should be used with the incorrect tape lengths so that the correct dimensions are obtained.

	Desired Dimension	Correct Tape Length	
5-12.	500.00 ft × 140.00 ft	100.08 ft	
5-13.	450.00 ft × 240.00 ft	99.96 ft	(*Ans.:* 450.18 ft × 240.10 ft)
5-14.	350.00 ft × 860.00 ft	100.07 ft	
5-15.	187.20 ft × 301.68 ft	99.93 ft	(*Ans.:* 187.33 ft × 301.89 ft)
5-16.	80.00 m × 120.00 m	29.96 m	
5-17.	200.00 m × 340.50 m	30.03 m	(*Ans.:* 199.80 m × 340.16 m)

5-18. It is desired to lay off a horizontal distance equal to 704.40 ft with a tape that is 49.90 ft long (and not 50.00 as indicated on the scale). What should the recorded distance be?

5-19. A 50-ft woven tape is used to set the corners for a building. If the tape is actually 50.12 ft long, what should the recorded distances be if the building is to be 182.00 ft by 97.50 ft?   (*Ans.:* 181.56 ft × 97.27 ft)

5-20. A distance is measured through rough country and is found to be 3476.2 ft. If on the average a plumb bob is used every 50 ft with a probable error of ±0.02 ft, what is the probable total error in the whole distance?

5-21. Rework Problem 5-20 if the plumb bob is used every 30 ft on the average and the distance measured is 1792.3 ft.   (*Ans.:* ±0.15 ft)

For Problems 5-22 to 5-25, distances were measured with 100-ft steel tapes and their average temperatures estimated. From these values and the standardized tape lengths (100.000 ft at 68°F), determine the correct distances measured.

	Recorded Distance (ft)	Average Tape Temperature @ Time of Measurement	
5-22.	2269.70	18°F	
5-23.	843.86	108°F	(*Ans.:* 844.08 ft)
5-24.	1772.10	98°F	
5-25.	2201.60	8°F	(*Ans.:* 2200.74 ft)

5-26. A steel tape which has a length of 100.000 ft at 68°F is to be used to lay off a building with the dimensions 230.00 ft by 410.00 ft.
   (a) What should be the tape readings if the tape temperature is 18°F at the time of the measurement?
   (b) Repeat part (a) if the tape temperature is 108°F.

For Problems 5-27 to 5-29, distances were measured with 30-m steel tapes and their average temperatures estimated. From these values and the standardized tape length (30.000 m at 20°C), determine the correct distances measured.

	Recorded Distance (m)	Average Tape Temperature @ Time of Measurement	
5-27.	240.31	40°C	(*Ans.:* 240.37 m)
5-28.	675.88	8°C	
5-29.	1 032.76	32°C	(*Ans.:* 1 032.90 m)

5-30. A distance was measured at a temperature of 8°F and was found to be 2166.68 ft. If the tape has a standardized length of 99.96 ft at 68°F, what is the correct distance measured? If the same distance were measured at a temperature of 108°F, what would be the probable value obtained, neglecting other errors?

5-31. A distance was measured with a 100-ft steel tape at a temperature of 98°F and found to be 2492.32 ft. The next winter the distance was remeasured with the same tape at a temperature of 18°F and was found to be 2493.73 ft. What part of the discrepancy between the two measurements should be caused by the temperature difference?   (*Ans.:* 1.30 ft)

5-32. A distance is measured as 2223.40 ft when the tape temperature is 28°F. If the same distance is measured again with the same tape but at a temperature of 98°F, what distance should be expected neglecting other errors? If the tape is 100.00 ft long at 68°F, what is the "true" distance measured?

5-33. A 100.00-ft tape is used to measure an inclined distance and the value determined is 3106.34 ft. If the slope is 6%, what is the correct horizontal distance obtained using the slope correction formula? (*Ans.:* 3100.75 ft)

5-34. A 200.00-ft tape is used to measure an inclined distance and the value determined is 4109.4 ft. If the slope is 4%, what is the correct horizontal distance? Use an appropriate slope correction formula.

5-35. Repeat Problem 5-34 if the measured value is 6210.60 ft and the slope is 5%.
(*Ans.:* 6202.84 ft)

5-36. A 30.00-m tape is used to measure an inclined distance and the value determined is 1 642.32 m. If the slope is 6%, what is the correct horizontal distance? Use an appropriate slope correction formula.

5-37. A tape weighing 2 lb (A = 0.006 sq. in.) has a length of 100.00 ft under a pull of 10 lb when supported for its full length. The tape is used to measure a line holding the tape on the ground with a pull of 35 lb. If the value obtained is 2686.45 ft, what should the correct distance be?   (*Ans.:* 2686.84 ft)

5-38. A tape weighing 2 lb (A = 0.006 sq. in.) has a length of 100.00 ft under a pull of 10 lb when supported for its full length. This tape is used to measure a line holding the tape on the ground with a pull of 35 lb. If the value obtained is 3102.68 ft, what should the correct distance be?

5-39. A 2-lb steel tape (with a cross-sectional area of 0.006 sq. in.) is 100.00 ft long when it is supported for its full length and subjected to a pull of 10 lb. What is the length of this tape when it is supported only at its ends and subjected to a pull of 40 lb?   (*Ans.:* 100.01 ft)

5-40. Repeat Problem 5-37 if the tape is supported at its ends and center line.

5-41. Rework Problem 5-38 if the tape is supported at its ends only and the pull is 20 lb.   (*Ans.:* 3101.57 ft)

5-42. Rework Problem 5-38 if the tape is supported at its ends and center line and if the pull is 30 lb.

5-43. A 30-m steel tape weighs 0.030 3 kg/m. Determine the sag effect if a tension of 10 kg is applied and the tape is supported at its ends only.

(Ans.: −0.010 m)

5-44. A tape weighing 0.910 kg and with a cross-sectional area of 3.87 mm² has a length of 30.00 m under a pull of 5 kg when supported for its full length. This tape is used to measure a line holding the tape on the ground with a tension of 12 kg. If the value obtained is 903.26 m and E = 2 050 000 kg/cm², what is the correct distance measured?

5-45. A tape weighing 0.910 kg and with a cross-sectional area of 3.87 mm² has a length of 30.00 m under a pull of 5 kg when supported for its full length. This tape is used to measure a distance with its ends only supported and E = 2 050 000 kg/cm² with a pull of 10 kg. If the value obtained is 1 168.32 m, what should the correct distance be?   (*Ans.:* 1 167.99 m)

5-46. A 100-ft tape weights 2 lb, has a cross-sectional area equal to 0.006 sq. in. and is 100.00 ft long at 68°F while subjected to a 10-lb pull and supported for its full length. A distance is measured in the field with this tape and found to be 838.49 ft. If the pull on the tape is 20 lb and if the tape is supported only at its ends, what is the correct distance measured?

In Problems 5-47 to 5-50, a 2-lb steel tape is 99.95 ft long at 68°F while supported for its full length with a 10-lb pull. It has a cross-sectional area of 0.006 sq. in. and $E = 29 \times 10^6$ psi. For these problems determine the correct value for each length assuming the tape is supported only at its ends.

	Recorded Distance (ft)	Tape Temperature (°F)	Pull on Tape (lb)	Slope	
5-47.	1697.32	108	30	4%	(*Ans.:* 1695.44 ft)
5-48.	3284.60	98	35	6%	
5-49.	1200.20	38	30	5%	(*Ans.:* 1197.78 ft)
5-50.	963.68	18	40	5%	

In Problems 5-51 to 5-53, a 0.910 kg steel tape is 30.03 m long at 20°C while supported for its full length with a 5-kg pull. It has a cross-sectional area of 3.87 mm² and $E = 2\ 050\ 000$ kg/cm². For these problems determine the correct value for each distance.

	Recorded Distance (m)	Tape Temperature (°C)	Pull on Tape (kg)	Slope	
5-51.	642.90	24	12	4%	(*Ans.*: 642.96 m)
5-52.	1 369.80	10	10	6%	
5-53.	459.32	28	12	5%	(*Ans.*: 459.18 m)

5-54. A 2-lb steel tape ($A = 0.006$ sq. in.) has a length of 100.00 ft at a temperature of 68°F when supported for its entire length and subject to an applied total tension of 10 lb.   With this tape a distance is measured as being 100.00 ft long under the following conditions:

(a) supported at its ends only

(b) total applied tension 30 lb

(c) alignment off by 2.00 ft (bent around tree on line at ₵)

(d) tape not horizontal by 4.00 ft

(e) average tape temperature 18°F

What is the correct distance measured?

6

Mistakes, Errors, and Miscellaneous Taping Operations

6-1 COMMON MISTAKES MADE IN TAPING

Some of the most common mistakes made in taping are described in this section, and a method of eliminating each is suggested.

Reading tape wrong.   A frequent mistake made by tapemen is reading the wrong number on the tape, for example, reading a 6 instead of a 9 or a 9 instead of a 6.   As tapes become older these mistakes become more frequent because the numbers on the tape become worn.   These blunders can be eliminated if tapemen develop the simple habit of looking at the adjacent numbers on the tape when readings are taken.

Recording numbers.   Occasionally the recorder will misunderstand a measurement that is called out to him.   In order to prevent this kind of mistake, he can repeat aloud the values, including the decimals, as he records them.

Missing a tape length.   It is not very difficult to lose or gain a tape length in measuring long distances.   The careful use of taping pins, previously described in Section 4-3, should prevent this mistake.   In addition, the surveyor can many times eliminate such mistakes by cultivating the habit of estimating distances by eye or by pacing or better by taking stadia readings whenever possible.

Mistaking end point of tape.   Some tapes are manufactured with the 0- and 100-ft points at the very ends of the tapes. Other tapes have them at a little distance from the ends. Clearly, tapemen should not make mistakes like these if they have taken the time to examine the tape before they begin to take measurements.

One-foot mistakes.   When a fractional part of a tape is being used at the end of a line, it is possible to make a one-foot mistake. Mistakes like these can be prevented by carefully following the procedure described for such measurements in Section 4-3. Also helpful are the habits of calling out the numbers and checking the adjacent numbers on the tape.

6-2 ERRORS IN TAPING

The following paragraphs present a brief discussion of the common taping errors. As these errors are studied it is important to notice that the effect of most of them is to cause the surveyor to get too much distance. If the tape is not properly aligned, not horizontal, or sags too much, if a strong wind is blowing the tape to one side, or if the tape has shortened on a cold day, it takes more tape lengths to cover the distance.

Alignment of tape.   A good rear tapeman can align the head tapeman with sufficient accuracy for most surveys, although it is more accurate to use the transit to keep the tape on line. In some cases it is necessary to use a transit when establishing new lines or when the tapemen are unable to see the ending point because of the roughness of the terrain. For this latter case it may be necessary to set up intermediate points on the line to guide the tapemen.

It is probable that most surveyors spend too much time improving their alignment, at least in proportion to the time they spend trying to reduce other more important errors. In taping a 100-ft distance the tape would have to be 1.414 ft out of line to cause an error of 0.01 ft. From this value it can be seen that for ordinary distances alignment errors should not be appreciable. As a matter of fact, experienced tapemen should have no difficulty in keeping their alignment well within a foot of the correct line by eye, particularly when the lines are only a few hundred feet or less in length.

Accidental taping errors.   Because of human imperfections tapemen cannot read tapes perfectly, cannot plumb perfectly, and cannot set taping pins perfectly. They will place the pins a little too far forward or a little too far back. These errors are accidental in nature and will tend to cancel each other somewhat. Generally, errors caused by setting pins and reading the tape are minor, but errors caused by plumbing may be very important. Their magnitudes can be reduced by increasing the care with which the work is done or by taping along slopes and applying slope corrections in order to avoid plumbing.

Tape not horizontal. If tapes are not held in the horizontal position, an error results which causes the surveyor to obtain distances that are too large. These errors are cumulative and can be quite large when surveying is done in hilly country. Here the surveyor must be very careful.

If a surveyor deliberately holds his tape along a slope, he can correct his measurement with the slope correction expression

$$C = \frac{v^2}{2s}$$

which was presented in Section 4-6. It might be noticed that if one end of a 100-ft tape is 1.414 ft above or below the other end, an error of

$$\frac{(1.414)^2}{200} = 0.01 \text{ ft}$$

is made. From this expression it can be seen that the error varies as the square of the elevation difference. If the elevation difference is doubled, the error quadruples. For a 2.828-ft elevation difference, the error made is

$$\frac{(2.828)^2}{200} = 0.04 \text{ ft}$$

Tape incorrect length. There important errors were discussed in Section 5-2 and must be given careful attention if good work is to be done. For a given tape of incorrect length, the errors are cumulative and can add up to sizable values.

Temperature variations. Corrections for variations in tape temperature were discussed in Section 5-3. Errors in taping caused by temperature changes are usually thought of as being cumulative for a single day. They may, however, be accidental under unusual circumstances with changing temperatures during the day and also with different temperatures at the same time in different parts of the tape. It is probably wiser to limit tape variations instead of trying to correct for them no matter how large they may be. Taping on cloudy days, early in the morning, or late in the afternoon or using invar tapes are effective means of limiting length changes caused by temperature variations.

Sag.   Sag effects (discussed in Section 5-4) cause the surveyor to obtain excessive distances.   Most surveyors attempt to reduce these errors by overpulling their tapes with a force which will stretch them sufficiently to counterbalance the sag effects. A rule of thumb used by many for 100-ft tapes is to apply an estimated pull of approximately 30 lb.   This practice is satisfactory for surveys of ordinary precision, but it is not adequate for those of high precision because the amount of pull required varies for different tapes, different support conditions, and so on. It is also difficult to estimate by hand the force being applied. A better method is to use a spring balance for applying a definite tension to a tape, the tension required having been calculated or determined by a standardized test to equal the normal tension of the tape.

Miscellaneous errors.   Some of the miscellaneous errors which affect the precision of taping are: (1) wind blowing plumb bobs, (2) wind blowing tape to one side causing the same effect as sag, and (3) taping pins not set exactly where plumb bobs touch ground.

6-3 MAGNITUDE OF ERRORS

To get some "feel" as to the effects of common taping errors, consider Table 6-1. In this table various sources of errors are listed together with the variations which would be necessary for each error to have a magnitude of ±0.01 ft when a 100-ft distance is measured with a 100-ft tape. The starred items (*) in the table may be greatly minimized if sufficient field data is obtained and appropriate mathematical corrections are applied.

TABLE 6-1 Errors of ±0.01 ft in 100-ft Measurements†

Source of Error	Magnitude of Error
1. Wrong length tape*	0.01 ft
2. Temperature variation*	15°F
3. Tension or pull variation*	15 lb
4. Sag*	7.5″ sag at center line rest of tape supported throughout
5. Alignment*	1.4 ft at one end
6. Tape not level*	1.4 ft difference in elevation
7. Plumbing	0.01 ft
8. Marking	0.01 ft
9. Reading tape	0.01 ft

†J. F. Dracup and C. F. Kelly, *Horizontal Control as Applied to Local Surveying Needs* (Falls Church, Virginia: American Congress on Surveying and Mapping, 1973), p. 16.

TABLE 6-2*

Source of Error	Magnitude of Error (ft)	Magnitude of Error Squared
Wrong length tape	±0.005	0.000025
Temperature variation (10°F)	±0.009	0.000081
Tension or pull variation (5 lb)	±0.003	0.000009
Plumbing (0.005 ft)	±0.005	0.000025
Marking (0.001 ft)	±0.001	0.000001
Reading tape (0.001 ft)	±0.001	0.000001
		Σ = 0.000142

Most probable error = $\sqrt{0.000142}$ = ±0.119 ft

Corresponding precision = $\dfrac{0.0119}{100}$ = $\dfrac{1}{8403}$

*J. F. Dracup and C. F. Kelly, *Horizontal Control as Applied to Local Surveying Needs* (Falls Church, Virginia: American Congress on Surveying and Mapping, 1973), p. 16.

For the discussion to follow it is assumed that a 100-ft steel tape is used to measure a distance of 100 ft. The ground is gently sloping and the entire length of the tape can be held horizontal at one time. In Table 6-2 several accidental errors and their magnitudes are given. At the bottom of the table the random error theory of Chapter 2 is used to estimate the magnitude of the most probable total error.

6-4 SUGGESTIONS FOR GOOD TAPING

If the surveyor studies the errors and mistakes that are made in taping, he should be able to develop a few rules of thumb that will appreciably improve the precision of his work. Following is a set of rules that have proved helpful in the field:

1. Tapemen should develop the habit of estimating by eye the distances which they are measuring because it will enable them to avoid most major blunders.
2. When reading a foot mark at some intermediate point on a tape, a tapeman should glance at the adjacent foot marks to be sure that he has read his mark correctly.
3. All points which the tapeman establishes should be checked. This is particularly true when the plumb bob has been used to set a point such as a tack in a stake.

4. It is easier to tape downhill whenever feasible. The rear tapeman can hold his tape end on the ground at the last point instead of having to hold his plumb bob over the point while the head tapeman is pulling against him, as would be the case in taping uphill.

5. If time permits (often it may not), distances should be taped twice, once forward and once back. Taping in the two different directions should prevent repeating the same mistakes.

6. Tapemen should assume stable positions when pulling the tape. This usually means feet widespread, leather thong wrapped around the hand (or use of taping clamp), standing to one side of the tape with arms in close to the body, and applying pull to the tape by leaning against it.

7. Since most taping gives distances that are too large, the surveyor can improve his work for ordinary surveys by pulling the tape very firmly, by estimating the smaller number when a reading seems to be halfway between two values, and perhaps even by setting taping pins slightly to the forward side.

6-5 MISCELLANEOUS TAPING OPERATIONS

In addition to the direct determination of the distance between two points, there are several miscellaneous operations that can be performed with a steel tape. They include:

Angle measurement. It is normally much wiser to measure angles with a transit or theodolite, but on some occasions such an instrument may not be available and a tape can be used to determine angle values within 5 to 10 minutes of arc. In Fig. 6-1 a convenient *horizontal* distance is measured down each line (say 100 ft), and the chord distance between points *a* and *b* is measured *horizontally*.

From trigonometry the following equation can be used to determine the magnitude of the angle α:

$$\sin \frac{1}{2}\alpha = \frac{\text{Chord length}}{200}$$

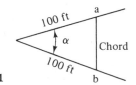

Figure 6-1

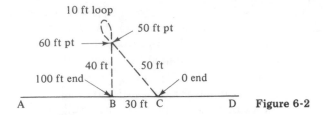

A   B   30 ft   C   D   **Figure 6-2**

Laying off right angles.   Occasionally a tape may be used for laying off a right angle (again assuming that a transit or theodolite is not readily available). This can be done quickly and precisely with a steel tape by working with the proportions of a 3-4-5 triangle, as shown in Fig. 6-2. It is assumed in this figure that a perpendicular is to be established at a point B on the straight line AD.

A distance of 30 ft is layed off from B to C, as shown in the figure. Then the 100-ft end of the tape is held at B and the tape is layed out so that a 10-ft loop is made from the 60-ft to the 50-ft marks and the 0-end is at B. When the tape is taut, the 60-ft and 50-ft marks will fall on a perpendicular to line AD passing through point B.

Erecting a perpendicular from a given point to an existing line.   Another use of steel tapes is the erection of a line from a given point perpendicular to an existing line. See Fig. 6-3, in which it is desired to erect a perpendicular line from point B to line AD. One method of doing this is to make use of the procedure just described for laying off right angles.

The point at which the perpendicular will hit line AD is estimated (C' in the figure). A right angle is layed off giving the line $B'C'$. This location is obviously in error, but it can be quickly adjusted by measuring the distance from B' to B and then measuring over the same distance from C' to C. Point C should then fall on the desired perpendicular BC. The right-angle procedure should again be used to check the perpendicularity of line BC.

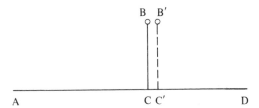

A   C C'   D   **Figure 6-3**

Figure 6-4

Taping obstructed lines.   The surveyor is continually faced with the problem of having something obstructing his work, for example, trees, buildings, and lakes.   Several methods of running straight lines where obstructions occur are described in the following paragraphs.

1. *Random Lines.*   It is desired to establish a straight line from A to B, but the surveyor cannot see from one point to the other, and he does not know the precise direction between them.   (See Fig. 6-4.)   A common method of handling the problem is for the surveyor to run convenient random straight lines working from the first point until he reaches the second one.   In other words, he starts at point A and runs a straight line in what seems to be the easiest way to proceed (that is fewest obstructions, reasonable slopes, and so on) in the general direction of B.   After some distance he sets a point (x in the figure) and proceeds to run another convenient straight line to y, and then to z, and so on until he reaches B.   It is necessary for him to tape each of these distances and measure the angles involved.   By a quick and convenient method called *latitudes and departures* (to be described in Chapter 14), he can compute the length and direction of the straight line AB and if necessary return to the field and establish the line on the ground.

2. *Offsetting.*   It is assumed that the surveyor is running a straight line.   He has already established points A, B, and C, but beyond C there is a tree along the line.   (See Fig. 6-5.)

    One method of handling the problem is to offset some convenient distance along the existing line, as at B and C.   The distances BB' and CC' are made with sufficient length to pass the tree (say 1 or 2 ft), and then a straight line is run through B' and C'

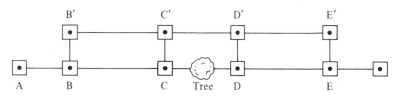

Figure 6-5

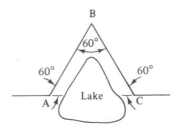

Figure 6-6

past the tree until points D' and E' are established. The offset distances $D'D$ and $E'E$ can then be established in order to get back on the original line. The surveyor, however, may choose to keep running the offset line until he encounters another tree at which time he can offset back to his original line. It should be obvious that the distances BC and DE should be fairly large and that the offset distances should be very short; otherwise the tape will not be too desirable for laying off the distance because of the problem laying off the 90°-angle with the tape.

3. *Equilateral Triangles.* A convenient method for getting around large obstacles such as lakes or large buildings involves using an equilateral triangle, as illustrated in Fig. 6-6. A 60°-angle is established at A and a sufficient distance is chained (AB) to pass the obstacle. Then a 60°-angle is layed off at B and the distance BC is taped equal to AB. Point C will be on the original line and the survey can continue. The distance desired, AC, will equal AB or BC.

4. *Other Triangles.* The student can easily think of several other precise methods of taping past an obstacle. For instance, in Fig. 6-7 a convenient angle α is laid off at B and a convenient distance l is taped. Then, as shown at C, the angle 2α is turned and the distance l is measured to point D. This puts the surveyor back on the original line AE. By measuring the angle α shown at D, he will be able to move on toward point E. From trigonometry, the distance BD equals $2l \cos \alpha$.

5. *Triangulation.* To obtain the distance from B to D across a large

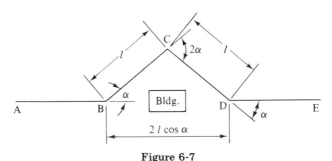

Figure 6-7

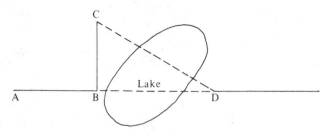

Figure 6-8

body of water (see Fig. 6-8), the distance *BC* can be layed off in a convenient direction, perhaps perpendicular to line *AB*, the distance taped, the angles at *B* and *C* measured, and the distance *BD* computed.   For checking, one can repeat the process from the other side of the lake. For better accuracy, it is desirable for the distance *BC* to equal at least one-half of the unknown distance *BD*. This type of problem is further discussed and numerically illustrated in Chapter 23.

6-6 TAPING PRECISION

The presentation of definite values that should be obtained for good, average, and poor taping is difficult because what is good under one set of conditions may be poor for another set of conditions. For example, a precision of 1/1000 is very poor when taping is done along a level road, but it may very well be satisfactory when the work is done through heavy underbrush in mountainous terrain.

Below are presented some supposedly reasonable precision values which should be expected under ordinary taping conditions. The average value is probably sufficient for most preliminary surveys and the good value is desirable for most other surveys.

Poor 1/1000

Average 1/3000

Good 1/5000

Taping can be done with precision much higher than 1/5000 if careful attention is paid to the reduction of the errors previously discussed in this chapter. Thus by carefully controlling tape tensions, precisely measuring tape temperatures, and applying corrections, and by using hand levels to keep tapes horizontal, or by taping along slopes and minimizing the use of plumb bobs but necessitating the measurement of the slopes and the application of the appropriate corrections, the surveyor will be able to tape with precisions of 1/10,000 and better.

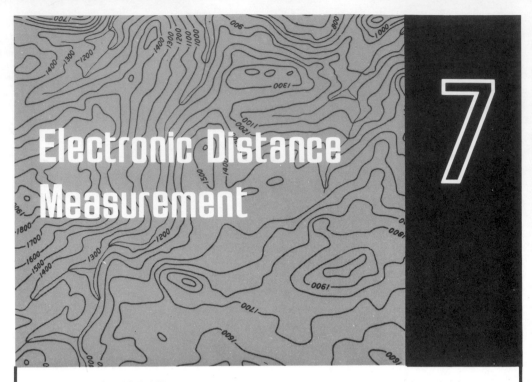

7-1 INTRODUCTION

Over the last 100 years there has been a constant but gradual improvement in the quality of surveying equipment. Not only have better tapes been manufactured but also better instruments for measuring elevations and directions have been produced. Within the past two decades the pace of improvement has quickened tremendously as the electronic distance-measuring devices have come of age.

Now there are available to the surveyor devices with which he can measure short distances of a few feet or long distances of many miles with extraordinary speed and precision. These devices save time and money and reduce the sizes of conventional survey parties. Furthermore, they may be used with the same facility where there are obstacles intervening such as lakes, canyons, standing crops, swampy and timbered terrain, hostile landowners or heavy traffic. Electronic distance-measuring devices have revolutionized distance measurement not only for geodetic surveys but also for ordinary land surveys.

Another important point is that most of these instruments automatically display direct readout measurements with the result that mistakes are greatly reduced. At the push of a button the operator may have the value shown in feet or in meters as desired. Normally the distances displayed are slope distances but in many of the newer instruments vertical angles are measured and horizontal distances computed and displayed.

Figure 7-1 HP3805A. An infrared EDM with which distances up to 1.6 km (1 mi) can be measured. Manufactured in U.S.A. by Hewlett-Packard Co. (*Courtesy Hewlett-Packard Co.*)

For the surveyor to fully understand the electronic measurement of distances, he or she would have to have a good background in physics and electronics. Fortunately, however, a person can readily use electronic distance-measuring equipment without really understanding very much about the physical phenomena involved. Though the equipment is expensive and complex, its actual operation is automatic and requires less skill than is needed with traditional instruments.

Electronic distance-measuring devices were not much used until the 1960s. The equipment was heavy, expensive, and required rather sophisticated maintenance. Nevertheless their tremendous advantages—the instant and precise measurement of distances up to 10 miles or more over forbidding terrain—inspired much research which has led to far better models. As a result the equipment has become lighter, cheaper and more maintenance-free. Generally the manufacturers do not recommend that the owners of EDM equipment service the units. They feel that more often than not worse problems will be created, resulting in even higher repair costs. It may be very practical, however, to call the company when an instrument problem occurs and describe the difficulty. They may be able to explain the necessary adjustment over the phone without the necessity of returning the equipment to the factory.

7-2 BASIC TERMS

Below are presented a few brief definitions which the reader will encounter in the sections to follow.

An *electronic distance-measuring device* is an instrument which transmits a carrier signal of electromagnetic energy from its position to a receiver located at another position. The signal is returned from the receiver to the instrument such that two times the distance between the two positions can be measured.

Visible light is generally defined as that part of the electromagnetic spectrum to which the eye is sensitive. It has a wavelength in the range of 0.4 to 0.7 μm (micrometers or microns).

Infrared light has frequencies below the visible portion of the spectrum; they lie between light and radio waves with wavelengths from 0.7 to 1.2 μm. Nevertheless, infrared light is generally put in the light-wave category because the distance calculations are made by the same technique.

An *electro-optical* instrument is one which transmits modulated light, either visible or infrared. It consists of a measuring unit and a reflector. The reflector consists of several so-called retrodirective glass cube corner prisms mounted on a tripod. The sides of the prisms are perpendicular to each other within very close tolerances.

Due to the perpendicular sides the prisms will reflect the light rays back in the same direction as the incoming light rays, hence the term retrodirective. Figure 7-2 illustrates this situation.

The trihedral prisms which are mounted on a tripod will reflect light back to the transmitting unit even though the reflector is out of perpendicularity with the light wave by as much as 20°. The number of prisms used depends on the distance to be measured and on visibility conditions. For short distances mirrors, reflectorized tape, and similar items will reflect the light beams satisfactorily.

A *laser* is one of several devices which produces a very powerful

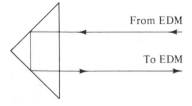

From EDM

To EDM

Figure 7-2 Reflection of visible or infrared light by a reflector.

single-color beam of light.   Low-intensity light waves are generated by the device and amplified into a very intense beam which spreads only slightly even over long distances.   The waves produced fall into the visible or into the infrared frequencies of the electromagnetic spectrum. Personnel must be informed about the necessary precautions to the eyes when working with lasers and they must strictly adhere to these safety requirements.

A *microwave* is an electromagnetic radiation which has an ultra-short wavelength or an ultrahigh frequency and lies in the region between infrared and short wave radio.

Radar makes use of the special properties of the extremely short, or micro, electromagnetic waves.   A radar set transmits these waves which are reflected or bounced back from objects in their path in the same manner that sounds bounce back from solid objects as echoes. The term *radar* was taken from the first letters of the words RAdio Detecting And Ranging.

7-3 THE FIRST EDMs

Radar was developed in the 1930s and was used very successfully in combat during World War II.   Distances could be measured almost instantaneously with fairly good accuracy and used to guide aircraft to targets or used by defenders to locate attacking aircraft.   As a result of radar's successful use, it is not surprising that geodesists began to study the possibility of applying radar principles to distance measurements in the years immediately following the war.

The first electronic distance measuring device developed for surveyors was the *Geodimeter* (the name being derived from the words GEOdetic DIstance METER).   It was conceived by a Swedish geodesist, Dr. Erik Bergstrand, around 1941, reduced to a workable device in 1947 and put on the market in 1952.   Today Geodimeters are in worldwide use.

Distance measurement with the Geodimeter is based on the accurate measurement of the time required for a visible light beam to travel from the instrument to a reflector (which acts as a mirror) and back to the instrument.   It is, therefore, necessary to have an unobstructed line of sight for this device to be used.   The reflector consists of one or more retroprisms which reflect the light rays to the source even if the prisms are not placed exactly perpendicular to the light rays.   The time is measured in terms of a phase difference between the sent and returning signals.   Depending on weather conditions the Geodimeter is useful for measuring distances up to a maximum of a few miles during daylight and up to 15 or 20 miles at night, although it has been used successfully for even greater distances.   Night measurements are a little better than day measurements because of the more stable atmospheric conditions. The average error with the Geodimeter is approximately 1 cm ± 1/500,000 the distance.

About a decade after the Geodimeter was developed the radio microwave devices were introduced. These included the Tellurometer, the Electrotape, and others. While the Geodimeter made use of light waves, these latter instruments measured the time required for microwaves to travel from the transmitter to a receiver at the other end of the line and back.

The microwave instruments have the advantage that they are not affected by light of any kind. Meterological conditions, however, are a source of error when these instruments are used for measuring long distances. Although measurements may be satisfactorily made when haze or mist or light rain or even light underbrush is present, there may be a problem when sights are taken across large bodies of water or over long stretches of highway. Such features have a tendency to reflect the microwave radiation, thus causing more than one path length between the instruments. This problem may be overcome by elevating the instruments above the ground as with towers or even with helicopters which are centered over stations by special optical devices called "hover sights."

Reflection of microwaves from ground or water surfaces is commonly called *ground swing*. Some surveyors when sighting from shore points across bodies of water have, instead of elevating their instruments, found it helpful to place the transmitter or master unit behind or partially blocked by a sand dune. This cuts off the bottom part of the radio wave while permitting the top part the necessary clearance across the water.[1] It is also possible to correct measurements for ground swing by taking 10 or 12 measurements each at a different carrier frequency and with uniform increments between the frequencies. The results are averaged or a graphical procedure is used to make the correction.

The Tellurometer was developed originally in 1954 by the Telecommunications Research Laboratory of the South African Council for Scientific and Industrial Research. T. L. Wadley conducted the work and his device was introduced in the United States in 1957. A short electromagnetic wave (a refined kind of radar) is sent from one unit, called the master, to another, called the remote instrument, where it is picked up and sent back to the master and the received signal compared with the transmitted signal. The Tellurometer, which can be used under any weather conditions day or night, has an inherent error

[1] R. W. Tomlinson and T. C. Burger, *Electronic Distance Measuring Instruments*, 2nd ed. (Washington, D.C.: American Congress on Surveying and Mapping, 1975), p. 7.

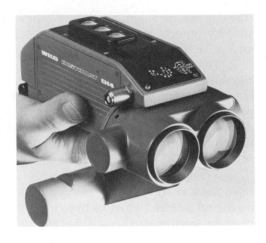

Figure 7-3 Wild Distomat D14. An infrared instrument with which distances up to 2 km (6600 ft) can be measured. One of the smallest and lightest EDMs on market. Measures horizontal and vertical angles, horizontal distances, elevation differences, and coordinate differences. Manufactured in Switzerland by Wild Herrbrugg Instruments, Inc. (*Courtesy Wild Herrbrugg Instruments, Inc.*)

of 1.5 cm ± 3/1,000,000 of the distance and is not affected by light of any kind.

The first EDM produced in the United States for the surveying profession was the *Electrotape*. It was produced by the Cubic Corporation of San Diego in 1958 and may be used to measure distances as large as 30 miles. The time required (again by a phase comparison technique) for a radio wave traveling at 186,000 miles/sec to travel to and from the other end of the line is measured. Identical instruments (called the Interrogator and the Responder) are set up at each end of the line and the distance between them can be measured from both ends. According to the manufacturer, its performance is independent of weather or light conditions such as day, night, rain, fog, or temperature extremes. Distances can be measured with an inherent error of 1 cm ± 1/300,000 of the distance.

The light wave and microwave devices permitted super work for medium and long-range distances up to 30 miles and soon became indispensable for organizations that needed to perform long-distance measurements with great precision, for example, the National Ocean Survey. But they did not really benefit the average American surveyor because they were very expensive and they did not provide sufficient precision for measurements under 1500 ft—yet that is where most of the average surveyor's work falls.

During the 1960s a solid-state device for generating infrared radiation permitted the development of relatively low-cost electronic distance-measuring devices that could be used for short-distance measurements of only a few feet and for longer distances up to 1 or 2 miles. The much higher frequency of infrared produces much narrower beams than those of radio microwaves, with the result that there are several advantages. More precise measurement can be made for short distances, there is less atmospheric interference and less trouble caused by reflection by objects along the path (as was the case with the wider beams), and reflection is permitted by a passive mirrorlike reflector.

The infrared EDMs operate with almost twice the wavelengths of the instruments which utilize visible light but the precisions obtained are approximately equal. The infrared diodes, however, have lower power transmission with the result that ranges are limited to maximum distances of approximately 1 or 2 miles. The infrared EDM units are light and compact and they yield precision equal to that of very careful taping even for short distances.

Nearly all of the EDM instruments manufactured in recent years have used as their carrier source either infrared or red laser light. Red laser and infrared carriers have a great advantage over the tungsten light used for earlier units. A filter can be used at the instrument which will permit only transmitted light to be passed. This means that the instruments can be used in daytime and direct sunlight and other reflected stray light (representing noise) can be prevented from entering the instruments.

7-4 CLASSIFICATION OF EDM INSTRUMENTS BY CARRIER SIGNALS

EDM instruments are often classed into three general groups by their carrier signals: e.g., the electro-optical instruments, the microwave instruments and the long-range radio-wave instruments. A few comments about each of these classes are made in the following paragraphs.

The *electro-optical instruments* include those devices which use visible light or invisible infrared light. Most of the newest short-range EDM instruments use infrared light.

Early EDM models such as Dr. Bergstrand's first Geodimeter made use of tungsten or mercury lamps. These instruments, however, had short operating ranges particularly during the day, because of the appreciable atmospheric scatter of the coherent light. Daytime range has been greatly increased by the use of the coherent light produced by gas lasers. Among the electro-optical instruments on the market are the Geodimeters, Cubitapes, Rangemasters, Microrangers, HPs (Hewlett-Packards), and others. Electro-optical instruments provide the most accurate EDM distance measurement because humidity has only negligible effects on the measurements.

A *microwave instrument* involves two electrical units, one being placed at each end of the line being measured. Each of the units, however, has a transmitter and a receiver along with a built-in communication system and the necessary measurement circuitry. These devices are involved with the time required for a radio microwave signal to travel from the transmitter to a receiver at the other end and back. The time is measured in terms of a phase difference between the sent and returning signals. Among the microwave instruments are the Tellurometer, Electrotape, Distameter and Microchain. The Tellurometers on the market today include electro-optical equipment as well as microwave devices.

The *long-range radio-wave instruments* make use of radio waves with which distances can be measured from 50 or 60 miles up to quite a few thousand miles. These devices are obviously useful for very large-scale work such as marine navigation, oceanographic surveys, location of oil-drilling platforms, etc. Some of the long-range radio-wave devices are the Autotape, Lambda, Omega, and others. Further discussion of these devices is beyond the scope of this text.

7-5 CLASSIFICATION OF EDMs BY THEIR OPERATIONAL RANGES

There are today many EDMs on the market and the list is constantly changing. They may be classed as short-range, intermediate-range, or long-range instruments. These approximate classifications are described in the paragraphs to follow:

1. *Short-range* EDMs are those which are capable of measuring distances up to a maximum of about 2 miles although some of them have maximum ranges of only a mile or even less. All of these instruments use infrared light generated by light-emitting diodes.

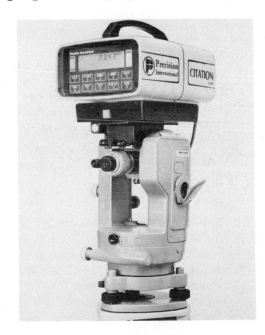

Figure 7-4 Citation CI-450. An infrared EDM with a range up to 4 km (2.5 mi). Displays horizontal distances and elevation differences after vertical angles are entered. Manufactured in U.S.A. by Precision International, Inc. (*Courtesy Wild Heerbrugg Instruments, Inc.*)

Figure 7-5 Topcon EDM Meter, Model DM-C3. An infrared instrument which can be used for measuring distances up to 3.2 km (2 mi). Displays horizontal distances and elevation differences after vertical angles are entered. Manufactured in Japan by Tokyo Optical Co., Ltd. (*Courtesy of Tokyo Optical Co., Ltd.*)

They are lighter, more compact, and more portable than are the longer range instruments making them very practical for the average land surveyor. Distances may generally be read directly while some reduction calculations are required for many of the longer range instruments.

2. *Intermediate-range* EDMs are those which can be used to measure distances of up to about 7 to 10 miles. They use lasers, microwaves, or tungsten or mercury lightwaves. The instruments are very useful for making very accurate control surveys of large areas such as those made by the National Geodetic Survey.

3. *Long-range* EDMs are capable of measuring distances as much as 30 or more miles. The carrier sources used have been tungsten light and laser light.

7-6 EDM EQUIPMENT AVAILABLE

Because EDM equipment is changing rapidly year by year it is not feasible to list and describe all of the different instruments available. As of this date (1982) there are over 30 commonly used models on the market. To learn more about this array of equipment the reader may very well like to obtain brochures from the various manufacturers or contact their representatives. For short-range instruments with which maximum distances up to about 2 miles can be measured, prices range from about $2500 to $10,000. For instruments with greater ranges up to 15 or 20 miles, prices generally will go above $10,000 and for those with ranges up to 40 or 50 miles or more, prices may exceed $50,000.

The following list includes some of the short-range EDM instruments commonly used by the average surveyor. The maximum distances for which they may be used depend on the instrument and vary from approximately 1600 ft up to about 10,000 ft (500 m to about 3050 m). Their weights vary from about 5 lbs up to about 44 lbs (2 kg to about 20 kg). The number or name of each of these instruments, the manufacturer and the radiation are shown in the list.

1. CD#6, MA 100—Tellurometer Division of Plessey Electronics, USA (infrared)
2. DI 3, DI 10—Wild, Switzerland (infrared)
3. DM 500, DM 2000—Kern, Switzerland (infrared)
4. Geodimeter 12—AGA, Sweden (infrared)
5. Geodimeter 76—AGA, Sweden (laser)
6. HDM 70—Cubic Western Data, USA (infrared)
7. HPs 3800, 3805, 3810—Hewlett-Packard, USA (infrared)
8. Micro-ranger—Keuffel & Esser, USA (infrared)
9. Reg Elta 14, SM 11—Zeiss, West Germany (infrared)
10. Citation CI-450—Precision International, USA (infrared)
11. Topcon DM-C2—Topcon Optical Co., Ltd., Japan (infrared)

7-7 BASIC EDM THEORY

The basic theory of EDM instruments is that distance equals time multiplied by velocity. Thus if the velocity of a radio or light wave and the time required for it to go from one point to another are known, the distance between the two points can be calculated. A very important relationship in the transmission of sound, radio, microwave, and light waves is given by the following expression:

$$V = f\lambda$$

where V is the velocity of propagation, f is the modulating frequency or number of cycles per unit of time of the energy, and λ is the wavelength or the distance traveled in one cycle. The unit of frequency is the hertz (Hz) which is equal to one cycle per second. A frequency of 10 MHz (or 10 megahertz) equals 10 million cycles per second. The wavelength λ is normally expressed in meters.

Both radio waves and light waves are electromagnetic. They have identical velocities in a vacuum (or space) equal to 299,792.458 ± 0.001 km/sec. These velocities are reduced in the atmosphere, however, because of density variations between atmospheric layers. In addition the velocities of radio and light waves are different from each other in the atmosphere. These are the velocities which need to be used in the calculations, and not laboratory values, since surveying measurements are of course made in the atmosphere. The ratio of the velocity V_0 of an electromagnetic wave in a vacuum divided by its velocity V in the atmosphere is called the *index of refraction* (n) and has a typical value of about 1.0003:

$$n = \frac{V_0}{V}$$

The velocity of the waves in the atmosphere is affected by the air's density. As a wave moves through the innumerable layers of the atmosphere it changes its velocity and direction because of refraction. To determine the value of the index of refraction, it is necessary to measure the temperature, humidity, and barometric pressure and then to substitute these values into an appropriate equation. Various publications have presented equations for computing the index. Though there are slight variations in the form of the equations and the results obtained, the differences are insignificant. The part of these equations involving relative humidity can be neglected for the electro-optical instruments as its effect is negligible. An index of refraction of about 1.0003 has been applied to the internal circuitry of the instruments. If the meteorological conditions existing at the time of the measurement yield an index appreciably different from this value, a correction should be made.

With EDM instruments the frequency of the electromagnetic waves can be precisely controlled but the velocity of those waves varies with temperature, pressure, and humidity. To make accurate measurements it is necessary to determine the meteorological conditions and make appropriate corrections. Usually these conditions are determined at each end of a line and averaged. For very large distances it may be desirable to determine these conditions at intermediate points as well.

As we have seen, a wave is transmitted from a transmitter to a reflector and returned. The total distance traveled is thus equal to two times the slope distance. When the wave returns it is detected and compared at the transmitter with the phase (a definite change with time, controlled by the frequency) of the original signal in order that the travel time can be determined.

To calculate the distance within an accuracy of ±0.5 ft would require that the time interval be measured to the nearest billionth of a second. Even with current equipment, such a time measurement is extraordinarily difficult to make. Yet the distance should be measured closer than ±0.5 ft, with the result that the time measurement problem is worse than stated. Fortunately, however, it is possible to solve the problem by making a very accurate phase measurement of the signal.

As previously stated, distance is measured with EDM devices by determining the time required for an electromagnetic wave to travel from a transmitter at one end of a line and to return from a reflector at the other end and back. This measurement is made indirectly by transmitting the wave to the reflector, and on its return it is detected and converted into an electric signal. This signal is a replica of the original waveform. It is, however, delayed in phase by an amount which is proportional to the time required for the wave to travel over the return path. Since the velocity of propagation of the wave can be accurately determined the distance can be calculated from the integral number of phase rotations plus the fraction of phase delay.[2]

7-8 TRAINING OF PERSONNEL

The procedures used for field measurements vary somewhat with the equipment from different manufacturers. As a result such procedures are not described here. Each manufacturer provides an equipment manual in which operating instructions are given. The instructions are quite simple and little training is needed before the units can be used.

Despite the simplicity of EDM equipment operation, however, it is advisable to have personnel trained as thoroughly as possible to get optimum results. No matter how fine and expensive the instruments used for a survey, they are of little value if they are not used knowledgeably. If a company spends $5,000 or $10,000 or more for EDM equipment but will not spend an extra few hundred dollars for detailed equipment training, the result will probably be poor economy.

A rather common practice among instrument companies is for one of their representatives to provide a few hours or even a day of training when an instrument is delivered. Such a short period is probably not sufficient to obtain the best results and to protect the large investment in equipment. Quite a few manufacturers offer a week-long training course. Participation in such a course is normally a wise investment and will yield long-range dividends. In addition, surveyors should consider attending workshops or short courses on EDM theory and practice given by the National Geodetic Survey and by manufacturers and universities.

[2]M. O. Schmidt and W. H. Rayner, *Fundamentals of Surveying*, 2nd ed., (New York: D. Van Nostrand Company, 1978), pp. 109–111.

Although EDM equipment is splendidly manufactured and its use can lead to very accurate surveys it is still necessary to take precautionary measures to minimize errors. The sources of errors in EDM work are the same as for other surveying work: personal, instrumental, and natural.

Personal errors are caused by such items as not setting the instruments or reflectors exactly over the points, not measuring instrument heights perfectly (see Section 7-12) and not accurately measuring weather conditions.

For precise distance measurement with an EDM device, it is necessary to center the instrument and reflector accurately over the end-points of the line. This topic of centering is discussed in detail in Chapter 12. Plumb bobs hanging on strings from the centers of instruments have long been used by surveyors for centering over points. They are still in constant use and have been employed by surveyors for countless centuries. For very precise work, however, the *optical plummet* is much to be preferred over the plumb bob. This is a special telescope device with which the surveyor can sight vertically from the center of the instrument to the point below. With the optical plummet, errors in centering can greatly be reduced (usually to a fraction of a millimeter). Its advantage over the plumb bob is multiplied when there is appreciable wind. The axis of the optical plummet must be periodically checked under lab conditions.

Instrumental errors are usually quite small if the equipment has been carefully adjusted and calibrated. Each of the EDM instruments, however, has a built-in error which varies from model to model. In general the more expensive the equipment, the smaller the error. The instrumental error is usually expressed as $\pm$ (some constant plus an additional value dependent on the distance measured D). For instance it might be $\pm(0.02 \text{ ft} + D/500,000)$. It can be seen that for short distances up to several hundred feet EDM equipment may not provide measurements as precise as those obtained by taping or with the subtense bar.

Whenever possible surveyors using microwave equipment should try to avoid high-voltage power lines, microwave towers, etc. Should they have to work in the vicinity of any of these structures, such information should be carefully noted in the field records since it may later be helpful should specifications not be met.

The *natural errors* present in EDM measurement are caused by variations in temperature, humidity, and pressure. Some EDM instruments automatically correct for atmospheric variables, for others it is necessary to "dial in" corrections into the instruments while for others it is necessary to make mathematical corrections. For instruments requiring adjustments, the manufacturers provide tables, charts, and explanations in their direction manuals as to how the corrections are to be made. For the microwave instruments it is necessary to correct for temperature, humidity and pressure while for the electro-optical instruments humidity can be neglected. (The humidity effect on microwaves is more than 100 times its effect on light waves.) Meteorological data should be obtained at each end of a line and sometimes at intermediate points and averaged if a higher precision is desired.

A new clean optical reflector probably doesn't scatter the light by more than about 2 arc seconds. After a few days or weeks of use, however, the presence of dirt scatters the light with the result that the divergence may be as large as 8 or more arc seconds.

Cleaning of the prisms is very involved and seldom can be done adequately by the average user. Soap and water are not satisfactory and the units should not be scrubbed since they may be scratched. A very lengthy cleaning process can be used involving first the blowing off of the dust followed by repeated applications of a light solvent such as acetone with clean unmedicated cotton balls. To adequately clean a retro-reflector probably takes 45 minutes to an hour and a few hundred cotton balls.

7 10 CALIBRATION OF EDM EQUIPMENT

It is important to periodically check EDM instrument measurements against the length of a National Geodetic Survey base line or other accurate standard. From the differences in the two values an *instrument constant* can be determined. This constant, which is a systematic error, enables the surveyor to make corrections to future measurements. Although a constant is furnished with the equipment, it is subject to change. This error results if the electrical center of the instrument does not coincide with the physical center of the instrument which is centered over the point. With the value so determined, a correction is applied to each subsequent measurement. In addition a record of the results and the dates when the checks were made should be kept in the surveyor's files in case of future legal disputes involving equipment accuracy.

There unfortunately seems to be a rather large percentage of practicing surveyors who think that EDM equipment can be continually used accurately without the necessity of calibration. These instruments, however, must be standardized periodically just as do other measuring devices. Both electro-optical and microwave equipment should be checked against an accurate base line at frequent intervals. The electronically obtained distances should be determined while

U.S. DEPARTMENT OF COMMERCE - NOAA
NOS - NATIONAL GEODETIC SURVEY
ROCKVILLE MD 20852 - MAY 27, 1981

CALIBRATION BASE LINE DATA
BASE LINE DESIGNATION: CLEMSON
PROJECT ACCESSION NUMBER: G16441

QUAD: N340824
SOUTH CAROLINA
PICKENS COUNTY

LIST OF ADJUSTED DISTANCES (APRIL 22, 1981)

FROM STATION N	ELEV.(M)	TO STATION N	ELEV.(M)	ADJ. DIST.(M) HORIZONTAL	ADJ. DIST.(M) MARK - MARK	STD. ERROR(MM)
0	228.600	150	229.071	149.9999	150.0006	.2
0	228.600	430	231.412	429.9949	430.0041	.4
0	228.600	1070	241.844	1069.9287	1070.0106	.6
150	229.071	430	231.412	279.9950	280.0048	.4
150	229.071	1070	241.844	919.9287	920.0174	.5
430	231.412	1070	241.844	639.9336	640.0186	.4

DESCRIPTION OF CLEMSON BASE LINE
YEAR MEASURED: 1981
CHIEF OF PARTY: WJR

THE BASE LINE IS LOCATED ABOUT 4.3 KM (2.7 MI) SOUTHEAST OF CLEMSON AND 2.9 KM (1.8 MI) WEST OF PENDLETON ALONG THE RIGHT-OF-WAY ON THE WEST SIDE OF UNITED STATES HIGHWAY 76 WHERE IT CROSSES THE ANDERSON-PICKENS COUNTY LINE. OWNERSHIP--MR. GEORGE WEATHERS, SOUTH CAROLINA HIGHWAY DEPARTMENT, PRE-CONSTRUCTION ENGINEER, POST OFFICE BOX 191, COLUMBIA, SOUTH CAROLINA 29202, TELEPHONE 803-758-3414.

TO REACH THE BASE LINE FROM THE SOUTH CAROLINA STATE HIGHWAY 93 OVERPASS ON UNITED STATES HIGHWAY 76 EAST OF CLEMSON, GO SOUTH ON HIGHWAY 76 FOR 2.6 KM (1.65 MI) TO NEW HOPE ROAD ON THE RIGHT AND THE 1070 METER POINT IN THE SOUTHWEST ANGLE OF THE INTERSECTION. TO REACH THE OTHER MARKS AND 0 METER POINT, CONTINUE SOUTH ON HIGHWAY 76 FOR 0.64 METER (0.4 MI) TO THE 430 METER POINT ON THE RIGHT, CONTINUE SOUTH FOR 0.32 KM (0.2 MI) TO A SIDE ROAD RIGHT AND THE 150 METER POINT IN THE SOUTHWEST ANGLE OF THE INTERSECTION, AND CONTINUE 0.16 KM (0.1 MI) SOUTH TO THE 0 METER POINT ON THE RIGHT ABOUT 0.9 M (3.0 FT) LOWER THAN THE HIGHWAY AND 21.5 M (70.5 FT) SOUTH OF THE ANDERSON COUNTY LINE.

THE 0 METER POINT IS A STANDARD NATIONAL GEODETIC SURVEY DISK STAMPED 0 1980, SET INTO THE TOP OF A ROUND CONCRETE MONUMENT 38 CM (15 IN) IN DIAMETER FLUSH WITH THE GROUND, LOCATED 40.9 M SOUTHEAST OF TELEPHONE JUNCTION BOX NUMBER 6, 21.5 M SOUTH OF ANDERSON COUNTY LINE SIGN, 21.5 M EAST OF WEST EDGE OF THE WOODS, 3.65 M WEST EDGE OF HIGHWAY 76, AND 1.15 M SOUTHEAST OF A METAL WITNESS POST.

THE BASE LINE IS A NORTH-SOUTH LINE WITH THE 0 METER POINT ON THE SOUTH END. IT IS MADE UP OF THE 0, 150, 430, AND 1070 METER POINTS WITH A POINT FOR THE CALIBRATION OF 100 FOOT TAPES SET SOUTH OF THE 0 METER POINT. ALL OF THE MARKS ARE SET ON A LINE PARALLEL TO THE HIGHWAY AND IN THE DITCH ON THE WEST SIDE OF THE ROAD. THIS BASE LINE IS NOT CONNECTED TO THE NATIONAL NOR THE LOCAL CONTROL NETWORKS.

THIS BASE LINE WAS ESTABLISHED IN CONJUNCTION WITH THE STATE OF SOUTH CAROLINA. FOR FURTHER INFORMATION, CONTACT THE DIRECTOR, SOUTH CAROLINA GEODETIC SURVEY, SOUTH CAROLINA DIVISION OF RESEARCH AND STATISTICAL SERVICES, OFFICE OF GEOGRAPHIC STATISTICS, 915 MAIN STREET, SUITE 203, COLUMBIA, SOUTH CAROLINA 29201. TELEPHONE 803-758-3604.

Figure 7-6

taking into consideration differences in elevations, meteorological data, etc.

The NGS compiles and publishes a calibration of base lines for each state showing location, elevations, horizontal distances, and other pertinent data.   Copies may be obtained by writing to NGS, National Ocean Survey, Rockville, MD 20852.   Copies are also on hand at each state's surveying society offices.   A sample copy of such data is shown in Fig. 7-6.

If a base line is not available, two points can be set up (as much as 5 miles apart for microwave equipment) and the distance between them measured.   A point can be set in between the other two and the two segmental distances measured.   (See Fig. 7-7.)   The sum of those two

Figure 7-7   A   B   C

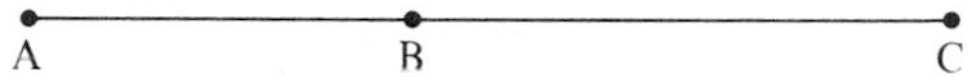

values should be compared with the overall length.   Should the three points not be in a line, angles will be needed to compute the components of the two segments to compare with the overall straight-line distance between the end-points.   The instrument constant can be calculated as follows—noting that the constant will be present in each of the three measurements:

$$\text{Instrument constant} = AC - AB - BC$$

It is also desirable to check barometers, thermometers, and psychrometers approximately once a month or more often if they are subject to heavy use.   These checks can usually be made with equipment which is available at most airports.

7-11 CONSTRUCTION LAYOUT

In construction work, a steel tape is normally used for laying out distances, since it is so convenient to handle.   This is particularly true for the average construction job where distances are relatively small.   When rather long lines are involved, EDM devices can be used advantageously.

One method of using EDM instruments for layout work is as follows.   An approximate distance can be established within a few feet by stadia.   An approximate point is set there as well as an additional one in line and on each side of the tentative point.   The distance to the tentative point can be measured by an EDM and the necessary correction made with a steel tape in line with the other two points.   The new point is checked with the EDM.   Some EDM devices are manufactured so they can be sighted on a moving reflector, thus simplifying the work.

It is to be remembered that horizontal distances are needed and slope corrections may have to be made.   Field layout work can be expedited appreciably if an EDM device is used which has an angle-measuring capability. Many such devices can measure vertical angles and compute and display horizontal distances.

7-12 COMPUTATION OF HORIZONTAL DISTANCES FROM SLOPE DISTANCES

All EDM equipment is used to measure slope distances. For most models the values obtained are corrected for the appropriate meteorological and instrumental corrections and then reduced to horizontal components. With many of the newer instruments, however, the computations are made automatically. As with taping along slopes, the horizontal values may be computed by making corrections with the slope correction formula (described in Section 4-6 of this text), by using the Pythagorean theorem, or by applying trigonometry. If the slope is quite steep, say greater than 10 to 15%, the slope correction formula (which is only approximate) should not be used.

To compute horizontal distances it is necessary either to determine the elevations at the ends of the line or to measure vertical angles at one or both ends. Example 7-1 illustrates the simple calculations involved when the elevations are known.

Example 7-1

A slope distance of 1654.32 ft was measured between two points with an EDM instrument. It is assumed that the atmospheric and instrumental corrections have been made. If the difference in elevation between the two points is 183.36 ft and if the heights of the EDM and its reflector above the ground are equal, determine the following:

(a) The horizontal distance between the two points using the slope correction expression.

(b) The horizontal distance between the two points using the Pythagorean theroem.

Solution

(a) *Using the slope correction expression*

$$C = \frac{v^2}{2s} = \frac{(183.86)^2}{(2)(1654.32)} = 10.22 \text{ ft}$$

$$h = 1654.32 - 10.22 = 1644.10 \text{ ft}$$

(b) *Using the Pythagorean equation (right triangles)*

$$h = \sqrt{(1654.32)^2 - (183.86)^2} = \textbf{1644.07 ft}$$

For Example 7-1 it was assumed that the distance measured was parallel to the ground or that is the heights of the EDM instrument and the reflector above the end-points were equal. If these values are not

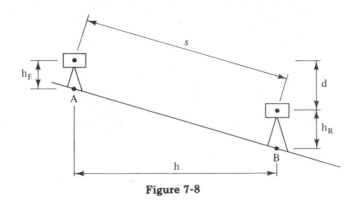

Figure 7-8

equal, that fact must be accounted for in the calculations. Such a situation is shown in Fig. 7-8 where h_E and h_R represent the heights of the instrument and the reflector respectively, d is the difference in elevation between the instrument and reflector, s is the slope distance determined, and h is the horizontal component desired. From this figure the value of d can be determined as follows:

$$d = \text{Elevation } A + h_E - (\text{Elevation } B + h_R)$$

Once d is determined the horizontal distance can be calculated as previously described in Example 7-1. If the Pythagorean theorem is used the results can be expressed in formula fashion as follows:

$$h = \sqrt{s^2 - [(\text{Elev } A + h_E) - (\text{Elev } B + h_R)]^2}$$

Example 7-2

A slope distance of 1836.42 ft was measured between two points with an EDM instrument. It is assumed that the necessary atmospherical and instrumental corrections have been made. The EDM is centered over one point with a ground elevation of 632.52 ft while the reflector is located over the end-point which has an elevation of 506.46 ft. If the heights of the EDM and the reflector above these points are respectively 4.80 ft and 5.20 ft, determine the horizontal distance between the two points.

Solution

$$h = \sqrt{(1836.42)^2 - [(632.52 + 4.80) - (506.46 + 5.20)]^2} = \textbf{1832.12 ft}$$

If, instead of measuring elevations and instrument heights, vertical angles are measured, the horizontal component may be calculated directly with trigonometry. For instance if the vertical angle α' shown in Fig. 7-9 between a horizontal and a line from the EDM to the center of the reflector is measured the horizontal distance will equal $s \cos \alpha'$. If, however, a theodolite is used to measure the true vertical angle be-

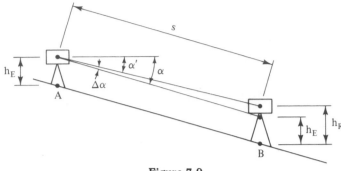

Figure 7-9

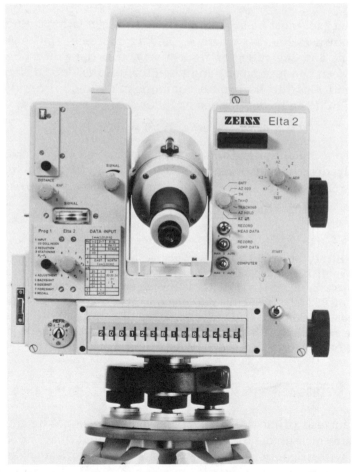

Figure 7-10 Zeiss Elta 2. An infrared tacheometer with which angles can be read to seconds and distances up to 4 km (2.5 mi). Manufactured by Zeiss in West Germany.

tween the two points A and B, represented by α in Fig. 7-9, it will be necessary to calculate the value of $\Delta\alpha$ and use it to correct the angle α. In the figure $\alpha' = \alpha - \Delta\alpha$. The value of $\Delta\alpha$ in seconds can be calculated with the following expression:

$$\Delta\alpha = \frac{(h_R - h_E)\cos\alpha}{s\,\sin 1''}$$

Some EDM instruments have the capability of vertical angle measurement, but with others it is necessary to set up a transit or theodolite to measure the angles. As a result sometimes the heights of the EDM, the reflector, the theodolite, and even the target may be different.

7-13 THE FUTURE FOR EDM EQUIPMENT

As always, it is difficult to predict the future but there seems to be a general trend in EDM usage toward the so-called *total station instruments*, also sometimes called *electronic tacheometers*. These instruments are combinations of theodolites with EDM units. Although several total station instruments are now on the market, they are beyond the pocketbook of the average surveyor. Presumably this will not always be the case. In addition to being used to measure slope distances and vertical angles, these instruments will compute and display horizontal distances. They will also probably be used to measure and display differences in elevation as well as directions. Moreover, they may also have some system capability for recording horizontal distances, elevations, and directions.

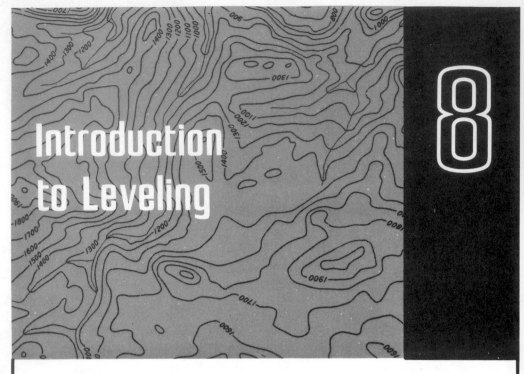

Introduction
to Leveling

8

8-1 IMPORTANCE OF LEVELING

The determination of elevations with a surveying instrument, which is known as *leveling*, is a comparatively simple but extremely important process.

The significance of relative elevations cannot be exaggerated. They are so important that one cannot imagine a construction project in which they are not critical. From terracing on a farm or the building of a simple wall to the construction of drainage projects or the largest buildings and bridges, the control of elevations is of the greatest importance.

8-2 BASIC DEFINITIONS

Below are presented a few introductory definitions which are necessary for the understanding of the material to follow. Additional definitions are presented as needed for the discussion of leveling in this chapter and the next. Several of these terms are illustrated in Fig. 8-1.

A *vertical line* is a line parallel to the direction of gravity. At a particular point it is the direction assumed by a plumb bob string if the plumb bob is allowed to swing freely.

A *level surface* is a surface of constant elevation that is perpendicular to a plumb line at every point. It is best represented by the shape that a large body of still water would take if it were unaffected by tides.

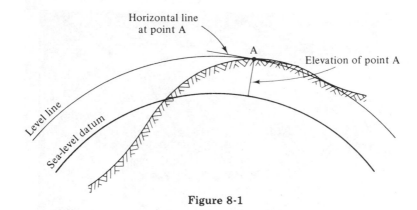

Figure 8-1

The *elevation* of a particular point is the vertical distance above or below a reference level surface (normally sea level) to the point in question.

A *level line* is a curved line in a level surface all points of which are of equal elevation. Every element of the line is perpendicular to gravity.

A *horizontal line* is a straight line tangent to a level line at one point.

8-3 REFERENCE ELEVATIONS OR DATUMS

For a large percentage of surveys, it is reasonable to use some convenient point as a reference or datum with respect to which elevations of other points can be determined. For instance, the surface of a body of water in the vicinity can be assigned a convenient elevation. Any value can be assigned to the datum, for example, 100 ft or 1000 ft, but the assigned value is usually sufficiently large so that no nearby points will have negative elevations.

In the past many assumed reference elevations were used in the United States, even for surveys of major importance. An assumed value may have been given to the top of a hill, or the surface of one of the Great Lakes, or the low-water mark of a river, or any other convenient point. These different datums created confusion and the current availability of the *sea-level datum* is a great improvement.

In the United States the sea-level datum is the value of mean sea level determined by averaging the hourly elevations of the sea over a long period of time, usually 19 years. Actually, the elevation of mean sea level is rising very slowly—probably because of the gradual melting of the polar ice caps and perhaps because of the erosion of ground surfaces. The change, however, is so slow that it does not prevent the surveying profession from using sea level as a datum.

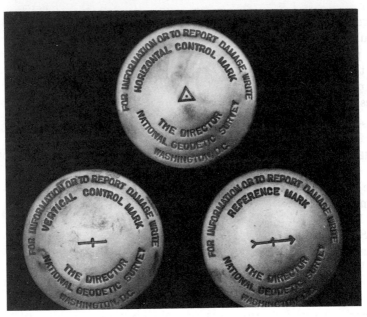

Figure 8-2 National Geodetic Survey brass disk marks used beginning in 1972 are designed to set in concrete or clamped to rods. (*Courtesy National Geodetic Survey.*)

In 1878, at Sandy Hook, NJ, the National Geodetic Survey (then the U.S. Coast and Geodetic Survey) began working on a transcontinental system of precise levels. In 1929, the agency made an adjustment of all first-order leveling (a term describing precise leveling defined in Section 9-11) in the U.S. and Canada, and established a datum for elevations throughout the country. This datum is referred to as the *National Geodetic vertical datum of 1929* or just the *sea-level datum of 1929*.

One of the most important tasks of the National Geodetic Survey and the U.S. Geological Survey is the establishment of a network of known elevations throughout the country. As a result of their work, eventually there will be no area in the country which will be far from a point whose elevation in relation to sea level is not known.

These two federal agencies have set up throughout the U.S. monuments whose elevations have been precisely determined. A monument, called a *bench mark*, is usually of concrete and has a brass disk imbedded in it, as illustrated in Fig. 8-2. For many years the practice was to actually record the elevations on the monuments. This, however, is no longer the case because the elevations of these monuments may be changed by frost action, earthquakes, and so on. The elevations are carefully checked at intervals and the surveyor who needs the elevation of a particular monument should contact the agency which set it, even though a value is given on the monument. More than half a million of these monuments have been placed in the United States and its possessions.

8-4 METHODS OF LEVELING

There are three general methods of leveling: spirit, trigonometric, and barometric. Although the surveyor is concerned almost entirely with spirit leveling, a brief description of each type is given below.

Spirit leveling, also called *direct leveling*, is the usual method of leveling. Vertical distances are measured in relation to a horizontal line and these values are used to compute the differences in elevations between various points. A spirit level (to be discussed in Section 8-5) is used to fix the line of sight of the telescope. This line of sight is the assumed horizontal line with respect to which vertical distances are measured.

Trigonometric leveling is leveling in which horizontal distances and vertical angles are measured and used to compute elevation differences. This method can be used for inaccessible points such as mountain peaks. For purposes of mapping, rough leveling, or preliminary surveys, the stadia method, which is actually a variation of trigonometric leveling, is very useful.

Barometric leveling involves the determination of elevations by measuring changes in air pressure. Although the decrease in air pressure with increasing elevations can be measured with mercurial barometers, these barometers are cumbersome and fragile and are impractical for surveying purposes. Instead, the light and sturdy, but less accurate aneroid barometers are used. These devices, commonly called *altimeters*, can be used to determine approximate elevations within 5 or 10 ft. Such precision is only sufficient for preliminary or reconnaissance work. They do nevertheless offer the advantage that approximate elevations over a large area can be quickly determined. If careful procedures and larger aneroid barometers are used, much better results can be obtained.

Barometer readings at the same elevation vary with local air pressure conditions and also are affected by temperature and humidity variations. Therefore, at least three barometers are needed for reasonably accurate elevation measurement. Ideally, one barometer is placed at a known elevation higher than that of the desired point and one is placed at a lower point of known elevation. All of the barometers are read and from the readings at the known elevations corrections can be made at the point at which the elevation is being determined.

8-5 LEVELS

A level consists of a high-powered telescope (20 to 40 diameters) with a spirit level attached to it in such a manner that when its bubble is centered, the line of sight is horizontal. The purposes of the telescope are to fix the direction of the line of sight and to magnify the apparent sizes of objects observed. The invention of the telescope is generally credited to the Dutch optician Hans Lippershey about 1607. In colonial times telescopes were too large for practical surveying and they were not used on surveying instruments until the end of the nineteenth century. Their use tremendously increases the speed and precision with which measurements can be taken. These telescopes have a vertical cross hair for sighting on points and a horizontal cross hair with which readings are made on level rods. In addition, they may have stadia hairs.

The telescope has three main parts: the objective lens, the eyepiece, and the reticle. The *objective lens* is the large lens located at the front or forward end of the telescope. The *eyepiece* is the small lens located at the viewer's end. It is actually a microscope which magnifies and enables the viewer to clearly see the image formed by the objective lens. The cross hairs and stadia hairs form a network of lines which is usually called the *reticle, reticule,* or just the *cross hairs.* The cross hairs are made from spider webs or fine wires, or they may consist of very fine lines etched in glass. A line drawn from the point of intersection of the cross hairs and the optical center of the objective is called the *line of sight* or *line of collimation* (the word *collimation* meaning lining up or adjusting the line of sight).

Some telescopes are said to be *external focusing.* Their objective lens is mounted on a sleeve which moves back and forth as the focusing screw is turned. *Internal focusing* is used for improved telescopes. These instruments have a lens which moves back and forth internally between the objective lens and the reticle.

The level tube is an essential part of most surveying instruments. The closed glass tube is precisely ground on the inside surface to make the upper half curved in shape so that the bubble is stabilized. If the tube were not so curved the bubble would be very erratic. The tube is filled with a sensitive liquid (usually a purified synthetic alcohol) and a small air bubble. The liquid used is stable and nonfreezing for ordinary temperature variations. The bubble rises to the top of the liquid against the curved surface of the tube. The tangent to the circle at that point is horizontal and perpendicular to gravity. The tube is marked with divisions symmetrical about its mid-point and thus when the bubble is centered, the tangent to the bubble tube is a horizontal line and is parallel to the axis of the telescope.

Nearly all of the older types of levels have two pairs of leveling screws, the pairs being placed directly opposite each other. In order to level the instrument, the level tube is alternately placed over opposite pairs of leveling screws. In each case the screws are turned until the bubble is centered. The process is repeated until it is possible to completely turn the telescope through 360° without movement of the bubble.

In centering the bubble, opposite screws are used and the screws are turned in toward the center $\supset\subset$ or in the opposite directions $\supset\subset$. In each case as the screws are turned, the bubble will follow the motion of the left thumb.

The levels used for precise leveling usually have only three leveling screws. This arrangement permits quicker and more precise leveling of the instrument. Any one of the three screws can be used alone for leveling, and the bubble will move toward any screw turned in a clockwise direction. The slight disadvantage of these levels is that if the three screws are all adjusted up or down, there will be a slight variation in the elevation of the telescope. This is not the case for adjustments with the four-screw level. A detailed discussion of setting up and leveling both the four-screw and three-screw instruments is presented in Section 8-7.

Some newer levels have only two leveling screws. They are constructed similar to the three-screw levels but one of the screws is re-placed with a fixed point. This means that when the instruments are set up and leveled at a given point they will be at a fixed elevation. If the instrument is moved slightly and releveled, it will return to the same elevation. To level the two-screw instrument, the telescope is turned until it is parallel to a line from the fixed point to one of the leveling screws. The instrument is leveled and then the telescope is turned to the other leveling screw; this instrument is again leveled and the process is repeated a few more times as needed.

Several types of levels are briefly discussed in the following paragraphs.

Wye level. This level has its telescope supported in wye-shaped supports and is held in place with curved clips. It is not commonly used today because other levels are more satisfactory. The telescope for the wye level can actually be removed and turned end for end for adjustment purposes. Although they are not as popular as the other levels, many of the old wye levels have more sensitive bubble tubes than do the average dumpy levels.

Dumpy level. Originally the dumpy level had an inverting eye-piece and as a result it was shorter (thus the name dumpy) than a wye level for the same magnification power. The dumpy level (Fig. 8-3) is used for almost all precise work and for most other leveling purposes as well. It is superior to the wye level in several important respects. Because the telescope is rigidly held in its supports together with the attached level tube, it is less likely to get out of adjustment than the wye level. This simpler and more dependable instrument has fewer movable parts than does the wye level and as a result it has fewer parts to become worn or get out of adjustment.

Builder's level. This level, sometimes called an *architect's level* or a *construction level*, is less expensive than the usual level. Its bubble tube is less sensitive and its telescope has less magnification power. In addition, it has a horizontal scale for measuring angles. It is generally used for those phases of construction in which great precision is not required, such as the establishment of grades for earthwork or the setting of forms for concrete.

Self-leveling or automatic level. The dumpy and wye levels are being replaced with the self-leveling and tilting levels. The self-leveling level has a small circular spirit level called a *bull's eye* level and three leveling screws. The bubble is approximately centered in the bull's eye and then the instrument itself automatically does the fine leveling.

The self-leveling level has a prismatic device called a *compensator* suspended on fine, nonmagnetic wires. When the instrument is approxi-

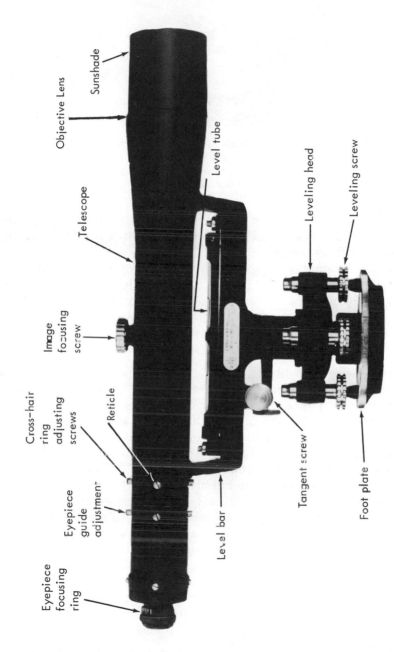

Objective Lens

Sunshade

Level tube

Telescope

Image focusing screw

Cross-hair ring adjusting screws

Reticle

Eyepiece guide adjustment

Eyepiece focusing ring

Level bar

Tangent screw

Leveling head

Leveling screw

Foot plate

Figure 8-3 Dumpy level. (*Courtesy Berger Instruments.*)

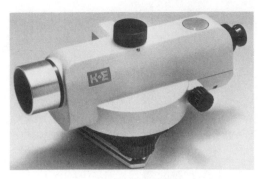

Figure 8-4 Eagle Automatic level for which a circular spirit level is centered once for each setup. It has a compensator that automatically levels the line of sight. Manufactured in U.S.A. by Keuffel & Esser Company. (*Courtesy of Keuffel & Esser Company.*)

mately centered the force of gravity on the compensator causes the optical system to swing almost instantaneously into a position such that its line of sight is horizontal.

The self-leveling level speeds up leveling operations and also is particularly useful where the ground is soft and/or when strong winds are blowing because the instrument automatically relevels itself when it is thrown slightly out of level. When the surveyor uses an ordinary level under these adverse conditions, he must constantly check the bubble to see that it remains centered.

Tilting level. A tilting level (Fig. 8-5) is one whose telescope can be tilted or rotated about its horizontal axis. The instrument can be leveled quickly and approximately by means of a bull's eye or circular type level and the telescope is moved through a small vertical angle by means of a tilting knob until the telescope is level.

The tilting level usually has a special arrangement of prisms which enables the user to center the bubble by means of a *split* or *coincident bubble*. The two halves of the bubble are actually the half ends of a single bubble and they will coincide when the bubble is centered. Manufacturers claim that split bubbles can be centered several times more precisely than the noncoincident types. Tilting levels are faster to use than are the wye and dumpy levels but slower than the self-leveling levels. Purchase prices of tilting levels fall in between those for dumpy and self-leveling levels.

The tilting features were formerly used primarily for precise government leveling, but they are being used more commonly for other work because they permit faster and more precise leveling.

Transit used as level. Although it is primarily used for angle measurement (as described in Chapter 12), the transit can also be used

Figure 8-5 Paragon tilting level. (*Courtesy Keuffel & Esser Co.*)

for leveling. The results are fairly precise but not as good as those obtained with standard levels because the telescope and bubble tube are shorter.

Laser levels. The laser (previously mentioned in Chapter 7 relating to EDM equipment) is effectively used today for several leveling operations. It is commonly used to create a known reference elevation or point from which construction measurements can be taken.

The lasers used for surveying and construction fall into two general classes: single-beam lasers and rotating-beam lasers. A single-beam laser projects a string line that can be seen on a target regardless of lighting conditions. The line may be projected in a vertical, horizontal, or inclined direction. The vertical line provides a very long plumb line, which builders have needed throughout history. Horizontal and inclined lines are very useful for pipe lines and tunnels. Section 20-16 of this text describes the use of lasers for pipe line construction.

A rotating-beam laser, which provides a plane of reference over open areas, can be rotated rapidly or slowly or can be stopped and used as a single beam. Today's rotating lasers are self-leveling and self-plumbing, thus providing both horizontal and vertical reference planes. The laser beam will not come on until the instrument is level. If the instrument is bumped out of position the beam shuts off and will not come back on until it's level again. Thus it is particularly advantageous when severe winds are occurring. The rotating beam can be set to provide a horizontal plane for leveling or a sloped plane as might be required for the setting of road or parking lot grades. It is precise for distances up to 1000 ft. This means that fewer instrument set-ups are required and, for construction jobs, it means that the laser can be set up some distance away and out of the way of construction equipment.

The laser is not usually visible to the human eye in bright sunlight and thus some type of detector is needed. The detector can either be a small hand-held or rod-mounted unit that may be moved up and down the level rod, or it may be an automatic detector. This latter device has an electronic carriage that moves up and down inside the rod and locates the beam.

Lasers can be very useful for staking pipe lines, parking lots, setting stakes to control excavation and fills; for topographic surveys; etc. A detailed discussion of lasers is provided in an article written by Tom Liolios.[1]

8-6 LEVEL RODS

There are many kinds of level rods available. Some are in one piece and others (for ease of transporting) are either telescoping or hinged. Level rods are usually made of wood and are graduated from zero at the bottom. They may be either *self-reading rods* that are read directly through the telescope or *target rods* where the rodman sets a sliding target on the rod and takes the reading directly. Most rods serve as either self-reading or as target rods.

Among the several types of level rods available are the *Philadelphia rod*, the *Chicago rod*, and the *Florida rod*. The Philadelphia rod, the most common one, is made in two sections. It has a rear section that slides on the front section. For readings between 0 and 7 ft, the rear section is not extended; for readings between 7 and 13 ft, it is necessary to extend the rod. When the rod is extended, it is called a *high rod*. The Philadelphia rod is distinctly divided into feet, tenths, and hundredths by means of alternating black and white spaces painted on the rod (as shown in Fig. 8-6).

The Chicago rod is 12 ft long and is graduated in the same way as the Philadelphia rod, but it consists of three sliding sections. The Florida rod is 10 ft long and is graduated in white and red stripes, each stripe being 0.10 ft wide. Also available for ease of transportation are tapes or ribbons of waterproofed fabric which are marked in the same way that a regular level rod is marked and which can be attached to ordinary wood strips. Once a job is completed, the ribbon can be removed and rolled up. The wood strip can be thrown away. The instrumentman can clearly read these various level rods through his

---

[1]Tom Liolios, *Lasers and Construction Surveying* (Wayne, Mich.: P.O.B. Publishing Co., August–September 1981, vol. 6, no. 6) pp. 38–41 and 63.

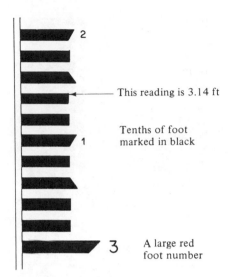

This reading is 3.14 ft

Tenths of foot
marked in black

A large red
foot number

Figure 8-6 The divisions of a Philadelphia rod.

telescope for distances up to 200 or 300 ft, but for greater distances he
must use a target. A target is a small red and white piece of metal (see
Figs. 8-7 and 9-7) attached to the rod. The target has a vernier (to be
discussed in Section 9-5) that enables the rodman to take a reading to
the nearest 0.001 ft.

If the rodman is taking the readings with a target and if the line
of sight of the telescope is above the 7-ft mark, it is obvious that he
cannot take the readings directly in the normal fashion. Therefore, the
back face of the rod is numbered downward from 7 to 13 ft. The target is set at a certain mark on the front face of the rod and as the back
section is pushed upward, it runs under an index scale and a vernier
which enables the rodman to take the reading on the front. Level rod
targets and verniers are discussed in Sections 9-5 and 9-6.

8-7 SETTING UP THE LEVEL

Before setting up the level the instrumentman should give some thought
to where he must stand in order to make his sights. In other words, he
will consider how to place the tripod legs so that he can stand comfortably between them for the layout of the work that he has in mind,
as shown in Fig. 8-8.

The tripod is desirably placed in solid ground where the instrument will not settle as it most certainly will in muddy or swampy areas.
It may be necessary to provide some special support for the instrument,

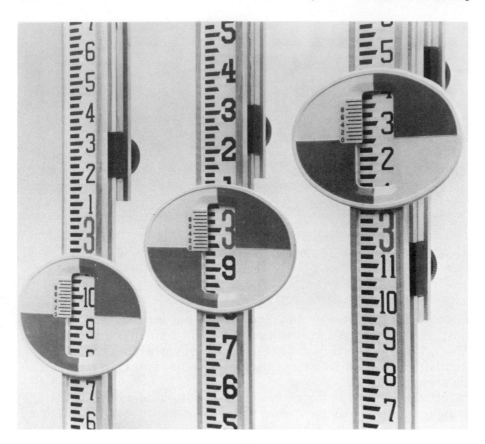

Figure 8-7 Level rods and targets.(*Courtesy Berger Instruments.*)

such as stakes or a platform. The tripod legs should be well spread apart and adjusted so that the footplate under the leveling screws is approximately level. The instrumentman walks around the instrument and pushes each leg firmly into the ground. On hillsides it is usually convenient to place one leg uphill and two downhill, as shown in Fig. 12-4.

After the instrument has been leveled as much as possible by ad-

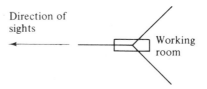

Figure 8-8

justing the tripod legs, the telescope is turned over a pair of opposite leveling screws if a four-screw instrument is being used. Then the bubble is roughly centered by turning that pair of screws in opposite directions to each other (as described in Section 8-5). The bubble will move in the direction of the left thumb. Next, the telescope is turned over the other pair of leveling screws and the bubble is again roughly centered. The telescope is turned back over the first pair and the bubble is again roughly centered, and so on. This process is repeated a few more times with increasing care until the bubble is centered with the telescope turned over either pair of screws. If the level is properly adjusted, the bubble should remain centered when the telescope is turned in any direction. It is to be expected that there will be a slight maladjustment of the instrument that will result in a slight movement of the bubble; however, the precision of the work should not be adversely affected if the bubble is centered each time a rod reading is taken.

The first step in leveling a three-screw instrument is to turn the telescope until the bubble tube is parallel to two of the screws (see 1 and 2 in Fig. 8-9). The bubble is centered by turning these two screws in opposite directions.

Next, the telescope is turned so that the bubble tube is perpendicular to a line through screws 1 and 2. The bubble is centered by turning screw 3 (see (Fig. 8-10).

These steps are repeated until the bubble stays centered when the telescope is turned back and forth.

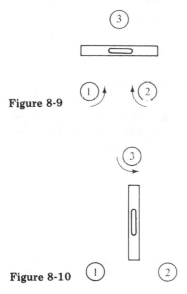

Figure 8-9

Figure 8-10

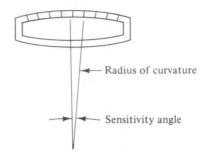

Radius of curvature

Sensitivity angle

Figure 8-11 Sensitivity of bubble tubes.

8-8 SENSITIVITY OF BUBBLE TUBES

The divisions on bubble tubes were at one time commonly spaced at $\frac{1}{10}$-in. intervals but today they are usually spaced at 2-mm intervals. The student often wants to know how much the rod readings will be affected if the bubble is off center by 1 or 2 or more divisions on the bubble tube when the readings are made. One way for him to find out is to place the level rod a certain distance from the level (say 100 ft) and take readings with the bubble centered; then with it one division off center; two divisions off center, etc.

The *sensitivity* of the bubble tube may be expressed in terms of the radius of curvature of the bubble tube. (See Fig. 8-11.) Obviously if the radius is large, a small movement of the telescope vertically will result in a large movement of the bubble. An instrument with a large radius of curvature is said to be sensitive. For very precise leveling very sensitive levels need to be used. The centering of the bubble for such instruments takes more time and therefore less sensitive instruments (which are more quickly centered) may be more practical for surveys of lesser precision.

Another way of expressing the sensitivity of the bubble tube is to give the angle through which the axis of the bubble tube must be rotated (usually given in seconds) to cause the bubble to move by one division on the scale (Fig. 8-11). If the movement of one division on the scale corresponds to $10''$ of rotation, the bubble is said to be a $10''$ bubble. For very precise levels, $0.25''$ or $0.5''$ bubbles or even more sensitive ones may be used. For ordinary construction leveling $10''$ or $20''$ instruments may be satisfactory.

8-9 CARE OF EQUIPMENT

Although surveying equipment is very precisely and delicately manu-factured, it can be very durable when properly used and maintained. In fact, these instruments may very well last a lifetime in the hands of a careful surveyor but a few seconds of careless treatment can result in severe or irreparable damage.

The long-lasting qualities of surveying equipment mean that the surveyor will occasionally have to use some rather antique levels and transits. This is not a disadvantage because a good old instrument will still be quite precise. Used transits and levels decline very little in value as the years go by. As a matter of fact their selling prices in dollars often remain stable for several decades. However, finding parts for older instruments is sometimes a problem.

Several suggestions are made in the following paragraphs concerning the care of the instruments used for leveling.

The level.   Before the level is removed from its box, the tripod should be set up in a firm position. The user should observe exactly how the level is held in the box so that after use he can return it to exactly the same position. After the instrument is taken from its box, it should be handled by its base when it is not on the tripod. It should be carefully screwed onto the tripod. The surveyor must not permit anything to interrupt him until this task is completed. Many levels have been severely damaged when careless surveyors began to attach them to their tripods but then allowed their attention to be diverted before tightening was completed.

If possible, levels should not be set up on smooth hard surfaces, such as building floors, unless the tripod points can be either set in indentations in the floor or firmly held in place by some other means, perhaps by triangular frames made for that purpose. Particular care should be exercised when the instruments are being carried inside buildings to avoid damage from possible collisions with doors, walls, or columns. In such locations, therefore, the level should be carried in the arms instead of on the shoulder. When working outside where it is normal to carry the instrument on the shoulder, the clamps should be left loose in order to allow the instrument to turn if it hits limbs or bushes.

A level should never be left unattended unless it is in a very protected location. If it is turned over by wind, cattle, children, or cars, the results will probably be disastrous. Some surveyors paint their tripod legs with bright colors. Such a scheme is particularly wise when work is being done near heavy traffic. Another reason for not leaving instruments unprotected is thieves. Surveyors' levels, transits, and EDM instruments can quickly and easily be sold for good prices. In addition the surveyor should protect his instruments as much as possible from moisture; but when the level does get wet, he should gently wipe dry everything except the lens. Moisture on the lens is usually permitted to evaporate because the lens can so easily be scratched while being dried. The lens should never be touched with anything other than a camel's-hair brush, or less desirably a soft silk handkerchief.   If the objective lens or the eyepiece lens becomes so dusty that it interferes with vision, it may be cleaned with a camel's-hair brush or with lens paper.

Leveling screws.   Perhaps the most common injury to levels by the novice surveyor is caused by applying too much pressure to the leveling screws.   If the instrument is in proper condition, these screws should turn easily and it should never be necessary to use more than the fingertips for turning them.

To emphasize the "light touch" that should be used with leveling screws, many surveying instructors require that their students carry out the following test: A leveling screw is loosened and a piece of paper is slipped underneath.   Then the screw is tightened just enough to hold the paper in place against a slight tug of the hand. The instructor then explains that leveling screws should never be turned with any greater force than is required to hold the paper.   This demonstration should make the student conscious of the need for care of these screws, which can become bound or even stripped if too much pressure is applied. If the threads are stripped, they usually have to be shipped back to the manufacturer for repair.

If the leveling screws do not turn easily, they may be cleaned with a solvent, such as gasoline, and the interior threads may be *very* lightly oiled with a light watch oil. When the level is taken indoors for storing or outdoors for use, its screws and clamps should be loosened because severe temperature changes may cause severe damage.

Level rod.   The level rod should never be dragged on the ground and its metal base should never be allowed to strike rocks, pavement, or other hard objects; such use will gradually wear away the metal base and will thereby cause leveling errors due to the change in the length of the rod itself.

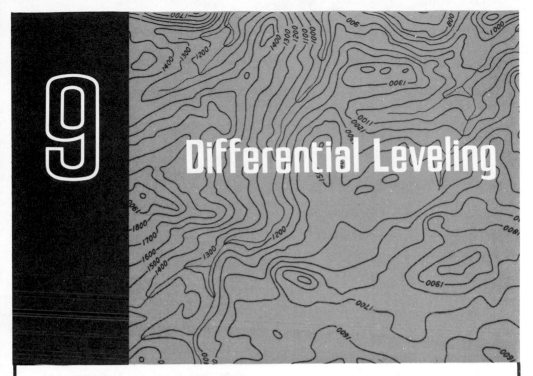

9 Differential Leveling

9-1 THEORY OF SPIRIT LEVELING

For an introductory description of direct or spirit leveling it is assumed that the surveyor has set up his instrument and has carefully leveled it. He then sights on the level rod held by the rodman on some point of known elevation (this sight is called a *backsight*, or BS). If he adds the backsight reading to the elevation of the known point, he will have the *height of the instrument* (HI), that is, the elevation of the line of sight of the telescope.

To illustrate this procedure, refer to Fig. 9-1 in which the HI is seen to equal $100.00 + 6.32 = 106.32$ ft.

If the HI is known, the telescope may be used to determine the elevation of other points in the vicinity by placing the level rod on some point whose elevation is desired and by taking a reading on the rod. Since the elevation of the point where the line of sight of the telescope intersects the rod is known (the HI), the rod reading called a *foresight* (FS) may be subtracted from the HI in order to obtain the elevation of the point in question.

The level may be moved to another area by using temporary points called *turning points* (TPs). The telescope is sighted on the rod held at a convenient turning point and a foresight is taken. This establishes the elevation of the point. Then the level can be moved beyond the TP and set up at a convenient location. A *backsight* is taken on the rod held at the turning point and the HI for the new location is established. This process can be repeated over and over for long distances.

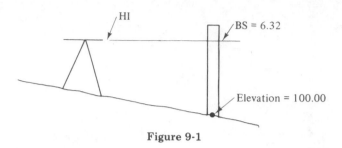

Figure 9-1

9-2 ADDITIONAL DEFINITIONS

A *bench mark* (BM) is a relatively permanent point of known elevation. It should be easily recognized and found and should be set fairly low in relation to the surrounding ground. It may be a concrete monument in the ground, a nail driven into a tree, an x mark in a concrete foundation, the top of a fire hydrant, or a similar object that is not likely to move. Bench marks that are to be permanent should be supported by structures which have completely settled and which extend below the frost line. The foundation of an old building usually meets these requirements very well. Parts of structures that must resist significant lateral forces such as retaining walls make rather poor bench marks. Such structures may move for years due to lateral earth pressure. Similar discussions can be made for the supporting foundations of poles and towers. Especially careful records should be made of bench marks because they may frequently be reused during the life of the job or for future work in the vicinity. They should be so completely and carefully described in the notes that another surveyor unfamiliar with the area can find and use them, perhaps years later.

A *turning point* (TP) is a temporary point whose elevation is determined during the process of leveling. It may be any convenient point on the ground, but it is usually wise to use a readily identifiable point such as a rock, a stake driven into the ground, a mark on the pavement, so that the level rod can be removed and put back in the same location as many times as required.

A *backsight* (BS) is a sight taken to the level rod held on a point of known elevation (either a BM or a TP) to determine the height of the instrument (HI). Backsights are also frequently referred to as *plus sights* because they are added to the elevations of points being sighted on to determine the height of the instrument.

A *foresight* (FS) is a sight taken to any point to determine its elevation. Foresights are often called *minus sights* because they are subtracted from HIs to obtain the elevations of points. *Notice that for any position of the instrument where the HI is known, any number of foresights may be taken to obtain elevations of other points in the area.*

The only limitations on the number of sights are the length of the level rod and the power of the telescope on the level. Obviously, one cannot sight on the level rod held at a point whose elevation is greater than that of the HI.

9-3 DIFFERENTIAL LEVELING

Differential leveling, which is the process of determining the difference in elevation between two points, is illustrated in Fig. 9-2 in which a line of levels is run from BM_1 to BM_2.

The instrumentman sets up his level at a convenient point and backsights on the level rod held on BM_1. This gives him his HI. The rodman moves to a convenient point (TP_1 in the figure) in the direction of BM_2. The instrumentman takes a foresight on the rod thus enabling him to compute the elevation of TP_1. He then moves the level to a convenient location beyond TP_1 and backsights on TP_1. This gives him his new HI. The rodman moves forward to a new location (TP_2), and so on. This procedure is continued until the elevation of BM_2 is determined.

The usual form for recording differential leveling notes is presented in Fig. 9-3 for the readings that were shown in Fig. 9-2. *The student would be wise to study these notes very carefully before he attempts leveling; otherwise he may become confused in recording his readings.*

In studying these notes he should particularly notice the math check. Since the backsights are positive and the foresights are negative, the surveyor should total them separately. The difference between these two totals must equal the difference between the initial and final elevations or else a math blunder has been made in the field book. *Since the math check is easy to make, there is no excuse for omitting it. For this reason, level notes are considered incomplete unless the check is made and shown in the notes.*

Another important point to observe in the notes is the careful description of the bench marks. Such a practice will enable the original

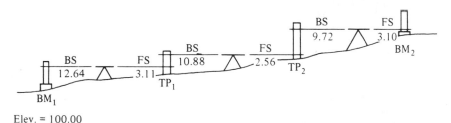

Figure 9-2

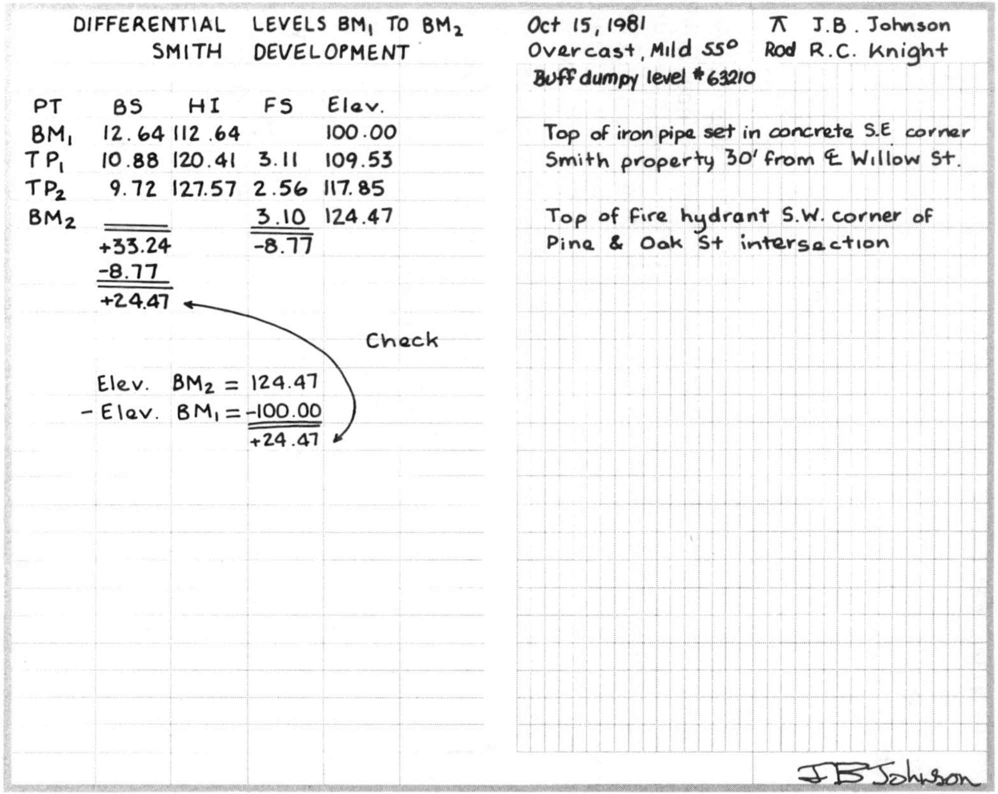

DIFFERENTIAL LEVELS BM₁ TO BM₂

SMITH DEVELOPMENT

PT	BS	HI	FS	Elev.
BM₁	12.64	112.64		100.00
TP₁	10.88	120.41	3.11	109.53
TP₂	9.72	127.57	2.56	117.85
BM₂			3.10	124.47

+33.24
-8.77
+24.47

Check

Elev. BM₂ = 124.47
- Elev. BM₁ = -100.00
+24.47

Oct 15, 1981   Π J.B. Johnson
Overcast, Mild 55°   Rod R.C. Knight
Buff dumpy level #6210

Top of iron pipe set in concrete S.E corner
Smith property 30' from E Willow St.

Top of fire hydrant S.W. corner of
Pine & Oak St intersection

J.B. Johnson

Figure 9-3. Differential level notes.

surveyor or another one to make use of these references while working in the area at a later date.

9-4 EARTH'S CURVATURE AND ATMOSPHERIC REFRACTION

Up to this point the discussion of leveling has assumed that when the instrument has been leveled its line of sight is a level line or a line of equal elevation. This is obviously not so and, in fact, the line of sight is perpendicular to a plumb line at one point only—at the instrument location. The student may then decide that the line of sight is a horizontal line, but this is not correct either.

When rays of light pass through air strata of different densities, they are refracted or bent downward. This means that in order to see an object on the ground some distance away, a person actually has to look above it. Refraction is greatest when the line of sight is near the ground or near bodies of water where temperature differences are large and, therefore, where large variations in air densities occur. Atmospheric refraction therefore varies with atmospheric conditions and has been known to be as large as 0.10 ft in a 200-ft distance. Under ordinary conditions, however, it is approximately equal to 0.093 ft in one mile and varies directly as the square of the horizontal distance.

Because of the earth's curvature, a horizontal line departs from a level line by 0.667 ft in one mile, also varying as the square of the horizontal distance. The effects of the earth's curvature and atmospheric refraction are represented in Fig. 9-4.

The combination of the earth's curvature and atmospheric refraction causes the telescope's line of sight to vary from a level line by approximately 0.667 minus 0.093 or 0.574 ft in one mile, varying as the square of the horizontal distance in miles. This may be represented in formula form as follows, where C is the departure of a telescope line of sight from a level line and M is the horizontal distance in miles:

$$C = 0.574M^2$$

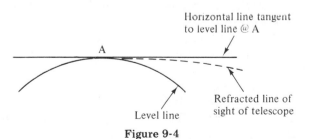

Figure 9-4

For a telescope reading on a rod 100 ft away, the reading would be in error by

$$(0.574) \left(\frac{100}{5280} \right)^2 = 0.000206 \text{ ft}$$

Similarly, a reading on a rod at a 300-ft distance would be in error by 0.00185 ft. At a 1000-ft distance the error would be 0.0206 ft.

If the backsight distance were exactly equal to the foresight distance for each setup of the instrument, it could be seen that errors caused by atmospheric refraction and the earth's curvature would cancel each other. Each of the readings would be too large by the same amount and since the same error would be added with the backsight and subtracted with the foresight, the net result would be to cancel them.

For surveys of ordinary precision, it is reasonable to neglect the effect of the earth's curvature and atmospheric refraction. The instrumentman might like to, by eye, make BS and FS distances approximately equal. For precise leveling, it is necessary to use more care in equalizing the distances. Pacing or even stadia may be used.

For SI units the correction for earth's curvature and atmospheric refraction is given by the following expression in which C is in meters and k is the distance in kilometers.

$$C = 0.0675 k^2$$

9-5 VERNIERS

A vernier is a device used for making readings on a divided scale closer than the smallest divisions on the scale. The vernier, which was invented by the Frenchman Pierre Vernier in 1620, is a short auxiliary scale attached to or moved along the divided scale.

Most of the targets used on level rods have verniers on them with which rod readings can be made to the nearest 0.001 ft. Figure 9-5 illustrates the construction and the reading of level rod verniers. The numbers on this particular rod are for 0.1 ft and 0.2 ft and thus the divisions in between are the 0.01 ft divisions. The vernier is shown to the right of the rod and is so constructed that 10 divisions on the vernier cover 9 divisions on the rod. Therefore, each division on the vernier is 0.009 ft.

In this figure the 0 or bottom mark on the vernier coincides with the 3.10 ft mark on the rod. It will be noted that the next division of the vernier falls $\frac{1}{10}$ of a division short of the next mark on the rod (or its 3.11 ft mark). The second division on the vernier falls $\frac{2}{10}$ of a division short of the 3.12 ft mark on the rod.

If the bottom of the vernier is moved up until the first division on the vernier coincides with the first division on the rod or the 3.11 ft mark, the bottom of the vernier will be located at 3.101 ft on the rod. In the same manner, if the vernier is moved up until the second division on the vernier coincides with the second mark on the main scale (3.12 ft), the bottom of the vernier will be located at 3.102 ft on the rod.

To read a level rod vernier, the bottom of the vernier (which is the center of the target) is lined up with the horizontal cross hair and the reading is determined by counting the number of vernier divisions until a division on the vernier coincides with a division on the rod. This reading is added to the last division on the rod below the bottom of the vernier. In Fig. 9-6 the rod reading at the bottom of the vernier is between 3.12 ft and 3.13 ft. The sixth division up on the vernier coincides with a division on the rod and so the rod reading is 3.12 plus the vernier reading 6 equals 3.126 ft. This also could be read as 3.18 – (6)(0.009) = 3.126 ft.

For the level rod and vernier described here, the smallest subdivision that can be read equals one-tenth of the scale division on the rod. Ten divisions on the vernier cover 9 divisions on the rod and the smallest subdivision that can be read with the vernier is

$$\frac{0.01}{10} = 0.001 \text{ ft}$$

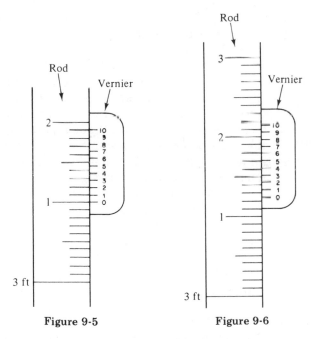

Figure 9-5   Figure 9-6

If another vernier is used for which 20 divisions of the vernier cover 19 divisions of the rod, the smallest subdivision that can be read is $\frac{1}{20}$ of the scale division or

$$\frac{0.01}{20} = 0.0005 \text{ ft}$$

From this information an expression can be written for the smallest subdivision that can be read with a particular vernier. Letting n be the number of vernier divisions, s the smallest division on the main scale, and D the smallest subdivision that can be read, the following expression can be written:

$$D = \frac{s}{n}$$

In surveying, one of the principal uses of verniers is for angle measurements and for this reason the subject of verniers is continued in Chapter 12 which deals with such measurements.

9-6 LEVEL ROD TARGETS

For long sights or for situations in which readings to the nearest 0.001 ft are desired, a level rod target may be used. Targets are small circular or elliptical pieces of metal approximately 5 in. in diameter painted red and white in alternate quadrants. They are clamped to the rods. As shown in Fig. 9-7, a vernier is part of the target.

The target is moved up or down as directed by the instrumentman until it appears to be bisected by the cross hair. At this point the horizontal cross hair coincides with the bottom of the vernier and the rodman takes the reading to the nearest 0.001 ft and this value is approximately checked by the instrumentman. An important point to remember about target readings is that although they are read to the nearest 0.001 ft, they can be no more accurate than the accuracy obtained in setting the target. The point is that for ordinary leveling "do not attach too much significance to readings taken to the nearest 0.001 ft." Although more accurate than those taken to the nearest 0.01 ft, the readings are probably not accurate to the third place. The reason is that the instrumentman just cannot signal the rodman to set the target on the rod as precisely as he can set the telescope cross hairs on the rod.

The precision of ordinary leveling can be increased somewhat by using targets as described in the preceding paragraph. Another way in which precision can be appreciably improved is by limiting the lengths of sights, perhaps to 100 ft or less, and by having the instrumentman estimate his rod readings through the telescope to the nearest 0.002 ft.

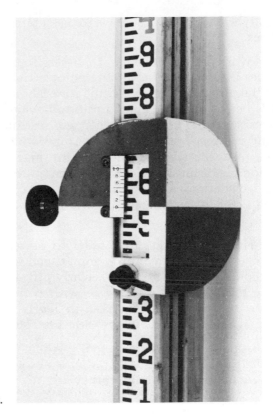

Figure 9-7. Level rod target.

9-7 COMMON LEVELING MISTAKES

The most common mistakes made in leveling are described in the following paragraphs.

Misreading the rod. Unless the instrumentman is very careful, he may occasionally read the rod incorrectly; for example, he may read 3.72 ft instead of 4.72 ft. This mistake most frequently occurs when the line of sight to the rod is partially obstructed by leaves, limbs, grass, rises in the ground, and so on. There are several ways to prevent such mistakes. The instrumentman should always carefully note the foot marks above or below the point where the horizontal cross hair intersects the rod. If he can't see those red foot marks, he may ask the rodman to slowly raise the rod until he sees a foot mark. To do this he may either call "raise for red" or give an appropriate hand signal as described in Section 9-12.

An excellent procedure for the instrumentman is to call out readings as he takes them. He can have the rodman, while still holding the rod properly, point to the reading with a pencil. Obviously, the pencil should coincide with the horizontal cross hair if the reading was taken correctly.   Another procedure is to use a target and have both men take readings.

Moving turning points.   A careless rodman causes serious leveling mistakes if he moves the turning points.   He holds the rod at one point while the instrumentman takes the foresight reading, and then while the level is being moved to a new position he puts the level rod down while he does something else.   If, when the instrumentman is ready for the BS, the rodman holds the rod at some other point, a serious mistake can be made because the new location may have an entirely different elevation.   Obviously, a good rodman prevents mistakes like this by using well-defined turning points or by clearly marking them with crayon (keel) if on pavement or by driving a stake or ax head into the earth.

Field note mistakes.   In order to prevent the recording of incorrect values, the instrumentman should call out the readings as he reads and records them.   This is particularly effective if the rodman is checking the readings with a pencil or with a target.   In order to prevent addition or subtraction mistakes in level notes, the math check described in Section 9-3 should be carefully followed.

Mistakes with extended rod.   When readings are taken on the extended portion of the level rod, it is absolutely necessary to have the two parts adjusted properly.   If they are not, mistakes will be made.

9-8 LEVELING ERRORS

A brief description of the most common leveling errors and suggested methods of minimizing them are presented in the next several paragraphs.

Level rod not vertical.   The instrumentman when sighting on the rod can see if the rod is leaning to one side or the other by means of the vertical cross hair in his telescope and, if necessary, he can signal the rodman to straighten up.   He cannot, however, usually tell if the rod is leaning a little toward or away from the instrument.   If the student thinks about this for a while, he will see that the smallest possible rod reading will occur when the rod is vertical. Many surveyors, therefore, have their rodman slowly "wave" the rod toward and away from the instrument and then record the smallest reading observed through the telescope.   A method used by some surveyors for ordinary leveling is to require each rodman (assuming that his stomach is not excessive) to hold the level rod so that it touches his nose and belt buckle, but this practice is not as satisfactory as waving the rod.   Some level rods, particularly those used for precise work, are equipped with individual circular levels which allow the rodman to plumb the rod by merely centering the bubble.   Other rods are equipped with conventional bubble tubes.   The use of rod levels or other methods of plumbing is preferable to waving the rod.   Waving the ordinary flat-bottom rod can cause small errors due to the rotation of the rod about its edges instead of about the center of its front face.

Setting of level rod.   It is essential to hold the level rod on firm definite points which will not settle and which are readily identifiable so that the rodman, if called away for some other work, may return to exactly the same spot.   If such convenient points are unavailable, it may be necessary to take turning points on the ordinary earth with the resulting possibility of settlement.   In order to minimize this possibility, the rodman should hold the rod on a stake driven in the ground, on the head of an ax stuck in the ground, or on some other similar base.

Mud, snow, or ice accumulation on base of rod.   If the rodman is not careful, he may easily have mud, snow, or ice sticking to the bottom of the rod.   This can cause severe errors in leveling.   He must learn not to drag the rod on the ground at any time and to keep the bottom of the rod carefully cleaned when there is snow, ice, or mud.

Rod not fully extended on high rod.   When the rear part of a Philadelphia rod is extended, it is called a high rod.   Such extension is necessary for readings from 7 to 13 ft.   Frequently, level rods have been damaged by letting the upper part of the rod slide down so rapidly that the blocks on the two sections are damaged.   The result is that the high rod readings may be in error and the rodman must carefully check the rod extension.

Incorrect rod length.   If a level rod is of incorrect length (and no rod is of perfect length), the rod readings will be in error.   If the length errors occur at the bottom of the rod they will theoretically be cancelled in the differential leveling process.   Errors due to a misfit in extending the rod, however, will not cancel if some of the readings are taken above the joint and some of them below.   This generally occurs when the surveyor is leveling up or down a slope.   For such cases he tends to read on the high part of the rod for downhill shots and on the low part for uphill shots.   Rod lengths should be checked periodically with a steel tape.

BS and FS distances not equal.   In Section 9-4 it was shown that if the lengths of backsights and foresights were kept equal for a particular setup, there would be no error caused by the earth's curvature and atmospheric refraction.   For ordinary work, it is sufficient to neglect or merely approximate by eye equal distances.   For more precise work, it is necessary to pace distances or even use stadia to keep BS and FS distances equal.

Errors due to instrument maladjustment are frequently more important than those due to atmospheric refraction and the earth's curvature. Particularly significant is the error produced if the axis of the bubble tube is not parallel to the line of sight of the telescope. (See Section 9-13.) If, however, BS and FS distances are kept equal, such errors will be greatly minimized.

Bubble not centered on level. If the bubble is not centered in the level tube when a reading is taken, the readings will be in error. It is suprising how easily this can happen. The instrumentman may sometimes brush against the instrument, the tripod legs may settle in soft ground, or the instrument may not be properly leveled or adjusted, with the result that when the telescope is turned the bubble does not stay centered. All of these factors mean that the instrumentman must be particularly careful. If he checks the bubble before and after readings are taken to be sure that it is centered, he will substantially reduce these errors.

Settling of level. In soft or swampy ground there definitely will be some settling of the tripod. Between the time of the backsight and the foresight readings there will be some settlement with the result that the foresight reading will be too small. Special care should be taken in selecting the firmest possible places to set up the instrument. In addition, as little time as possible should be taken between the readings (use two rodmen if possible). A further precaution in minimizing settlement errors is to take the foresight reading first on alternate setups.

Instrument out of adjustment. The adjustment of levels is discussed in Section 9-13 and is very important. The surveyor will with experience learn to constantly make simple checks to see that the instruments are adjusted properly.

Improper focusing of telescope (parallax). If a person looks at the speedometer of a car from different angles, he will read different values. This is due to *parallax*. If the indicator and the speedometer scale were located exactly in the same plane, parallax would be eliminated.

Sometimes when the surveyor is sighting through his telescope, he finds that if he moves his eye a slight distance from one side to the other, there is an apparent movement of the cross hairs on the image or the object seems to move. Again, this is parallax and it can cause appreciable errors unless it is corrected. The surveyor should carefully focus the objective lens until the image and the cross hairs appear to be exactly in the same place, that is, until they do not move in relation to each other when the eye is moved back and forth. The distorting effect of parallax will then be prevented.

Heat waves.   Sometimes heat waves are so intense that they cause large errors in rod readings. They may be so bad in the middle part of the day that work must be stopped until the waves subside. Heat wave errors can be minimized by reducing the lengths of sights. In addition, since the waves are worse near the ground, points should be selected so that the lines of sight are 3 or 4 or more feet above the ground.

Wind.   Occasionally, high winds cause accidental errors because the winds actually shake the instrument so much that it is difficult to keep the bubble centered. Shorter sight distances can lessen these errors.

9-9 SUGGESTIONS FOR GOOD LEVELING

After reading the lengthy list of mistakes and errors described in the last two sections, the novice surveyor may be as confused as the beginning golfer who is trying to remember 15 different things about his swing as he tries to hit the ball. In order to perform good leveling, however, remember the following few general rules:

1. Anchor tripod legs firmly.
2. Check to be sure that the bubble tube is centered before and after rod readings.
3. Take as little time as possible between BS and FS readings.
4. For each setup of the level, use BS and FS distances that are approximately equal.
5. Either provide rodmen with level rods that have level tubes (circular, conventional, etc.) with which the rods can be plumbed or have them wave the rods slowly toward and away from the instrument.

9-10 COMMENTS ON TELESCOPE READINGS

The instrumentman should learn to keep both eyes open when looking through the telescope. First, it is quite tiring to keep closing one eye all day to take readings. Second, it is convenient to keep one eye on the cross hair and the other eye open to locate the target.

If a person wears ordinary glasses for magnification purposes with no other corrections, he will not need to wear his glasses while looking through the telescope. The adjustment of the lens will compensate for the eye trouble.

9-11 PRECISION OF DIFFERENTIAL LEVELING

This section presents as a guide the approximate errors which should result in differential leveling of different degrees of precision. It is assumed that levels of average condition and in good adjustment are used. *Rough leveling* pertains here to preliminary surveys in which readings are taken only to the nearest 0.1 ft and in which sights of up to 1000 ft may be used. In *average leveling*, rod readings are taken to the nearest 0.01 ft, and BS and FS distances may be approximately balanced by eye, particularly when leveling on long declines or upgrades, and sights up to 500 ft may be used. It is probable that 90% of all leveling falls into this category. In *excellent leveling*, readings are made to the nearest 0.001 ft, BS and FS distances are approximately equalized by pacing, and readings are taken for distances no greater than 300 ft. For *precise leveling*, the procedures to be described in Section 10-3 are followed.

The average errors will probably be less than the values given here.[1] In these expressions M is the number of miles leveled and the values resulting from the expressions are in feet.

1. Rough leveling $\pm 0.4\sqrt{M}$
2. Average $\pm 0.1\sqrt{M}$
3. Excellent $\pm 0.05\sqrt{M}$
4. Precise $\pm 0.02\sqrt{M}$

For instance, if differential leveling is done over a route of 6 miles, the maximum error resulting from surveying of average precision should not exceed $\pm 0.1\sqrt{6} = \pm 0.24$ ft. These values are given for leveling done under ordinary conditions. Surveyors who frequently work in very hilly parts of the country might have some difficulty in maintaining these degrees of precision.

If very rough leveling is done using the hand level, the maximum error can be limited to approximately $3.0\sqrt{M}$.

The surveyor often encounters the terms first-order leveling, second-order leveling and so on. A few remarks are made here to explain these terms which may otherwise be confusing to the reader. More information on this topic and related subjects is presented in Chapter 23.

The Federal Geodetic Control Committee of the U.S. Department of Commerce has established three distinct orders of accuracy for

---

[1] This information on maximum probable errors is as given in *Elementary Plane Surveying* by Raymond E. Davis (New York: McGraw-Hill Book Company, 1955), pp. 82–83.

TABLE 9-1 Standards of Classification—Vertical Control (Federal Geodetic Control Committee)

	First-Order		Second-Order		Third-Order
	Class I	Class II	Class I	Class II	
Recommended uses	Basic framework of the National Network and metropolitan area control. Regional crustal movement studies. Extensive engineering projects. Support for subsidiary surveys.		Secondary framework of the National Network and metropolitan area control. Local crustal movement studies. Large engineering projects. Tidal boundary reference. Support for lower order surveys.	Densification within the National Network. Rapid subsidence studies. Local engineering projects. Topographic mapping.	Small-scale topographic mapping. Establishing gradients in mountainous areas. Small engineering projects. May or may not be adjusted to the National Network.
Relative accuracy between directly connected points or bench marks (standard error)	$0.5 \ \text{mm} \sqrt{K}$	$0.7 \ \text{mm} \sqrt{K}$	$1.0 \ \text{mm} \sqrt{K}$	$1.3 \ \text{mm} \sqrt{K}$	$2.0 \ \text{mm} \sqrt{K}$

vertical control or leveling. These are, in descending order: first-order, second-order and third-order with the first two orders subdivided into classes I and II. Table 9-1 is a copy of this classification system.[2] It will be noted in the table that first- and second-order leveling is applicable to the basic control for national geodetic surveys and to the secondary framework of those systems respectively. Second-order leveling and third-order leveling apply to large and small engineering projects respectively. In the table the letter K represents the distance between bench marks in kilometers.

9-12 HAND SIGNALS

For all types of surveying it is essential for the various personnel to keep close communication with each other. Very often calling back and forth is completely impractical because of the distances involved or because of noisy traffic or earth moving machinery in the area. In the absence of walkie-talkies, therefore, a set of hand signals that is clearly understood by everyone involved is often a necessity. Considering the material presented earlier in this chapter, they are particularly useful when the level rod target is being used.

The instrumentman should remember that he has a telescope with which he can observe the rodman; the rodman, however, cannot see the instrumentman nearly so clearly. As a result he must be very careful to give clear signals to the rodman. Following are the most commonly used hand signals (see also Fig. 9-8):

Plumb the rod. One arm is raised above the head and moved slowly in the direction that the rod should be leaned (Fig. 9-8a).

Wave the rod. The levelman holds one arm above his head and moves it from side to side (Fig. 9-8b).

High rod. To give the signal for extending the rod, the instrumentman holds his arms out horizontally and brings them together over his head (Fig. 9-8c).

Raise the rod. Sometimes for very short sights the red foot marks will not fall within the telescope's field of view and with the "raise for red" signal the instrumentman asks that the rod be raised a little so he can determine the correct foot mark. He holds one arm straight forward, with palm up, and raises it a short distance (Fig. 9-8d).

[2]*Classification, Standards of Accuracy, and General Specifications of Geodetic Control Surveys*, February 1974 (Rockville, Md.: U.S. Department of Commerce), p. 3.

(a) Plumb the rod.   (b) Wave the rod.   (c) High rod.

(d) Raise for red
(side view).   (e) All right.   (f) Pick up the
instrument.   (g) Raise the target.

(h) Lower the target.   (i) Clamp the target.

Figure 9-8

All right.   The arms are extended horizontally and waved up and down (Fig. 9-8e).

Pick up the instrument.   The party chief may give this signal when a new setup of the instrument is desired. The hands are raised quickly from a downward position as though an object is being lifted (Fig. 9-8f).

Raise the target.   If one hand is raised above the shoulder with the palm visible, it means to raise the target (Fig. 9-8g). If a large movement is needed, the hand is moved abruptly, but if only a small movement is needed, the hand is moved slowly.

Lower the target.   Lowering the hand below the waist means to lower the target (Fig. 9-8h).

Clamp the target.   The instrumentman, keeping one arm horizontal, moves his hand in vertical circles. This means to clamp the target (Fig. 9-8i).

9-13 ADJUSTMENT OF DUMPY LEVELS

A level is manufactured with great care and precision, but it must be
checked periodically to see that it is in proper adjustment. The instrument will get out of adjustment and parts will become worn and loose
fitting even though it may be handled very carefully. If it is handled
roughly, the problem will be magnified. This section is devoted to
those adjustments that can be made by the surveyor in the field. The
surveyor should constantly check his instruments to see that they are
properly adjusted. It is a good habit to check at least one of the relations discussed below each day that the instrument is used.

In the following paragraphs are listed the relations that should
exist in a dumpy level, the procedure that should be used to see if they
do exist, and the adjustments necessary if they don't. Figure 9-9 illustrates the terms that will be used in the following discussion:

The bubble tube.   When the instrument is leveled, the axis of
the bubble tube should be perpendicular to the vertical axis of the
instrument.

In order to check to see if this relation is present, the instrument is
carefully leveled over opposite pairs of leveling screws. Then the telescope is turned through $180°$, that is, end to end over the same pair of
screws. If the bubble moves, the desired relation does not exist. The
amount of movement of the bubble equals twice the error involved.
The reason for the doubled error is discussed in Section 9-14 under the
heading "Principle of Reversion." This can be corrected by moving the

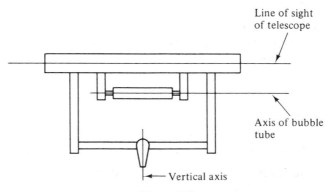

Figure 9-9

bubble halfway back to center by using the adjusting screws at the ends of the bubble tube. Once this is done, the test should be run again to see that the proper correction has been made.

The cross-hair ring.   The horizontal cross hair should lie in a plane perpendicular to the vertical axis of the level. If it does not meet this requirement, an error will occur with each reading. That error will tend to cancel if the same location on the cross hair is used each time for making readings. Furthermore the error will theoretically be zero if all readings are taken at the center of the reticle or that is if the vertical hair is centered on the rod each time.

In order to make this test, the level is set up and the telescope sighted on a sharply defined point with one end of the horizontal cross hair. Then the telescope is moved slowly about the vertical axis of the instrument (that is, the telescope is moved horizontally) to see if the cross hair remains on the point. If it does not, it is necessary to rotate the cross-hair ring. This is done by slightly loosening the four adjacent capstan screws on the ring and turning the ring slightly either with light pressure of the fingers or by tapping the ring lightly with a pencil. Then the test is run again and the process repeated until the desired relation exists.

Line of sight.   The line of sight of the telescope should be parallel to the axis of the bubble tube.

The "two-peg test" is commonly used to see whether or not this relation exists. The instrument is carefully set up at a convenient point. Stake A is driven in the ground, say 150 ft from the instrument, in one direction and stake B is driven in the ground the same distance on the other side of the instrument. Then, as shown in Fig. 9-10, readings a and b are taken on the rods. The correct difference in elevation between the two points equals $a - b$.

Next, the level is moved to one of the stakes, say B, and is placed so that the telescope will swing within approximately $\frac{1}{2}$ in. of the level rod held on the stake. The instrument cannot be focused on the rod that close, but if the instrumentman looks through the telescope backward to the rod (he will not be able to see the cross hairs), he will be able to see a very short segment of the rod. By holding a pencil at the center of this segment, he can take a very precise reading. This value

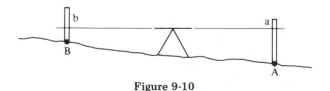

Figure 9-10

Figure 9-11

(labeled c in Fig. 9-11) will be correct even if the line of sight of the telescope is not parallel to the bubble tube axis.

The instrumentman can then take a reading on the rod at point A. If the instrument is properly adjusted, the difference in readings c and d will equal the difference between readings a and b. If reading d does not check, the instrument is out of adjustment. The correct value of d can be calculated and the instrument adjusted with the top and bottom cross-hair ring capstan screws. One is loosened and the other tightened, or vice versa, until the horizontal cross hair is on the desired reading. For example, if a = 6.42 and b = 3.32, the difference in elevation from A to B is 3.10. If reading c is 5.04, then d should equal 8.14 for an instrument in proper adjustment. The effects on the earth's curvature and atmospheric refraction have been neglected for this discussion.

The desired relations and tests have been given a particular order so that any adjustments made will have the least possible affect on other desired relations. When the series of tests is completed, it will be necessary to run through the series again (unless no adjustments were necessary) to see if the adjustments have affected other relations. For an instrument in extremely poor adjustment, it may be necessary to repeat the series several times.

9-14 PRINCIPLE OF REVERSION

Adjustments of surveying instruments are frequently checked by the principle of reversion. The instrument is inverted or turned 180° in its position with the result that the error in question is doubled and is thus more obvious. As an illustration of this principle, it is assumed that the bubble for a level is carefully centered as previously described. With the bubble centered and the telescope over a pair of leveling screws, the telescope is rotated through 180°. If the bubble does not remain centered, the axis of the bubble tube is not perpendicular to the vertical axis of the instrument. As will be described in the paragraphs to follow, the error has been doubled. As a result the bubble should be moved halfway back to the centered position by raising or lowering one end of the capstan screws at the end of the tube.

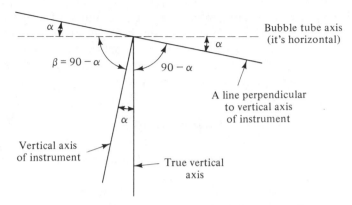

Figure 9-12. Initial position (bubble centered).

If the axis of the bubble tube is not perpendicular to the vertical axis of the instrument, the centering of the bubble will not put the vertical axis of the instrument in a truly vertical position. This can be seen in Fig. 9-12 where it is shown that if the bubble tube axis is not perpendicular to the vertical axis of the instrument by an angle α the instrument vertical axis will be located an angle α from the correct vertical axis.

In Fig. 9-12 the smaller angle from the vertical axis of the instrument to the bubble tube is designated as β and equals $90 - \alpha$. If the telescope is now rotated $180°$ horizontally the angle β will also rotate $180°$. The axis of the bubble tube will move away from its original horizontal position by an angle 2α as shown in Fig. 9-13. In order to correct the error, it will be necessary to raise or lower one end of the

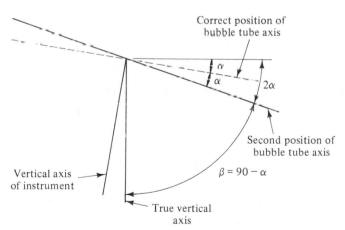

Figure 9-13. Second position (telescope rotated $180°$ from initial position).

bubble tube with the capstan screws at the ends of the tube until the bubble is halfway back to the center position. Then the axis of the bubble tube will be correctly perpendicular to the vertical axis of the instrument. If the bubble is centered, it will stay centered as the telescope is rotated around the vertical axis.

PROBLEMS

In Problems 9-1 to 9-3, complete and check the following sets of level notes.

9-1.

Station	BS	HI	FS	Elevation
BM_1	6.49			100.00
TP_1	8.21		6.48	
TP_2	9.06		5.03	
TP_3	10.34		3.96	
BM_2			1.09	

(*Ans.:* BM_2 = 117.54)

9-2.

Station	BS	HI	FS	Elevation
BM_1	6.34			342.60
TP_1	5.49		3.21	
TP_2	6.96		2.09	
TP_3	4.44		6.11	
TP_4	3.12		5.84	
BM_2			6.66	

9-3.

Station	BS	HI	FS	Elevation
BM_1	6.042			496.360
TP_1	5.874		5.620	
TP_2	4.926		4.114	
BM_2			6.764	
TP_3	5.482		3.626	
BM_3			1.942	

(*Ans.:* BM_3 = 503.382)

In Problems 9-4 to 9-6, set up and complete differential level notes for the information shown in the accompanying illustrations. Include the customary math checks.

9-4.

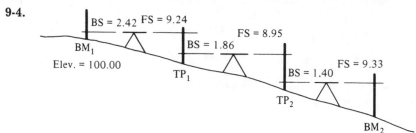

9-5.

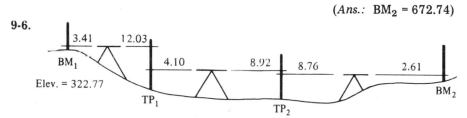

Elev. = 640.42

(*Ans.:* BM$_2$ = 672.74)

9-6.

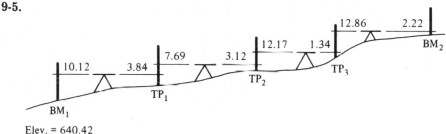

In Problems 9-7 to 9-9, the accompanying illustrations represent plans for differential leveling. The values shown on each line represent the sights taken along those
lines. Prepare and complete the necessary field notes for this work. (Y = instrument setup.)

9-7.

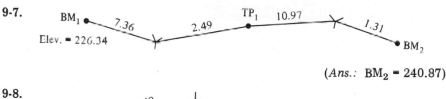

(*Ans.:* BM$_2$ = 240.87)

9-8.

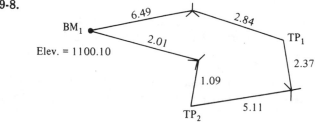

9-9.

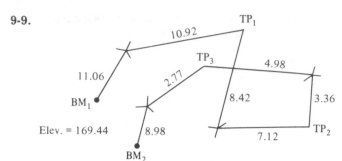

(*Ans.:* BM$_2$ = 163.05)

In Problems 9-10 to 9-14, the following level rod readings are given in the order in which they were taken. In each case the first reading is taken on BM$_1$ and the last reading is taken on BM$_2$, the point whose elevation is desired. Set up the differential level notes, including the customary math check. The elevation of BM$_1$ is given under each problem number.

Problem 9-10	Problem 9-11	Problem 9-12	Problem 9-13	Problem 9-14
100.00	256.68	1569.12	741.18	268.90
3.82	6.10	10.24	2.11	9.62
9.71	9.37	2.33	7.48	3.46
3.86	6.49	10.98	1.09	8.49
8.53	6.22	3.38	8.32	2.90
5.09	4.33	12.52	2.52	7.16
6.24	8.12	4.03	9.42	2.04
6.42	3.16	10.39	2.96	5.88
5.83	9.72	3.52	6.87	3.11
	4.14		2.26	
	3.89		9.82	

(*Ans. 9-11:*
BM$_2$ = 243.58)   (*Ans. 9-13:*
                         BM$_2$ = 710.21)

9-15. In running a line of levels from BM$_1$ (elevation 1242.40) to BM$_2$, the following readings were taken in the order given: 4.68, 3.24, 4.10, 2.88, 6.10, 3.47, 9.66 and 6.84. Set up and complete the level notes including the math check.   (*Ans.:* BM$_2$ = 1250.51)

9-16. In running a line of levels from BM$_1$ (elevation 162.11) to BM$_2$, the following readings were taken in the order given: 9.87, 3.98, 8.89, 4.09, 10.87,

5.55, 8.22, 5.29, 8.90, and 6.65. Set up and complete the level notes including the math check.

9-17. A line of levels was run into a mine shaft. All of the points (BMs and TPs) were located in the shaft ceiling and readings were taken by inverting the level rod. Complete the resulting level notes shown, including the math check.   (*Ans.:* BM_2 = 324.01)

Station	BS	HI	FS	Elevation
BM_1	9.33			342.18
TP_1	7.96		5.33	
TP_2	9.11		4.87	
TP_3	7.20		2.31	
BM_2			2.92	

9-18. Repeat Problem 9-16, assuming that all points (BMs and TPs) were located in the top of a tunnel and were taken by inverting the level rod.

9-19. Compute the combined effect of the earth's curvature and atmospheric refraction for distances of 100 ft; 200 ft; 500 ft; 2000 ft; 10 miles.
(*Ans.:* 0.0002 ft, 0.0008 ft, 0.0051 ft, 0.0824 ft, 57.4 ft)

9-20. A BS of 3.89 ft is taken on a level rod at a 100-ft distance, and a FS of 7.11 ft is taken on the rod held 1000 ft away. What is the error caused by the earth's curvature and atmospheric refraction? What is the correct difference in elevation between the two points?

9-21. In differential leveling from BM_1 to BM_2, the BS and FS distances for readings were as follows: BS 300 ft, FS 100 ft, BS 600 ft, FS 100 ft, BS 450 ft, FS 300 ft, BS 400 ft, and FS 200 ft. What is the error in the elevation of BM_2 caused by atmospheric refraction and the earth's curvature?
(*Ans.:* +0.014 ft)

9-22. Two towers A and B are located on flat ground and their bases have equal elevations above sea level. A person on tower A whose eye level is 50 ft above the ground can just see the top of tower B which is 90 ft above the ground. How far apart are the towers?

9-23. What BS or FS distances for an instrument setup will cause an error due to the earth's curvature and atmospheric refraction equal to 0.005 ft? 0.02 ft? 0.10 ft?   (*Ans.:* 493 ft, 986 ft, 2204 ft)

9-24. A man whose eye level is 5.0 ft above the ground is standing by the ocean. He can just see the top of a 120-ft lighthouse across the water. Neglecting tidal and wave effects, how far away is the lighthouse?

9-25. A surveyor is going to take a 10-mile sight across a lake from the top of one tower to a target on the top of another tower. It is desired to keep the line of sight 12 ft above the lake surface. At what equal heights above the shoreline should the instrument and the target be located?   (*Ans.:* 26.35 ft)

9-26. Calculate the error involved in the following level rod readings if a 13.0-ft rod is assumed to be 6 in. out of plumb at its top.
(a) A BS of 12.100 ft.
(b) A FS of 3.800 ft.

9-27. A line of levels is run from point A (elevation 842.46 ft) to point B to determine its elevation. The value so obtained is 614.86 ft. If a check of the rod after the work is done reveals that the bottom 0.04 ft has been worn away, what is the correct elevation of point B if there were 15 instrument setups?

(*Ans.:* 614.86 ft)

9-28. The smallest divisions on a level rod are $\frac{1}{10}$ of a foot. If 20 divisions on the vernier cover 19 divisions on the main scale, what is the least value readable?

9-29. The least divisions on a level rod are tenths of a foot. Describe a vernier which will enable the surveyor to read the rod to the nearest $\frac{1}{50}$ of a foot.

(*Ans.:* 5 divisions on vernier covering 4 divisions on rod)

9-30. For a level rod which is graduated to hundredths of a foot, design a target vernier so that it can be read to the nearest 0.002 ft.

9-31. Repeat Problem 9-29 except that the rod to be used will be read to the nearest 0.005 ft.   (*Ans.:* 20 divisions on vernier covering 19 divisions on rod)

9-32. The two-peg test is used to check to see if the line of sight of the telescope is parallel to the bubble tube axis. The instrument is set up halfway between points A and B and rod readings on A and B are respectively 6.46 and 7.09 ft. The level is moved very close to point A and readings of 5.10 ft on A and 5.67 ft on B are taken. What should this last reading on B equal for the instrument to be in proper adjustment?

9-33. A two-peg test is run as described in Problem 9-32. With the instrument halfway between points A and B the readings on those two points are respectively 1.923 and 2.162 m. With the level located near point A the reading on A is 1.524 m while on B it is 1.802 m. What should the last reading on B equal for the instrument to be in proper adjustment?   (*Ans.:* 1.763 m)

10-1 RECIPROCAL LEVELING

A useful method for leveling across wide, deep ravines, wide rivers, or other large bodies of water is called *reciprocal leveling*. We note that long sights are required across these features, and for such sights, errors will be large because of the earth's curvature and atmospheric refraction. In addition, rod reading errors and errors caused by imperfect instrument adjustment can be appreciable on these shots. Reciprocal leveling is especially devised in order to reduce these errors.

For this description, reference is made to Figs. 10-1 and 10-2 in which the elevation of point A on one side of a river is known and the elevation of point B on the other side is desired.

The instrument is set up very close to A, as shown in Fig. 10-1, and a BS is taken on a level rod held at A after which a foresight is taken on a level rod at B. Because the distance is large and the errors are appreciable, several FS readings are taken and averaged.

The instrument is moved to a point near B, as shown in Fig. 10-2; several BS readings are taken on A and averaged, after which a FS reading is taken on B.

When the instrument is near A, the average FS reading on B (a -sight) contains the appreciable error. When the instrument is near B, the average BS reading on A (a +sight) is the one that has the appreciable error. If the difference in elevation between points A and B, determined in each of the cases, is averaged, the large errors should

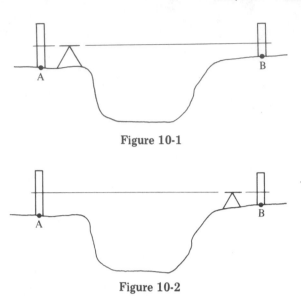

Figure 10-1

Figure 10-2

theoretically be removed, as shown in the following expressions where D is the difference in elevation between the two points. In these expressions it is assumed that the averages of the long sights have already been calculated.

$$D = \frac{[(BS) - (FS_{avg} \text{ including error})] + [(BS_{avg} \text{ including error}) - (FS)]}{2}$$

$$D = \frac{(\text{sum of backsights}) - (\text{sum of foresights})}{2}$$

If two leveling rods are used so that no appreciable time elapses between BS and FS readings, the precision of the work can be improved. In addition, the use of two levels with which simultaneous observations can be made will reduce the effects of atmospheric refraction variations.

10-2 ADJUSTMENTS OF LEVEL CIRCUITS

Levels over one route. If a surveyor runs a line of levels from a bench mark in order to set and establish the elevations of several bench marks some distance away, he will have to tie back into the original bench mark (or to some other bench mark). If he does not do this, he will not be able to check against serious discrepancies in his work. If he were, however, to level back to the starting bench mark, he would in all probability (no matter how careful his methods) obtain a value different from his starting one.   As a result he would have to adjust proportionately the measured elevations for the new bench marks which he set along the route.

The majority of the errors occurring in leveling are accidental, that is, there is just as much chance of getting too large a value with a particular reading as there is of getting too small a value.   It has previously been shown that the probable total error in a series of accidental errors tends to vary as the square root of the number of chances for the error to occur.   In leveling, therefore, the probable total error varies as the square root of the number of setups of the instrument.

For this discussion, it is assumed that the number of setups is the same in any one mile as in any other mile and thus the total probable error varies as the square root of the distance.

It can be shown that corrections for such errors should be proportional to the square of the probable errors, or, in this case, proportional to $(\sqrt{\text{distance}})^2$ and thus proportional to the distance.   It follows that the logical correction to the measured elevation of a particular point in a level circuit should be to the total correction as the distance to that point from the beginning is to the total circuit distance.

For this discussion, the line of levels shown in Fig. 10-3 is considered.   The surveyor starts from BM_1 with its known elevation and establishes BM_2, BM_3, and BM_4 and levels back to BM_1.   The observed elevation for each of the bench marks is shown in Table 10-1 with the calculations of their most probable elevations.

Levels over different routes.   If several lines of levels are run over different routes from a common beginning point to a common ending point where it is desired to establish a bench mark, it is obvious that different results will be obtained.   It is desired to obtain the most probable elevation of this new point.

As previously mentioned, leveling errors vary approximately as the square of the distance leveled.   Thus, in comparing different lines of levels to the same point, the shorter a particular route, the greater will be the importance or the weight given to its results.   In other words, the

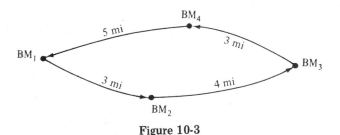

Figure 10-3

TABLE 10-1

Point	Distance from BM$_1$ (mi)	Observed Elevation	Correction	Most Probable Elevation
BM$_1$	0	200.00	0.00	200.00
BM$_2$	3	209.20	-0.06	209.14
BM$_3$	7	216.44	-0.14	216.30
BM$_4$	10	211.86	-0.20	211.66
BM$_1$	15	200.30	-0.30	200.00

shorter a line, the more accurate its results should be. Thus, the weight of an observed elevation varies inversely as the length of its line.

For the example shown in Fig. 10-4, route a is run for 4 mi from BM$_1$ in order to establish BM$_2$ where an elevation of 340.21 ft is obtained. A second route labeled b is also run from BM$_1$ to BM$_2$ over a distance of 2 mi and a measured elevation of 340.27 ft is obtained.

Because the second route is half as long as the first one, twice as much weight should be given to its observed value. From this information the most probable elevation of BM$_2$ can be obtained with the following expression:

$$\frac{(1)(340.21) + (2)(340.27)}{3} = 340.25 \text{ ft}$$

The preceding calculation is perfectly acceptable, but when several routes are involved and the numbers are not as simple, the calculations might be more convenient to handle as shown in Table 10-2. Here the weight given to each route equals the reciprocal of its length in miles.

Another example, somewhat more complicated, is presented in Table 10-3 and is shown in Fig. 10-5. Again, BM$_1$ has a known elevation and it is desired to set and establish the elevation of BM$_2$. To accomplish this objective, four different routes are run from BM$_1$ to BM$_2$. The lengths of each of these routes and the values of the elevations obtained for BM$_2$ are given in the table and the most probable elevation of BM$_2$ is determined.

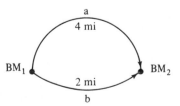

Figure 10-4

TABLE 10-2

Route	Length (mi)	Measured Elevation of BM_2	Measured Elevation -340.00	Weight of Route	Weighted Difference
a	4	340.21	0.21	$\frac{1}{4} = 0.25$	$\frac{0.25}{0.75} \times 0.21 = 0.07$
b	2	340.27	0.27	$\frac{1}{2} = 0.50$	$\frac{0.50}{0.75} \times 0.27 = 0.18$
				$\Sigma = 0.75$	$\Sigma = 0.25$

Most Probable Elevation of BM_2 = 340.25 ft

TABLE 10-3

Route	Length (mi)	Measured Elevation of BM_2	Measured Elevation -106.00	Weight of Route	Weighted Difference
a	1	106.50	0.50	$\frac{1}{1} = 1.00$	$\frac{1.00}{1.93} \times 0.50 = 0.26$
b	3	106.44	0.44	$\frac{1}{3} = 0.33$	$\frac{0.33}{1.93} \times 0.44 = 0.08$
c	2	106.52	0.52	$\frac{1}{2} = 0.50$	$\frac{0.50}{1.93} \times 0.52 = 0.13$
d	10	106.36	0.36	$\frac{1}{10} = 0.10$	$\frac{0.10}{1.93} \times 0.36 = 0.02$
				$\Sigma = 1.93$	$\Sigma = 0.49$

Most Probable Elevation of BM_2 = 106.49 ft

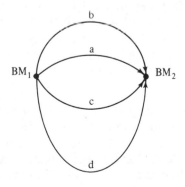

Figure 10-5

For a detailed discussion of level net adjustments the reader is referred to pages 157–167 of the book *Surveying for Civil Engineers*, 2nd ed., by the late Phillip Kissam, McGraw-Hill, Inc., 1981.

10-3 PRECISE LEVELING

The term *precise leveling* is usually applied to the leveling practiced by the National Geodetic Survey. The purpose of this section is not to describe in detail how precise leveling is performed but rather to indicate methods by which the average surveyor working with ordinary equipment can substantially raise the precision of his work. He can apply one or more of the procedures presented in the following paragraphs with improvements in his leveling for each one.

Miscellaneous. Several practices that improve the precision of leveling were mentioned in Chapter 9. These include the following: setting tripod legs firmly in solid ground, allowing the least time possible between BS and FS readings for each setup, using clearly marked TPs, limiting sight distances to 300 ft or less, careful plumbing of level rods, and avoiding leveling during strong winds and during severe heat waves. Another useful practice is to keep lines of sight at least 2 ft above the ground because of atmospheric refraction effects.

Shading the level. If a level is used on warm days in the direct sunlight, the result may be unequal expansion of different parts of the level with consequent errors. For instance, if one end of a bubble tube becomes warmer than the other end, the bubble will move toward the warmer end. This problem may be greatly minimized by shading the instrument from the direct rays of the sun with an umbrella while observations are being made and while the instrument is being moved from one point to another. Although the umbrella may make the surveyors more comfortable, that is not its purpose (see Fig. 10-6).

Three-wire leveling. For three-wire leveling, it is assumed that the level is equipped with stadia hairs in addition to its regular cross hairs. All three cross hairs are read to the nearest 0.001 ft by the rodman who uses a target. The average of the three readings is taken as the correct reading. It is helpful to balance BS and FS distances very well. This is readily done by reading the interval between the stadia hairs.

Precise leveling rods. For ordinary leveling, the usual level rod is satisfactory, but precision can be improved by using a rod treated in some manner against contraction and expansion in order to minimize the effects of changes of temperature and humidity. The preferable solution involves the use of a graduated invar tape which is independent of the main part of the rod so that it is free to slide in grooves on each

Figure 10-6 Precise leveling observations. (*Courtesy National Geodetic Survey.*)

side of the rod if the rod changes in length. Precise leveling rods are equipped with a thermometer and a bull's-eye level or a pair of ordinary levels placed at 90° to each other for plumbing the rod. The rods used by the National Geodetic Survey, are marked in a black and white checkerboard fashion so that the telescope cross hair will always fall on a white space where its position can be precisely estimated. Furthermore these rods have the characteristic of being clearly visible in both light and dark conditions.

Double-rodded lines.   The use of two level rods and two sets of TP, BS, and FS readings may improve leveling precision a little. Two sets of notes are kept, and the observed elevations of the points in question are averaged. When this procedure is used, it is desirable to use TPs for the two lines which have elevation differences of at least 1 ft or more so that the possibility of making the same 1 ft mistakes in

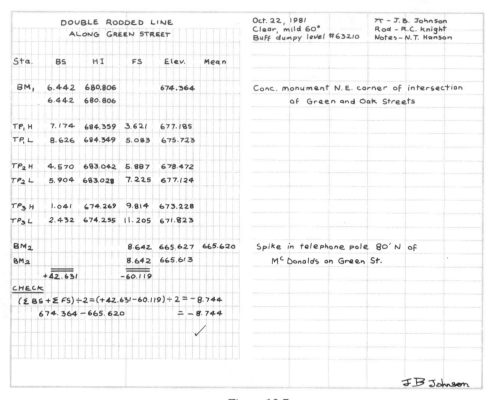

Sta.	BS	HI	FS	Elev.	Mean
BM₁	6.442	680.806		674.364	
	6.442	680.806			
TP₁ H	7.174	684.359	3.621	677.185	
TP₁ L	8.626	684.349	5.083	675.723	
TP₂ H	4.570	683.042	5.887	678.472	
TP₂ L	5.904	683.028	7.225	677.124	
TP₃ H	1.041	674.269	9.814	673.228	
TP₃ L	2.432	674.255	11.205	671.823	
BM₂			8.642	665.627	665.620
BM₂			8.642	665.613	

DOUBLE RODDED LINE
ALONG GREEN STREET

Oct. 22, 1981
Clear, mild 60°
Buff dumpy level #63210

π – J. B. Johnson
Rod – R.C. Knight
Notes – N.T. Hanson

Conc. monument N.E. corner of intersection
of Green and Oak Streets

Spike in telephone pole 80' N of
Mc Donalds on Green St.

+42.631   −60.119

CHECK

$(\Sigma BS + \Sigma FS) \div 2 = (+42.631 - 60.119) \div 2 = -8.744$

$674.364 - 665.620 \qquad = -8.744$

✓

J. B. Johnson

Figure 10-7

both lines is reduced. In order to minimize systematic errors, rods may be swapped between the lines at convenient intervals, for example, on alternate instrument setups. Double-rodded lines are also useful when it is necessary to run a quick set of differential levels in order to establish an elevation when there is not sufficient time to check the work by returning to the initial point or to some other point whose elevation has previously been established.

An example set of notes for a double-rodded line is shown in Fig. 10-7. In these notes the turning points are listed as H or L (highline or lowline) to identify the BS and FS values on the two different lines.

10-4 PROFILE LEVELING

For purposes of location, design, and construction it is necessary to determine elevations along proposed routes for highways, canals, railroads, water lines, and other similar projects. The process of determining a series of elevations along a fixed line is referred to as *profile leveling*.

Profile leveling consists of a line of differential levels with a series of intermediate shots taken during the process. The instrument is set up at a convenient point and a BS taken to a point of known elevation in order to determine the HI. If a bench mark of known elevation is not available, it is necessary to either establish one by differential levels from an existing BM or to set one up and give it an assumed elevation. (This latter procedure might not be satisfactory for some projects, for example, those involving water.)

After the HI is established, a series of foresights are taken along the center line of the project. These readings are referred to herein as *intermediate foresight readings* (IFS). Other surveyors may call them either *ground rod readings* or just rod readings. These readings are taken at regular intervals, say 50 to 100 ft, and at points where sudden changes in elevations occur, such as at the tops and bottoms of river banks, edges and center lines of roads and ditches, and so on. In other words, shots are taken where necessary in order to give a true picture of the ground surface along the route.

When it is no longer possible to continue with the IFS readings from the instrument position, it is necessary to take a FS to a TP and move the instrument to another position from which another series of readings can be taken. A portion of a typical set of profile level notes is shown in Fig. 10-8. The author has included a partial math check for the differential leveling part of these notes. The check includes the BS and FS values from the elevation of BM_{78} to the HI elevation just beyond TP_2.

It will be noted that when surveys are made along fixed routes such as these, the distances from the starting points are indicated by stationing. The usual practice is to set stakes along the center line of the project at regular intervals, say 50 to 100 ft. Points along the route with even multiples of 100 ft are referred to as *full stations* as 0 + 00, 1 + 00, 2 + 00, and so on. Intermediate stations are referred to as *plus stations.* For instance, a point 234.65 ft from the beginning station would be designated as 2 + 34.65.

It is usually unnecessary to take the intermediate foresight readings with as high a degree of precision as that needed for the regular FS values. If the surface is irregular, as in a field or a forest, readings to the nearest hundredth of a foot are unnecessary and perhaps a little misleading. They are usually taken to the nearest tenth of a foot, as in the example presented in Fig. 10-8. When surfaces are smooth and regular, as along a paved highway, it is not unreasonable to take the IFS readings to the nearest 0.01 ft.

During profile leveling it is usually very wise to set a series of BMs because they can be very useful at a later date, for example, when grades are being established for construction. These control points should be set a sufficient distance from the proposed project center

FOR PROFILE LEVELING HIGHWAY
PROPOSED CONGAREE

Oct 29, 1981 π - J.B. Johnson
Overcast, Mild 55° Rod - R.C. Knight
Buff dumpy level # 65210 Notes - N.T. Hanson

Sta.	BS	HI	FS	IFS	Elev.	
BM 78	3.11	103.11			100.00	Brass monument 208'N of sta 0+00
0+00				8.6	94.5	
+50				7.3	95.8	
1+00				5.9	97.2	
+50				5.8	97.3	
2+00				6.1	97.0	
+50				7.3	95.8	
TP₁	4.73	105.67	2.17		100.94	Stone
3+00				3.7	102.0	
+22.7				3.2	102.5	Top of ditch
+24				8.7	97.0	Bottom of ditch
+35.6				8.6	97.1	Bottom of ditch
+38				5.5	100.2	Top of ditch
+50				5.3	100.4	
4+00				7.0	98.7	
+50				8.0	97.7	
TP₂	6.07	110.68	1.06		104.61	Stump
5+00				11.9	98.8	
+50				10.0	100.7	

```
      +13.91
      -3.23     Elev. BM 78 = 100.00
      +10.68    Elev. HI    = 110.68
              Checks +10.68
```

J.B.Johnson

Figure 10-8

line so that they hopefully will not be obliterated during construction operations. When the profile is completed, it is necessary to check the work by tying into another BM or by running a line of differential levels back to the beginning point.

10-5 PROFILES

The purpose of profile leveling is to provide the information necessary to plot the elevation of the ground along the proposed route. A *profile* is the graphical intersection of a vertical plane, along the route in question, with the earth's surface. It is absolutely necessary for the planning of grades for roads, canals, railroads, sewers, and so on.

Figure 10-9 shows a typical profile of the center line of a proposed highway. It is plotted on profile paper especially made for this purpose, and having horizontal and vertical lines printed on it to represent distances both horizontally and vertically. It is common to use a vertical scale much larger than the horizontal one (usually 10:1) in order to make the elevation differences very clear. In this figure a horizontal scale of 1 in. = 200 ft and a vertical scale of 1 in. = 20 ft are used. The plotted elevations for the profile are connected by freehand because it is felt that the result is a better representation of the actual ground shape than would be the case if the points were connected with straight lines.

The author has sketched trial grade lines in the figure for the proposed highway. The major purpose of plotting the profile is to allow the establishment of grade lines. On an actual job, different grade lines are tried, until the cuts and fills balance sufficiently within reasonable haul distances with grades that are not too excessive.

For highway projects, it is normally convenient to plot the plan and profile on the same sheet. The top half of the paper is left plain so that the plan can be plotted above the profile. The plan is not shown in Fig. 10-9.

10-6 CROSS SECTIONS

Cross sections are lines of levels or short profiles made perpendicular to the center line of a project. They provide the information necessary for estimating quantities of earthwork. There are two general types of cross sections: the ones required for route projects such as roads and the ones required for borrow pits. This section is devoted to the first of these two types; the latter is discussed in Section 19-2.

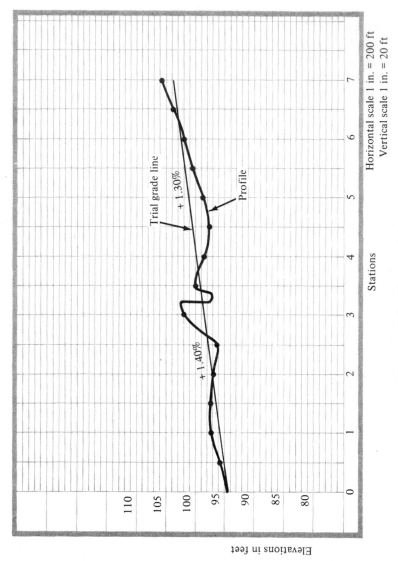

Horizontal scale 1 in. = 200 ft
Vertical scale 1 in. = 20 ft

Figure 10-9 Profile and trial grade lines.

For route surveys, cross sections are taken at regular intervals such
as the 50- or 100-ft stations and at sudden changes in the center-line
profile. There is a tendency among surveyors to take too few sections,
particularly in rough country. To serve their purpose, the sections must
extend a sufficient distance on each side of the center line so that the
complete area to be affected by the project is included. Where large
cuts or fills seem probable, greater distances from the center line should
be sectioned.

The necessary elevations can be determined with a regular level,
with a hand level, or with a combination of both. Sometimes a hand
level is held or fastened on top of a stick or board about 5 ft tall (called
a Jacob staff) and is used to determine the needed elevations. Elevations are usually taken at regular intervals as say 25, 50, 75 ft, and so
on, on each side of the route center line as well as at points where
significant changes in slopes or features (as streams, rocks, etc.) occur.

In Fig. 10-10 the level or Jacob staff is set up at the center line of
the project. The height of the telescope of the instrument above the
ground (HI in the figure) is measured and found to be 5.1 ft. The rodman is sent out 15 ft to the left and a FS reading of 11.2 is taken. The
difference in elevation from the center line to the point in question is
thus −6.1 ft. Then the rodman moves out 18 more ft to a point 33 ft
from the center line. The FS reading is 11.5 ft and the difference in
elevation is −6.4 ft. If the difference in elevation becomes too large for
the level rod height or if in going uphill the elevation differences are
larger than the HI of the instrument, it will then be necessary to take
one or more turning points to obtain all the necessary readings.

A convenient form of recording cross-section notes is shown in
Fig. 10-11. This figure includes profile notes on the left-hand page and
cross-section notes on the right-hand page. On the cross-section page,
the numerators are the differences in elevations from the center-line
stations to the points in question; the denominators are the distances

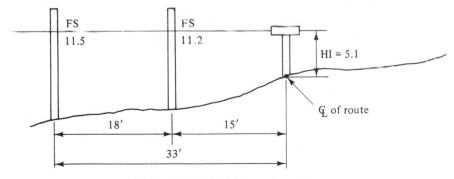

Figure 10-10 Taking cross sections.

Figure 10-11 Profile and cross-section notes.

CROSS SECTIONS FOR ROAD 10A

Nov 5, 1981
Overcast, Mild 50°
Buff Dumpy Level # 63210

𝛑 - J.B.Johnson
Rod - R.C.Knight
Notes - N.T.Hanson

STA	BS	HI	FS	IFS	ELEV	Left			℄	Right
+87				6.2	276.4	$+\frac{0.7}{35}$	$-\frac{0.4}{15}$		$-\frac{0.2}{18}$	$+\frac{0.1}{35}$
+50				6.3	276.3	$+\frac{0.4}{35}$	$+\frac{0.3}{15}$		$-\frac{0.4}{18}$	$-\frac{0.6}{35}$
13+00				6.9	275.7	$+\frac{0.5}{35}$	$+\frac{0.2}{16}$		$+\frac{0.2}{15}$	$+\frac{0.2}{33}$
TP	7.52	282.58	4.50		275.26					
+50				4.9	274.7	$-\frac{0.3}{35}$	$-\frac{1.4}{25}$	$-\frac{0.8}{12}$	$+\frac{0.2}{16}$	$+\frac{0.5}{33}$
12+00				5.1	274.5	$-\frac{8.2}{38}$	$-\frac{3.1}{33}$	$-\frac{2.3}{15}$	$+\frac{0.2}{17}$	$+\frac{0.8}{33}$
+50				5.2	274.4	$-\frac{9.0}{39}$	$-\frac{4.6}{33}$	$-\frac{3.0}{18}$	$\frac{0.0}{20}$	$\frac{0.0}{33}$
+23				5.2	274.4	$-\frac{7.3}{37}$	$-\frac{6.0}{33}$	$-\frac{4.0}{15}$	$\frac{0.0}{15}$	$+\frac{0.8}{33}$
11+00				5.3	274.3	$-\frac{6.3}{40}$	$-\frac{6.4}{33}$	$-\frac{6.1}{15}$	$+\frac{0.3}{15}$	$+\frac{0.1}{33}$
BM₅	5.87	279.56	3.31		273.69	Nail in root 12" pecan 90' left sta 10+50				

J.B.Johnson

from the center line. The surveyor must record carefully the signs for these numbers. A plus sign is given to the points that are higher than the center line and a minus sign to those below.

Many surveyors plot their level notes running up the page from the bottom. Such a form is quite logical in that, as the surveyor faces forward along the center line, the area to the right of the center line is shown on the right-hand side of the page and the area to the left is shown on the left-hand side.

PROBLEMS

10-1. In reciprocal leveling across a large ravine the following sets of rod readings were taken. Point A on one side has a known elevation of 319.66 ft; B is a point on the other side whose elevation is desired. Determine the elevation of point B.

Instrument near A: BS on A = 5.64; average FS on B = 8.46.

Instrument near B: average BS on A = 4.38; FS on B = 7.14.

(*Ans.:* 316.87 ft)

10-2. Reciprocal leveling across a river between points A and B gives the values shown. If the elevation of point A is 3050.67 ft, what is the elevation of point B?.

Instrument near A: BS = 3.42; FS readings = 6.84, 6.88, 6.83.

Instrument near B: BS readings = 4.02, 3.98, 4.00; FS = 7.12.

10-3. Adjust the following unbalanced level circuit.

Point	Distance from BM$_1$ (miles)	Observed Elevation (ft)
BM$_1$	0	919.40
BM$_2$	3	906.84
BM$_3$	9	910.17
BM$_4$	15	932.18
BM$_1$	18	919.76

(*Ans.:* BM$_2$ = 906.78 ft)
(*Ans.:* BM$_3$ = 909.99 ft)
(*Ans.:* BM$_4$ = 931.88 ft)

10-4. Adjust the unbalanced level circuit shown in the accompanying table.

Point	Distance from BM$_1$ (miles)	Observed Elevation (ft)
BM$_1$	0	811.16
BM$_2$	3	823.38
BM$_3$	7	831.77
BM$_4$	9	840.86
BM$_5$	14	820.08
BM$_1$	22	810.50

10-5. A line of levels was run to set the elevation of several bench marks. The following values were obtained:

Point	Distance from BM$_1$ (miles)	Observed Elevation (ft)
BM$_1$	0	834.30
BM$_2$	2	819.62
BM$_3$	5	801.38
BM$_4$	7	817.70
BM$_1$	11	834.16

(*Ans.:* BM$_2$ = 819.65 ft)
(*Ans.:* BM$_3$ = 801.44 ft)
(*Ans.:* BM$_4$ = 817.79 ft)

Is the closure satisfactory for excellent leveling ($\pm 0.05 \sqrt{M}$) as described in Chapter 9? If so, adjust the elevations for all of the bench marks.

10-6. A closed loop of differential levels was run to establish the elevations of several bench marks with the following results:

Distances (ft)		Observed Elevations (ft)	
A to B	6000	BM$_A$	419.66
B to C	5000	BM$_B$	423.76
C to D	3000	BM$_C$	431.09
D to E	4800	BM$_D$	422.16
E to A	2700	BM$_E$	416.28
		BM$_A$	419.85

Is the closure satisfactory for average leveling ($\pm 0.10 \sqrt{M}$) from Chapter 9? If so, adjust the elevations of all the bench marks.

10-7. BM_1 has a known elevation and it is desired to establish the elevation of BM_2 by running differential levels over three different routes from BM_1, as shown in the accompanying table. What is the most probable elevation of BM_2?

Route	Length (miles)	Measured Elevation of BM_2 (ft)
a	3	808.42
b	6	808.38
c	12	808.29

(*Ans.:* 808.39 ft)

10-8. Several lines of levels are run over different routes from BM_1 in order to set BM_2 and establish its elevation. The lengths of these routes and the value of the elevations determined are shown in the accompanying table. Determine the most probable elevation of BM_2.

Route	Length (miles)	Measured Elevation of BM_2 (ft)
a	1	291.16
b	3	291.42
c	7	291.54
d	12	291.02

10-9. Set up and complete the level notes for a double rodded line from BM_{11} (elevation 1642.324) to BM_{12}. In the following rod readings H refers to high route and L to low route: BS on BM_{11} = 7.342; FS on TP_1 H = 2.306; FS on TP_1 L = 4.107; BS on TP_1 H = 9.368; BS on TP_1 L = 11.162; FS on TP_2 H = 3.847; FS on TP_2 L = 5.111; BS on TP_2 H = 8.339; BS on TP_2 L = 9.619; FS on TP_3 H = 4.396; FS on TP_3 L = 5.448; BS on TP_3 H = 7.841; BS on TP_3 L = 8.896; FS on BM_{12} = 3.329.

(*Ans.:* 1661.342)

10-10. Complete the following set of profile level notes:

Station	BS	HI	FS	IFS	Elevation
BM$_1$	2.42				863.82
TP$_1$	3.04		11.41		
0 + 00				10.8	
1 + 00				9.7	
2 + 00				9.0	
2 + 70.5				6.4	
2 + 76.2				6.3	
3 + 00				8.2	
TP$_2$	1.03		4.21		
4 + 00				12.2	
5 + 00				9.7	
5 + 55.0				12.1	
5 + 65.2				11.8	
6 + 00				5.2	
BM$_2$			3.66		

(*Ans.:* BM$_2$ = 851.03 ft)

10.11 The center line for a proposed highway has been staked. With an HI of 820.24 ft the following intermediate foresights are taken at full stations beginning at Sta. 0 + 00: 1.2, 3.6, 4.1, 6.0, 6.9, 7.5, 8.6, 9.3, 10.2, 9.8, 9.2, 8.7, 8.1, 7.9, 8.0, 8.3, 8.8, and 9.4. Plot the profile for these stations.

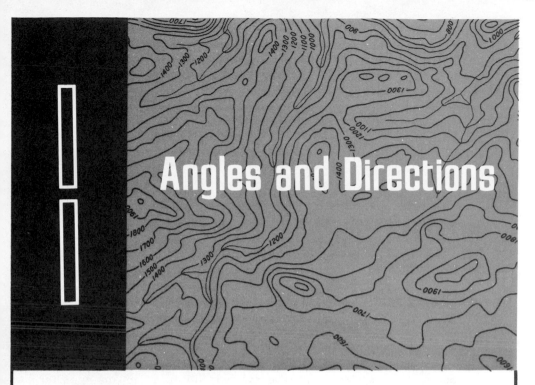

Angles and Directions

11-1 MERIDIANS

In surveying, the direction of a line is described by the horizontal angle that it makes with a reference line or direction. Usually this is done by referring to a fixed line of reference which is called a *meridian*. There are three types of meridians: true, magnetic, and assumed. A *true meridian* is the direction of a line passing through the geographic north and south poles and the observer's position, as shown in Fig. 11-1. True directions are obtained by sighting on the sun or on one of the numerous stars whose astronomical position is known (the North Star, or Polaris, is the most common). The subject of true directions is discussed in Chapter 18. A *magnetic meridian* is the direction taken by the magnetized needle of a compass at the observer's position; an *assumed meridian* is an arbitrary direction taken for convenience.

True meridians should be used for all surveys of large extent and, in fact, are desirable for all surveys of land boundaries. They do not change with time and can be reestablished decades later. Magnetic meridians have the disadvantage of being affected by many factors some of which vary with time. Furthermore there is no precise method available for establishing what magnetic north was years ago in a given locality. An assumed meridian may be used satisfactorily for many surveys which are of limited extent. The direction of an assumed meridian is usually taken roughly in the direction of a true meridian.

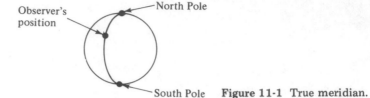

Figure 11-1 True meridian.

These meridians have a severe disadvantage—the problem of reestablishing their direction if the points of the survey on the ground are lost.

11-2 UNITS FOR MEASURING ANGLES

Among the methods used for expressing the magnitude of plane angles are the sexagesimal system, the centesimal system, and the methods using radians, and mils. These systems are briefly described in the following paragraphs.

Sexagesimal system. In the United States as in many other countries the sexagesimal system is used in which the circumference of circles is divided into 360 parts or degrees. The degrees are further divided into minutes and seconds. Thus an angle may be written as $36°27'32''$. The National Geodetic Survey as it adopts the metric system in 1983 plans to continue using the sexagesimal system.

Centesimal system. In some countries, particularly in Europe, the centesimal system is used in which the circle is divided into 400 parts called grads (or grades). It will be noted that $100^g = 90°$. The grads are each divided into 100 centigrads or 100^c (also called centesimal minutes) and the centigrads are each divided into 100 decimilligrads or 100^{cc} (also called centesimal seconds). The centesimal system is a decimal system which is of some advantage when mathematical calculations are being made. An angle may be expressed as 122.3968^g. The first pair of digits beyond the decimal represents the centigrads while the second pair of digits beyond the decimal represents the decimilligrads. Occasionally the centesimal minutes and seconds are written with primes and double primes. For instance the preceding angle could be written as $122^g39'68''$.

The radian. Another measure of angles frequently used particularly for calculation purposes is the radian. One radian is defined as the angle subtended at the center of a circle by an arc length exactly equal to the radius of the circle. This definition is illustrated in Fig. 11-2.

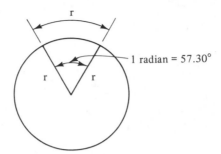

Figure 11-2 The radian.

The circumference of a circle equals 2π times its radius r and thus there are 2π radians in a circle. Therefore one radian equals

$$\frac{360°}{2\pi} = 57.30°$$

(One grad is equal to 0.01571 radian.)

The mil.   Another system of angular units divides the circumference of the circle into 6400 parts or mils. This particular system of angle measurement is primarily used in military science.

11-3 AZIMUTHS

A common term used for designating the direction of a line is the *azimuth*. The azimuth of a line is defined as the clockwise angle from the north end or the south end of the reference meridian to the line in question. For ordinary plane surveys, azimuths are generally measured from the north end of the meridian. This will be the case for the problems presented in this text. Astronomers and the National Geodetic Survey, however, work with the southern end of the meridian. In 1983 or 1984 the NGS plans to start using north azimuths.

The magnitude of an azimuth can fall anywhere from 0° to 360°. The north azimuths of several lines AB, AC, AD are shown in Fig. 11-3. The values are respectively 60°, 172°, and 284°.

Every line has two azimuths (forward and back). Their values differ by 180° from each other, depending on which end of the line is being considered. For instance, the *forward azimuth* of line AB is 60° and its *back azimuth*, which may be obtained by adding or subtracting 180°, equals 240°. Similarly, the back azimuths of lines AC and AD are 352° and 104°, respectively.

Azimuths are referred to as true, magnetic, or assumed, depending

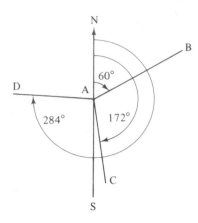

Figure 11-3

on the meridian used. The type of meridian being used should be clearly indicated.

11-4 BEARINGS

Another method of describing the direction of a line is to give its *bearing*. The bearing of a line is defined as the smallest angle which that line makes with the reference meridian. It cannot be greater than 90°. In this manner, bearings are measured in relation to the north or south ends of the meridian and are placed in one of the quadrants so that they have values of NE, NW, SE, or SW.

In Fig. 11-3 the bearing of line *AB* is N60°E, that of *AC* is S8°E, and that of *AD* is N76°W. As with azimuths, it will be noticed that every line has two bearings, depending on which end of the line is being considered. For instance, the bearing of line *BA* in Fig. 11-3 is S60°W.

Depending on the reference meridians being used, bearings may be true, magnetic, or assumed. Thus, it is important, as it is for azimuths, to indicate clearly the type of reference meridian being used.

11-5 THE COMPASS

Man has been blessed on earth with a wonderful direction finder, the magnetic poles. The earth's magnetic field and the use of the compass have been known to navigators and surveyors for many centuries. In fact, before the sextant and transit were developed, the compass was the only means by which the surveyor could measure angles and directions.

For many centuries there has been a legend which says that the compass was originally developed by a Chinese emperor who had to fight a battle in a heavy fog. The story goes that he invented a chariot which always pointed toward the south and he thereby was able to locate his enemy. Actually, no one knows who first developed the compass and the honor is claimed by the Greeks, the Italians, the Finns, the Arabs, and a few others. Regardless of its inventor, it is known that the compass was available to sailors during the Middle Ages.

The magnetic poles are not points but oval areas that are located not at the geographic poles but approximately 1300 miles away. To-day, the magnetic north pole is located approximately in the northern part of Canada's Baffin Island. The compass needle lines up with magnetic north and in most places this means the needle points slightly east or west of true north, depending on the locality. The angle between true north and magnetic north is referred to as the *magnetic declination*. (Navigators call it the *variation* of the compass.) The magnetic lines of force in the northern hemisphere are also inclined downward from the horizontal toward the magnetic north pole. The magnetized needles of compasses are counterbalanced with a little coil of brass wire on their south ends against the *dip* of the needles at the north end.

Not only are the magnetic fields not located at the geographic poles of the earth, but magnetic directions are also subject to several variations: long-term variations, annual variations, daily variations, as well as variations caused by magnetic storms, local attractions, and so on.

11-6 VARIATIONS IN MAGNETIC DECLINATION

The angle of declination at a particular location is not constant but varies with time. For periods of approximately 150 years there is a gradual unexplainable shift in the earth's magnetic fields in one direction after which a gradual drift occurs in the other direction to complete the cycle in the next 150-year period. This variation, called the *secular variation*, can be very large and it is quite important in checking old surveys whose directions were established with a compass. There is no known method of predicting the secular change and all that can be done is to make observations of its magnitude at various places around the world. Records kept in London for several centuries show a range of magnetic declination from $11°E$ in 1580 to $24°W$ in 1820.[1] The

---

[1] R. C. Brinker and P. R. Wolf, *Elementary Surveying*, 6th ed. (New York: Thomas Y. Crowell, Inc., 1977), p. 159.

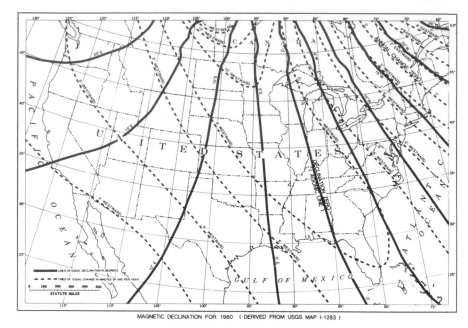

MAGNETIC DECLINATION FOR 1980 (DERIVED FROM USGS MAP I-1283)

Figure 11-4 Magnetic declination for 1980 (derived from USGS map I-1283). (*Courtesy U.S. Geological Survey, Department of the Interior.*)

time period between extreme eastern and western declinations varies with the locality. It can be as short as 50 years or even less and as long as 180 years or more.

Maps are available which provide the values of magnetic declination throughout the United States. The chart shown in Fig. 11-4 was prepared from observations at more than 8500 stations throughout the country. On this chart points of equal declination are connected with lines called *isogonic lines*. For some parts of the country magnetic declinations are zero and the lines connecting them are called *agonic lines*. It is thought that this chart provides magnetic declinations within about 15′ of their correct values for more than half of the country. For some extreme cases, however, they may be in error by as much as 1°.

In addition to secular variations in the magnetic declinations, there are also annual and daily variations of lesser importance. The *annual variations* usually amount to less than a 1-minute variation in the earth's magnetic field. Daily there is a swing of the compass needle through a cycle, causing a variation of as much as approximately one-tenth of one degree. The magnitude of this *daily variation* is still so small in comparison to the inaccuracies with which the magnetic compass can be read that it can be neglected.

11-7 LOCAL ATTRACTION

The direction taken by a compass needle is affected by magnetic attractions other than that of the earth's magnetic field. Fences, underground pipes, reinforcing bars, passing cars, nearby buildings, iron ore deposits under the earth's surface, and other steel or iron objects may have a considerable effect on compass readings. In addition, the effect of power lines, particularly because of variations in voltage, may be so great that compasses are useless in some areas. Even the surveyor's wristwatch, pen, belt buckle, taping pins, or steel tape may have a distorting effect on compass readings.

On many occasions the surveyor may not realize that the magnetic bearings he reads with the compass have been affected by local attractions. In order to detect local attraction, he must read both forward and back bearings for each line to see if they correspond reasonably well. To read a forward bearing, the surveyor sights along a line in the direction of the next point. In order to read a back bearing for the same line, he moves to the next point and sights back to the preceding point. If the two readings vary significantly from each other, local attraction has probably been the cause. This topic is continued in Section 11-9.

11-8 READING BEARINGS WITH A COMPASS

Compass readings may be used for surveys in which speed is important and when only limited precision is required. In addition, they may be used for checking more precise surveys or for rerunning old property lines that were originally run with compasses. They may also be used for preliminary surveys, rough mapping, timber cruising, and checking angle measurements.

For many years the surveyor's compass (see Fig. 11-5) was used for determining directions. This instrument was originally set up on a single leg called a Jacob staff. Later, tripods were used. The reader should particularly notice the folding upright sight vanes or peep sights which were used for alignment. These sights were fine slits running nearly the whole length of the vanes. This obsolete instrument (now a collector's item on the antique circuit) was formerly used for laying off old land boundaries. Today it is occasionally used when attempting to reestablish old property lines that were originally run with the same type of instrument.

In order to read a bearing with a surveyor's compass or with the compass on a transit, the surveyor sets up his instrument at a point at

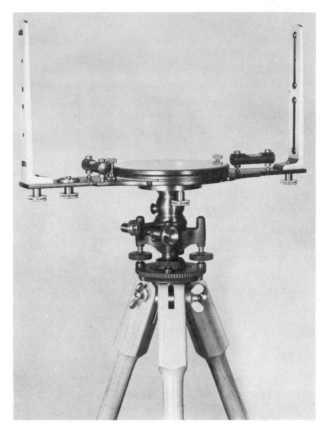

Figure 11-5 Surveyor's compass. (*Courtesy of Teledyne Gurley.*)

one end of the line whose bearing is desired, releases the needle clamp, and sights down the line to a point at the other end of the line. He then reads the bearing on the circle at the north end of the needle and reclamps the compass needle. When the compass is unclamped, a jeweled bearing at the center of the needle rests on a sharp pivot point, allowing the needle to swing freely. It is important to take care of the point and keep it from becoming dull and causing a sluggish needle. This is avoided by reclamping the needle, that is, by raising the needle off the pivot when the compass is not in use in order to minimize wear.

The condition or sensitivity of a compass needle can easily be checked by drawing the needle out of position with a piece of magnetic material such as a knife blade. If when the magnetic material is removed the needle returns closely to its original position, it's in good condition. If the needle is sluggish and does not return closely to its original position, the pivot point can be resharpened with a stone—a rather difficult task and it's probably better to have it replaced.

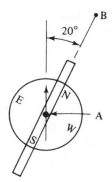

Figure 11-6 The telescope is over point A and sighted toward B. The bearing is N20°E. Notice that the scale turns with the telescope, but the needle continues to point toward magnetic north.

In Fig. 11-6 the instrument is assumed to be located at point A and the telescope is sighted toward point B so that bearing AB can be determined. It will be noticed that whichever way the telescope is directed the needle points toward magnetic north and bearings are read at the north end of the needle. For this to be possible, the positions of E and W must be reversed on the compass. Otherwise, when the telescope is turned to the NE quadrant, the needle would be between N and W on the compass. This fact can be seen in Fig. 11-6.

Sometimes in handling a compass, its glass face will become electrically charged because of the friction created by rubbing it with the hands or a piece of clothing. When this happens, the needle may be attracted to the undersurface of the glass by the electric charge. This charge may be removed from the glass either by breathing on it or by touching the glass at several points with wet fingers.

Many compasses, for example the ones on transits, are equipped so that they may be adjusted for magnetic declinations; that is, the scale or circle under the compass which is graduated in degrees may be rotated in relation to the telescope by the amount of the declination. In this way the compass may be used to read directly approximate true bearings even though declinations may be quite large.

In rerunning old surveys that were originally done with compasses it is frequently necessary to try to reestablish some obliterated lines with the compass. In such cases it is absolutely necessary (if many years have elapsed) to estimate changes that have occurred in magnetic declinations.

For new surveys, it is usually desirable to work with true bearings. This practice provides a permanence to surveys and greatly simplifies the work of future surveyors. A true bearing may be determined for one line (probably from observations of the sun or Polaris), and the true bearings for the other lines may be calculated from the angles measured with the transit or theodolite to the nearest minute or closer. Of course, if there are a great many lines involved, it is well to recheck true directions every so often during the survey. (Magnetic bearings may well be used for present-day surveys by reading the magnetic bearing for one line and by computing the bearings for the other sides from the measured angles, as was the case for the true bearings.)

Despite the increasing use of true directions, it is a good practice to read magnetic bearings for each line. If mistakes are made in angle measurements, the magnetic bearings may frequently be used to determine where the mistakes occurred.

Compasses may only be read to approximately the nearest 30 minutes and perhaps estimated to the nearest 15 minutes. Transits and theodolites may be read to the nearest minute or closer. In addition, compass readings may be decidedly affected by local magnetic attractions. As a result, *the precision obtained is quite limited, and compasses are becoming more nearly obsolete each year.*

11-9 DETECTING LOCAL ATTRACTION

When magnetic bearings are read with a compass, local attraction can frequently be a problem; thus all readings should be carefully checked. This is normally accomplished by reading the bearing of each line from both of its ends. If the forward and back bearings differ by 180°, there probably is no appreciable local attraction. If the bearings do not differ by 180° local attraction is present and the problem is to discover which bearing is correct.

For this discussion, consider the traverse of Fig. 11-7. It is assumed that the forward and back bearings were read for line AB as being S81°30'E and N83°15'W, respectively. Obviously local attraction is present at one or both ends of the line.

One method of determining the correct value is to read bearings from points A and B to a third point as perhaps C in the figure and then to move to that third point and read back bearings to A and B. If forward bearing AC agrees with back bearing CA, this shows that there is no appreciable local attraction at either A or C. Therefore the local attraction is at B and the correct magnetic bearing from A to B is S81°30'E.

It should be noted that, at a particular point, local attraction will draw the needle a certain amount from the magnetic meridian. Thus all readings taken from that point should have the same error due to local attraction (assuming the attraction is not a changing value as say volt-

A   C

      B   D   **Figure 11-7**

age variations in a power line), and the angles computed from the bearings at that point should not be affected by the local attraction.[2]

11-10 TRAVERSE AND ANGLE DEFINITIONS

Before proceeding with a discussion of angle measurement, several important definitions concerning position and direction need to be introduced. In surveying, the relative positions of the various points (or stations) of the survey must be determined. Generally, this is done by measuring the straight line distances between the points and the angles between those lines.

A *traverse* may be defined as a series of successive straight lines that are connected together. They may be *closed*, as are the boundary lines of a piece of land, or they may be *open*, as for highway, railroad, or other route surveys. Several types of angles are used in traversing and these are defined in the following paragraphs.

An *interior angle* is one that is enclosed by the sides of a closed traverse (see Fig. 11-8).

An *exterior angle* is one that is not enclosed by the sides of a closed traverse (see Fig. 11-8).

An *angle to the right* is the clockwise angle between the preceding line and the next line of a traverse. In Fig. 11-9 it is assumed that the surveying party is proceeding along the traverse from A to B to C, and so on. At C the angle to the right is obtained by sighting back to B and measuring the clockwise angle to D.

A *deflection angle* is the angle between the extension of the preceding line and the present one. Two deflection angles are shown in Fig. 11-10. There the necessity for designating them as being either right (R) or left (L) should be noted. In each of these cases it is as-

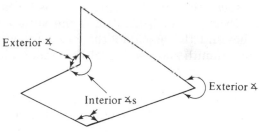

Figure 11-8

[2]C. B. Breed and G. L. Hosmer, *Elementary Surveying, 11th ed.* (New York: John Wiley & Sons, Inc., 1977), pp. 27–28.

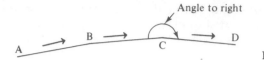

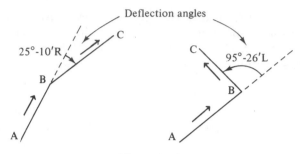

Figure 11-9

Figure 11-10

sumed that traversing is proceeding from *A* to *B* to *C*, and so on. The use of deflection angles permits easy visualization of traverses and facilitates their representation on paper. In addition the calculations of successive bearings or azimuths is very simple.

11-11 TRAVERSE COMPUTATIONS

If the bearing of one side of a traverse has been determined and the angles between the sides have been measured, the bearings of the other sides can easily be calculated. Actually several possible methods of solving such a problem can be used but regardless of which procedure is chosen, preparation of a careful sketch of the known data is required. Once the sketch is made, the required calculations are obvious.

   One way to solve most of these problems is by making use of deflection angles. Example 11-1 illustrates the situation in which the bearing of one line and the angle to the next line are given and it is desired to find the azimuth and bearing of that second line.

Example 11-1

For the traverse shown in Fig. 11-11, the bearing of side *AB* is given as well as the interior angle at *B*. Compute the north azimuth and the bearing of side *BC*.

Solution   A sketch is made of point *B* showing the north–south and east–west directions; the location of *AB* is extended past *B* (as shown by the dotted line), and the deflection angle at *B* is determined (see Fig. 11-12). With the value of this angle known the azimuth and the bearing of side *BC* are obvious.

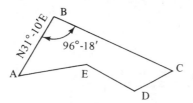

Figure 11-11

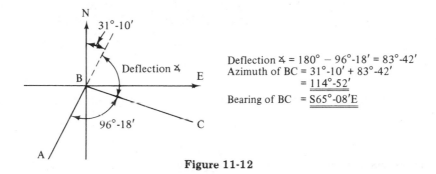

Deflection ⋊ = $180° - 96°\text{-}18' = 83°\text{-}42'$
Azimuth of BC = $31°\text{-}10' + 83°\text{-}42'$
$= \underline{114°\text{-}52'}$

Bearing of BC   = $\underline{\underline{S65°\text{-}08'E}}$

Figure 11-12

If the bearings of two successive lines are known, it is easy to compute the angles between them. A sketch is made, the deflection angle is calculated, and the value of any angle desired becomes evident. Example 11-2 presents the solution of this type of problem.

Example 11-2

For the traverse shown in Fig. 11-13 the bearings of sides AB and BC are given. Compute the interior angle at B.

Solution   A sketch is made (Fig. 11-14), the deflection angle from line AB extended to line BC is calculated, and any needed angle is immediately obvious.

At the end of this chapter are given several exercise problems dealing with angles, bearings, and azimuths. Before proceeding with the chapters that follow, the reader should be sure that he can handle all of these problems. He or she should be able to compute azimuths of lines from their bearings, and vice versa, calculate deflection angles from bearings, work from angles to bearings and bearings to angles, and be

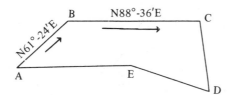

Figure 11-13

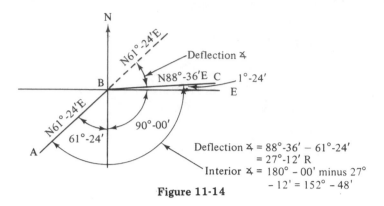

Figure 11-14

able to solve the magnetic declination problems discussed in Section 11-12.

11-12 MAGNETIC DECLINATION PROBLEMS

The reader may become confused when first attempting to calculate the true bearing of a line from its magnetic bearing for which the magnetic declination is known. A problem presenting similar difficulties is the calculation of the magnetic bearing of a line today when its magnetic bearing at some time many years ago is known, as are the entirely different magnetic declinations at the two times. These problems, however, may easily be handled if the student remembers one simple rule: *Make a careful sketch of the data given.* Example 11-3 presents the solution of a magnetic declination problem.

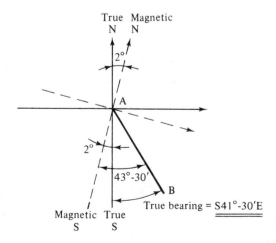

Figure 11-15

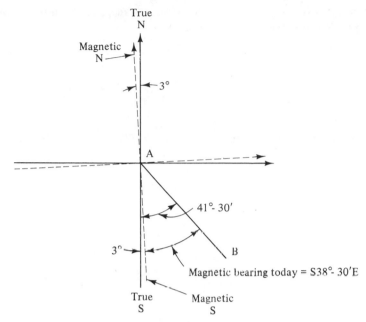

True
N

Magnetic
N

3°

A

41°-30'

3°

B

Magnetic bearing today = S38°-30'E

True
S

Magnetic
S

Figure 11-16

Example 11-3

The magnetic bearing of line AB was recorded as S43°30'E in 1888. If the declination was 2°00'E, what is the true bearing of the line? If the declination is now 3°00'W, what is the magnetic bearing of the line today?

Solution   A sketch is made (Fig. 11-15) in which the true directions (N, S, E, W) are shown as solid lines and the magnetic directions are shown as dotted lines. Magnetic north is shown 2°00'E or clockwise of true north and line AB is shown in its proper position 43°30'E of magnetic south. Once the sketch is completed, the true bearing of the line is obvious.

To determine the magnetic bearing of the line today another sketch is made (Fig. 11-16) showing the true bearing of the line and the present declination. From this completed sketch the magnetic bearing of the line today is obvious.

PROBLEMS

11-1. (a) Convert 36.4285^g to sexagesimal units.   (*Ans.:* 32°47'08")
      (b) Convert 61°17'15" to centesimal units.   (*Ans.:* 68.0972^g)

11-2. A circular arc has a radius of 420.00 ft and a central angle of 41°16'30". Determine the central angle in radians and the arc length.

11-3. Three lines have the following north azimuths: 132°16', 237°33', and 306°57'. What are their bearings?

(*Ans.:* S47°44'E, S57°33'W, N53°03'W)

11-4. Determine the north azimuths for sides *AB*, *BC*, and *CD* in the accompanying sketch for which the bearings are given.

11-5. Calculate the north azimuth for sides *OA*, *OB*, *OC*, and *OD* in the accompanying figure.

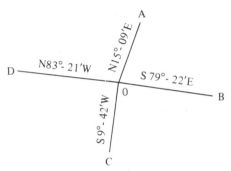

(*Ans.:* 15°09′, 100°38′, 189°42′, 276°39′)

11-6. Find the bearings of sides *BC* and *CD* in the accompanying figure.

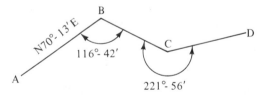

11-7. Compute the bearings of sides *BC* and *CD* in the accompanying figure.

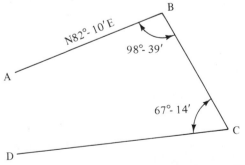

(*Ans.:* S16°29′E, N83°43′W)

11-8. What are the bearings of sides *CD*, *DE*, *EA*, and *AB* in the accompanying figure?

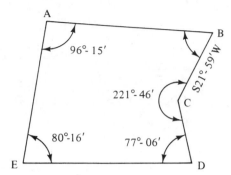

11-9. Determine the angles *AOB*, *BOC*, and *DOA* for the figure of Problem 11-5.

(*Ans.:* 85°29′, 89°04′, 98°30′)

11-10. Compute the value of the interior angles at *B* and *C* for the figure shown.

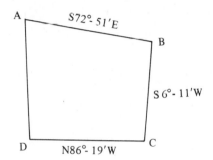

In Problems 11-11 to 11-14, compute all of the interior angles for each of the figures shown.

11-11.

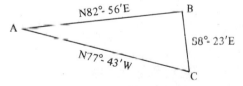

(*Ans.:* A = 19°21′, B = 91°19′, C = 69°20′)

11-12.

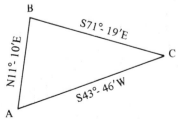

11-13.

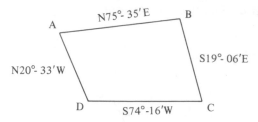

(*Ans.:* A = 83°52′, B = 94°41′, C = 86°38′, D = 94°49′)

11-14.

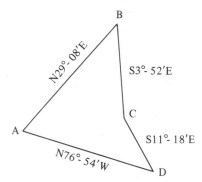

11-15. Compute the deflection angles for the traverse of Problem 11-12.

(*Ans.:* A = 147°24′R, B = 97°31′R, C = 115°05′R)

11-16. Compute the deflection angles for the traverse of Problem 11-14.

11-17. From the data given, compute the missing bearings.

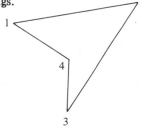

  (a) 1–2 = _____
  (b) 2–3 = _____
  (c) 3–4 = N3°18′W
  (d) 4–1 = _____
  (e) Interior ∢ @ 1 = 51°37′
  (f) Interior ∢ @ 2 = 35°56′
  (g) Interior ∢ @ 4 = 226°14′

(*Ans.:* 1–2 = N78°51′E, 2–3 = S42°55′W, 4–1 = N49°32′W)

11-18. From the data given, compute the missing bearings.

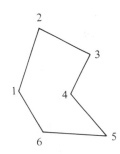

  (a) 1–2 = N27°58′E
  (b) 2–3 = _____
  (c) 3–4 = S38°16′W
  (d) 4–5 = _____
  (e) 5–6 = N77°43′W
  (f) 6–1 = _____
  (g) Interior ∢ @ 1 = 122°16′
  (h) Interior ∢ @ 2 = 80°15′
  (i) Interior ∢ @ 4 = 239°57′

11-19. For the accompanying figure, compute the following:
    (a) deflection angle at B (*Ans.:* $74°32'$R)
    (b) interior angle at B (*Ans.:* $105°28'$)
    (c) bearing of line CD (*Ans.:* S$23°46'$E)
    (d) north azimuth of DA (*Ans.:* $298°51'$)

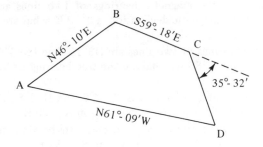

11-20. For the figure shown, compute the following:
    (a) bearing of line AB
    (b) interior angle at C
    (c) north azimuth of line DE
    (d) deflection angle at B

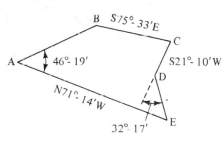

11-21. For the figure shown, compute the following:
    (a) the deflection angle at B (*Ans.:* $93°46'$R)
    (b) the bearing of CD (*Ans.:* S$54°09'$W)
    (c) the north azimuth of DE (*Ans.:* $162°48'$)
    (d) the interior angle at E (*Ans.:* $66°02'$)
    (e) the exterior angle at F (*Ans.:* $254°04'$)

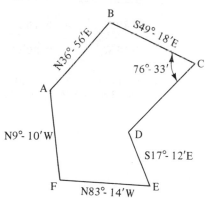

11-22. The following values are deflection angles for a closed traverse: (a) $116°17'R$; (b) $86°56'R$; (c) $107°23'R$; (d) $81°19'L$; and (e) $130°43'R$. If the bearing of side CD is $S85°51'W$, compute the bearings of the other sides.

11-23. The magnetic north azimuth of a line is $306°30'$ while the magnetic declination is $4°15'W$. What is the true azimuth of the line?   (*Ans.:* $302°15'$)

11-24. At a given place the magnetic bearings of two lines are $N46°32'E$ and $S61°38'E$. If the magnetic declination is $3°10'E$, what are the true bearings of the lines?

11-25. The magnetic bearings of two lines are $N6°15'E$ and $S86°45'E$. If the magnetic declination is $8°30'W$, what are the true bearings of the lines?
$$(Ans.: \text{ N}2°15'\text{W, N}84°45'\text{E})$$

11-26. The true bearings of two lines are $S87°14'W$ and $N78°52'E$. Compute their magnetic bearings if the magnetic declination is $6°00'E$.

11-27. Change the following true bearings to magnetic bearings for a $5°30'E$ magnetic declination: $N3°16'E$, $N4°18'W$, and $S86°37'E$.
$$(Ans.: \text{ N}2°14'\text{W, N}9°48'\text{W, N}87°53'\text{E})$$

11-28. The magnetic north azimuth of a line was $132°15'$ in 1890 when the magnetic declination was $6°30'E$. If the magnetic declination is now $1°45'W$, determine the true azimuth of the line and its magnetic azimuth today.

11-29. In 1860 the magnetic bearing of a line was $S85°30'E$ and the magnetic declination was $3°30'E$. Compute the magnetic bearing of this line today if the magnetic declination is now $3°W$. What is the true bearing of this line?   (*Ans.:* True = $S82°00'E$, Magnetic = $S79°00'E$)

From the information given in Problems 11-30 to 11-33, determine the true bearing of each line and its magnetic bearing today.

	Magnetic Bearing	Magnetic Declination in 1905	Magnetic Declination Today
11-30.	$S59°30'W$	$4°30'W$	$2°30'E$
11-31.	$S87°15'E$	$6°15'E$	$3°15'W$
11-32.	$N3°30'W$	$5°30'E$	$9°15'E$
11-33.	$N4°30'E$	$7°45'W$	$3°30'W$

(*Ans. 11-31:* True = $S81°00'E$, Magnetic today = $S77°45'E$)
(*Ans. 11-33:* True = $N3°15'W$, Magnetic today = $N0°15'E$)

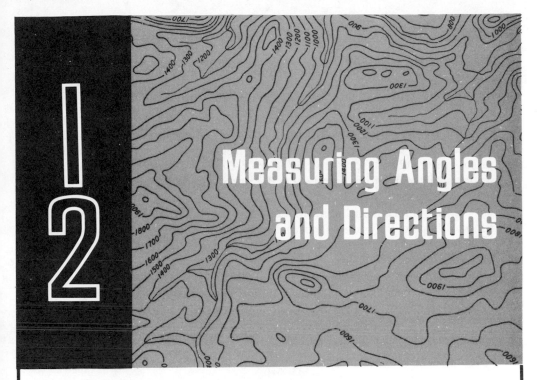

Measuring Angles and Directions

12-1 INTRODUCTION

The instruments used for measuring horizontal and vertical angles are divided into two groups—transits and theodolites—but the distinction between the two is not clear. Originally, both these instruments were called theodolites. The source of the term theodolite is not known with certainty. In any case, the theodolite is an instrument used for measuring horizontal and vertical angles. The first theodolites were manufactured with long telescopes that could not be inverted, but as time passed some were made with shorter telescopes that could be inverted end for end or "transited," and they were called transits.

Today most instruments have telescopes that can be inverted; thus the original distinction between the two no longer applies and to a large extent what they are called is a matter of local usage. It seems to be the convention that the more precise and optically read instruments are called theodolites. Several European companies have long been producing superb, optically read instruments with which American firms have been unable to compete.

The earlier theodolites used verniers and micrometer microscopes for reading angles, but modern theodolites have optical systems with which the user may obtain both horizontal and vertical angles through an eyepiece located near the telescope. Optical theodolites have several features quite distinct from normal transits: The scales are lighted for night use; there are three leveling screws; and there are no telescope levels and no compasses. Furthermore, they not only have enclosed optical systems for reading angles, but they also have optical systems for centering the instruments over points.

Actually, theodolites have horizontal and vertical circles for angle measurements, as do transits, but the circles are made of glass instead of metal. Light passes through the glass circles, and with the aid of glass prisms the readings from the circles are reflected into the eyepiece. The upper scale seen in the eyepiece is the vertical circle and the lower scale is the horizontal circle.

Some instruments combine features of both the transit and the theodolite and they are called *optically read transits*. They have glass circles as do theodolites but angles are read by means of glass verniers and magnifying eyepieces.

Most American surveyors use transits, although theodolites have several advantages over transits (higher precision for less money and fewer chances of making mistakes). The result is that they have re-

Figure 12-1 Transit. (*Courtesy Teledyne Gurley.*)

placed transits for some types of work and their use is constantly increasing in the U.S.   The first theodolite has been credited to the Danish astronomer Roemer, who in 1690 constructed the instrument for the observation of stars.   It was nearly a century later before it was used for surveying work.[1]   Actually, the modern transit as it is known today is a development of the nineteenth century.

The surveyor's transit (Fig. 12-1) is a versatile instrument that may be used to measure vertical and horizontal angles, prolong straight lines, perform leveling of a precision almost equivalent to that obtained with levels, determine magnetic bearings, and measure distances by stadia. This universal instrument is used for land surveying, mapping, construction surveys, and astronomical observations.   It has long been recognized as the American surveyor's most important instrument.

The transit is typically used to measure angles to the nearest one minute of arc, but better instruments are available (at somewhat higher costs) with which angles may be read to the nearest 30, 20, 15, or even 10 seconds.   Theodolites with which angles may be read to the nearest one second are common, but some European theodolites are available with which angles may be read to the nearest 0.2 seconds.   Sections 12-2 through 12-11 are devoted to transits while Sections 12-12 through 12-16 are concerned with theodolites.

12-2 PARTS OF THE TRANSIT

The various parts of a transit and their use may best be learned by handling and working with them.   For this reason, only a brief introduction to these parts is presented here.   Detailed discussions of their operations and manipulations are given in the following sections of this chapter.   The transit consists of three fundamental parts: the alidade, the horizontal circle, and the leveling head assembly.   See Fig. 12-2 for a picture of a transit and these parts.

The *alidade*, or upper plate, is the top part of the instrument and includes a circular cover plate that is rigidly connected to a solid conical shaft called the *inner spindle*.   It also includes the telescope (with a magnification power of approximately 18 to 28 diameters), the telescope level tube, the vertical circle (which is mounted on the telescope for measuring vertical angles), two plate leveling tubes (one parallel to the telescope and one perpendicular to it), two pairs of verniers for reading the horizontal circle, and probably a compass.   The telescope

---

[1] A. R. Legault, H. M. McMaster, and R. R. Marlette, *Surveying* (Englewood Cliffs, N.J.: Prentice-Hall, Inc., 1956), p. 83.

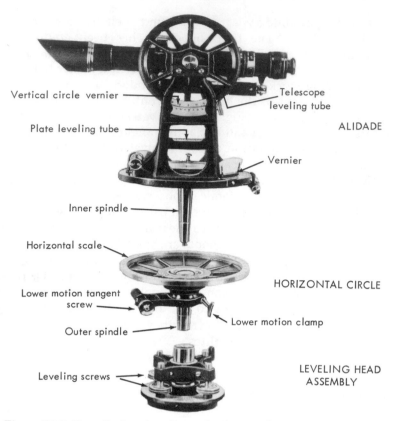

Figure 12-2 Transit showing three fundamental parts. (*Courtesy of Keuffel & Esser Co.*)

rotates up or down on a pair of side shafts which are called *trunnions*. This word was originally used to refer to the same side shafts on cannons.

The *horizontal circle* or lower plate is the scale with which horizontal angles are measured. It is usually graduated in degrees and halves of degrees and numbered every 10 degrees clockwise and counterclockwise. The underside of the horizontal circle is attached to a hollow, vertical, tapered spindle called the *outer spindle* into which the inner spindle fits. In addition, there is a clamp (called the *lower motion clamp*) that is used to permit or prevent rotation of the outer spindle.

The *leveling head assembly* consists of the four leveling screws and the footplate. The leveling screws are set into cups in order to prevent them from scoring the footplate. Included in the assembly is a device that permits the transit to be moved laterally from $\frac{1}{4}$ to $\frac{3}{8}$ in. without movement of the tripod.

12-3 CARE OF THE TRANSIT

Although the general rules given here for taking care of the transit are closely related to the ones previously given for the level, the subject is of sufficient importance to warrant some repetition. The major rule, as for the level, is "don't drop the instrument" because very serious damage will almost surely result. The following list presents other important items to remember in caring for the transit:

1. When the transit is being transported in a vehicle, the instrument should be held in the lap or kept in its box.
2. Hold the transit in the arms and not on the shoulder when carrying it inside a building.
3. Place the tripod with its legs well apart and sunk firmly in the ground.
4. Do not place the instrument on a smooth, hard surface such as a concrete slab unless some provision (such as a triangular frame) is made to keep the tripod legs from slipping.
5. Do not turn its screws tightly. If more than fingertip force is needed to turn them, the instrument either needs cleaning or repairs. Overtightening may lead to appreciable damage to the instrument.
6. Never leave an instrument unattended because it may be upset by wind, vehicles, children, farm animals, or it may be stolen.
7. If precipitation occurs, put the dustcap over the objective lens. In addition, it is a good idea to carry a waterproof silk bag with which to cover the transit.
8. If the instrument gets wet, let it dry because wiping may damage the finish as well as the scale graduations, and the lens may even be scratched.
9. If the lenses get dusty, clean them only with a camel hair brush, lens paper, or, less desirably, with a soft silk rag.
10. The working parts of a transit are lubricated with graphite and thus only a *very very* slight amount of oil or none at all should be used on the instrument. Oil tends to collect dust and it also becomes sticky at low temperatures.

12-4 SETTING UP THE TRANSIT

In order to measure angles, the transit is set up over a definite point such as an iron pin or a tack in a stake. For centering purposes, a plumb bob is suspended from a hook or chain beneath the instrument.

Of course, many modern transits are equipped with optical plumbing devices that permit very quick and accurate centering. These so called optical plummets were previously discussed in Section 7-9.

The transit is leveled in very much the same way that the level is, but there is one exception. Because the transit has two plate bubble tubes at right angles to each other, the instrumentman has only to turn the telescope until the axis of each of these tubes is parallel to a line through opposite leveling screws. He may then proceed with the leveling operation without having to turn the telescope as he works from one pair of leveling screws to the other.

When both bubble tubes are properly centered, the telescope may be turned through 180° to see if the bubbles remain centered. If the bubbles move, the instrument is out of adjustment, but if the movement is slight, for example, less than one division, the instrumentman is wasting time in trying to recenter the bubble every time he moves the telescope.

In order to center the transit over a particular point, the instrumentman places the tripod so that the plumb bob hangs approximately over the point. In doing this he tries to keep the leveling head approxi-

Figure 12-3 Transit set on a steep slope. Notice the manner in which the tripod is set with one leg uphill and two legs downhill.

mately level. He roughly levels the instrument to see where the plumb bob will be. If it is an appreciable distance off center, he picks up the tripod without moving the relative position of the legs and moves the instrument in the desired direction. He repeats the rough leveling process after which he may have to move the instrument again. When he gets the plumb bob fairly close to center, he may move it closer and closer by pressing one or two tripod legs more firmly into the ground. (Some surveyors use transits that have adjustable leg tripods that permit the lengthening or shortening of the legs as required to center the plumb bob.)

Finally, when the plumb bob is within approximately $\frac{1}{4}$ to $\frac{3}{8}$ in. of its proper position, two adjacent leveling screws are loosened. This loosens all four of the leveling screws and the head of the instrument may be pushed or slid over the required distance. When the plumb bob is centered, the instrumentman retightens the screws, relevels the instrument, and checks to see if the plumb bob remains centered. If it does not, he may have to repeat this last part of the procedure.

12-5 READING TRANSIT VERNIERS

In Section 9-5, the reading of level rod verniers was discussed. In this section the subject is continued but in relation to angle measurements. The verniers used for reading horizontal and vertical angles are identical in principle with target rod verniers. In other words, a person who can use the level rod vernier can use a transit vernier just as well.

The graduated scale on the lower plate is provided with two sets of numbers, one increasing in a clockwise direction from 0° to 360° and the other increasing from 0° to 360° in a counterclockwise direction. For reading these scales, two pairs of verniers (A and B) are provided. These are located on opposite sides of the upper plate, the A vernier being located below the eyepiece for convenience. There are two A verniers and two B verniers because one vernier is needed at each location for the clockwise numbers and one for the counterclockwise numbers. These verniers are located in windows in the cover plate covered with glass.

For most angle measurements, the A vernier is the only one used, but for more precise work, both A and B verniers are used and the mean of the readings is used. The purpose of taking the two sets of readings is to reduce instrumental errors caused by imperfections in the scale and the vernier.

For the average transit scale, the smallest divisions are one-half degrees and the verniers are so constructed that 30 divisions on the vernier cover 29 divisions on the main scale. As a result, the smallest subdivision which can be read is

$$D = \frac{s}{n} = \frac{\frac{1}{2}^\circ}{30} = 1'$$

Pictures of two transit verniers, together with example readings, are illustrated in Fig. 12-4. When the telescope has been turned in a

STYLES OF GRADUATIONS

The circles and verniers of transits are graduated in various ways. The usual styles of graduation and the method of numbering the horizontal circle and vernier are shown below.

GRADUATED 30 MINUTES READING TO ONE MINUTE
DOUBLE DIRECT VERNIER

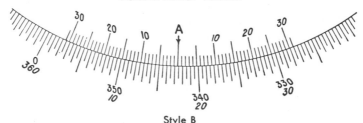

Style B

Style B represents the usual graduation of the horizontal circle of a transit with its vernier, as furnished on K & E Transits Nos. P5081C and P5136. This is an ordinary double direct vernier, reading from the center to either extreme division (30). The circle is graduated to half degrees, and the vernier (from 0 to 30) comprises 30 divisions; therefore, the value of one division on the vernier is 30 minutes ÷ 30 = 1 minute.

The figure reads $17^\circ + 25' = 17^\circ 25'$ from left to right and $342^\circ 30' + 05' = 342^\circ 35'$ from right to left.

GRADUATED 20 MINUTES READING TO 30 SECONDS
DOUBLE DIRECT VERNIER

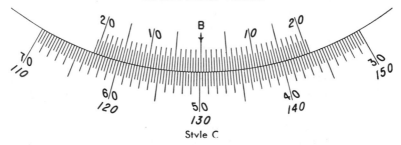

Style C

Style C represents the graduation and vernicr of a 6¼ in. transit such as No. P5085C. This is also a double direct vernier, reading from the center to either extreme division (20). The circle is graduated to 20 minutes or 1200 seconds and there are 40 divisions in the vernier; consequently, the value of one division on the vernier is 1200 seconds ÷ 40 = 30 seconds.

The figure reads $130^\circ 00' + 9' 30'' = 130^\circ 9' 30''$ from left to right, and $49^\circ 40' + 10' 30'' = 49^\circ 50' 30''$ from right to left.

Figure 12-4 (*Courtesy Keuffel & Esser Co.*)

clockwise direction, the numbers increasing in that direction will be read on the horizontal scale. Similarly, the verniers whose numbers are increasing in the same direction (clockwise) will be used. The letter A at the zero degree mark signifies which of the verniers is being used.

In order to reduce the possibility of mistakes, it is wise to make a rough estimate by eye of the fractional part of a circle division involved to check against the vernier reading. This practice will greatly reduce the number of mistakes occurring in angle measurement. Some transits are equipped with attached magnifying glasses to aid the instrument-man in reading the verniers. If this is not the case, a hand magnifying glass or reading glass should be used.

A magnifying glass if not permanently attached to the transit or theodolite is conveniently kept on a string around the user's neck. If he keeps it in his pocket, he will often have to fumble around to find it, or he may take it home and fail to bring it back the next day, or he could leave it lying on the instrument and might drop it.

Other transits are available in which the horizontal scale can be read to the nearest 30 seconds or less. Different instruments have different arrangements of verniers and the instrumentman should carefully study them before he uses the instrument. Vertical circles normally have verniers with which vertical angles can be read to the nearest minute.

12-6 POINTING THE INSTRUMENT

The surveyor should point at his targets and not aim at them. Aiming, which is excessive pointing at a target, causes eyestrain and makes the target look as though its moving. The most reliable sighting is usually the first trial when the target first appears to be aligned with the cross hairs. When aligning the cross hairs on the target with the tangent screw or slow motion screw, it is desirable that the last movement be in a clockwise direction. Such a procedure prevents the occurrence of small errors due to slack in the slow motion screw.

All pointings should be made with the target close to the intersection of the vertical and horizontal cross hairs. Figure 12-5 shows some of the common cross-hair arrangements for transits and theodo-

Figure 12-5 Some cross-hair arrangements.

lites. The black dot in each case indicates the recommended position of the target.

In order to measure an angle, the telescope is pointed at one target and then turned to another. If, as is often true, the targets in question are represented by tacks in wooden stakes and if they are reasonably close to the instrument, it is often possible to leave the tacks sticking up where they may be seen with the telescope. If this is not feasible, a nail or taping pin may be held on top of the tack. When the stakes, iron pins, or whatever are too low for sighting, a plumb bob may be held over the point and the telescope lined up with the string. Sometimes a card or paper is attached to the string so that it will be more visible.

For longer sights, it is necessary to use a larger item for sighting, usually a range pole. The alternate feet of these wooden, metal, or fiberglass poles are painted red and white in order to make them more visible. The rodman should stand directly behind the rod, and when the bottom of the rod is not visible to the instrumentman, the rodman must be sure that the rod is vertical. If the rodman is needed at some other point, he may stick the rod with its pointed end into the ground behind the point. Where the sight distances are not too large, he may set the pole so that it leans across the point and tie the string of a plumb bob around the pole so that the plumb bob hangs over the point.

12-7 MEASUREMENT OF HORIZONTAL ANGLES

For this discussion it is assumed that a transit is located at point A, as shown in Fig. 12-6, and it is desired to measure the angle between the lines AB and AC.

Both the upper motion and the lower motion clamps are released and the horizontal scales are adjusted by turning the instrument on its spindle with the hands until the A vernier is near zero. The clamps are then tightened and the upper motion tangent screw is turned until the vernier reading is zero. The lower motion is released and the telescope is sighted close to a range pole or other object at point B after which the lower motion is clamped and the line of sight is pointed precisely

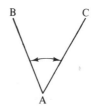

Figure 12-6

on point B with the lower motion tangent screw. The reading at the A vernier should still be zero.

In order to turn the angle, the upper motion is released and the telescope pointed toward C. The fine adjustment of the line of sight to C is made by clamping the upper motion and turning the upper motion tangent screw after which the angle is read at the A vernier as previously described.

12-8 CLOSING THE HORIZON

Before the student attempts to use the transit to measure horizontal angles for an actual traverse, he probably needs a good practice session to be sure that he fully understands the operation of his instrument. One excellent exercise is referred to as "closing the horizon" or "measuring the angles about a point."

The transit is set up at a convenient point, and several taping pins are stuck in the ground at convenient distances from the transit. The angles between the taping pins are measured, the scale being set back to zero for each measurement. When the work is completed, the angles are added together to see if their sum equals $360°$. Once the student is able to do this correctly, he will have learned how to use properly the upper and lower motions of the transit.

Sample notes for this field exercise are shown in Fig. 12-7. For practice in reading the compass and as a rough check on the horizontal angles, the bearings of the lines from the transit to the chaining pins are recorded and the angles between them are computed.

Closing the horizon is useful for checking the measurement of any angle. If the surveyor has measured one or more angles for some position of the instrument, he can set the scale to zero and measure the angle required to complete the circle. If he adds the angles together to see if the total is $360°$, he has a very good check on his work.

12-9 MEASURING ANGLES BY REPETITION

All surveyors on more occasions than they would care to admit will make mistakes in measuring angles. After they have measured an angle they would like to be as sure as possible that no mistakes were made so they will not have to return and repeat the measurement. Usually, it is much easier to prevent mistakes than it is to figure out where they occurred at a later time. A method that nearly always eliminates mistakes in angle measurement is that of *measuring angles by repetition*. After an angle is measured, the lower motion clamp and the lower

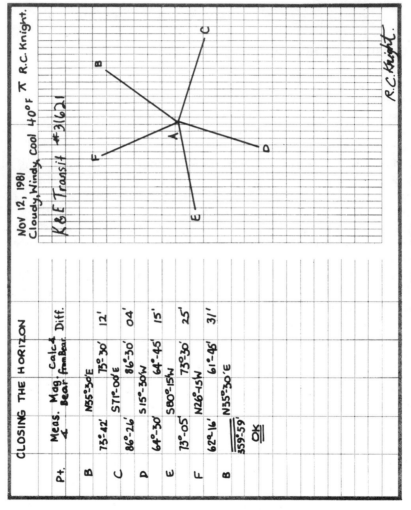

Figure 12-7

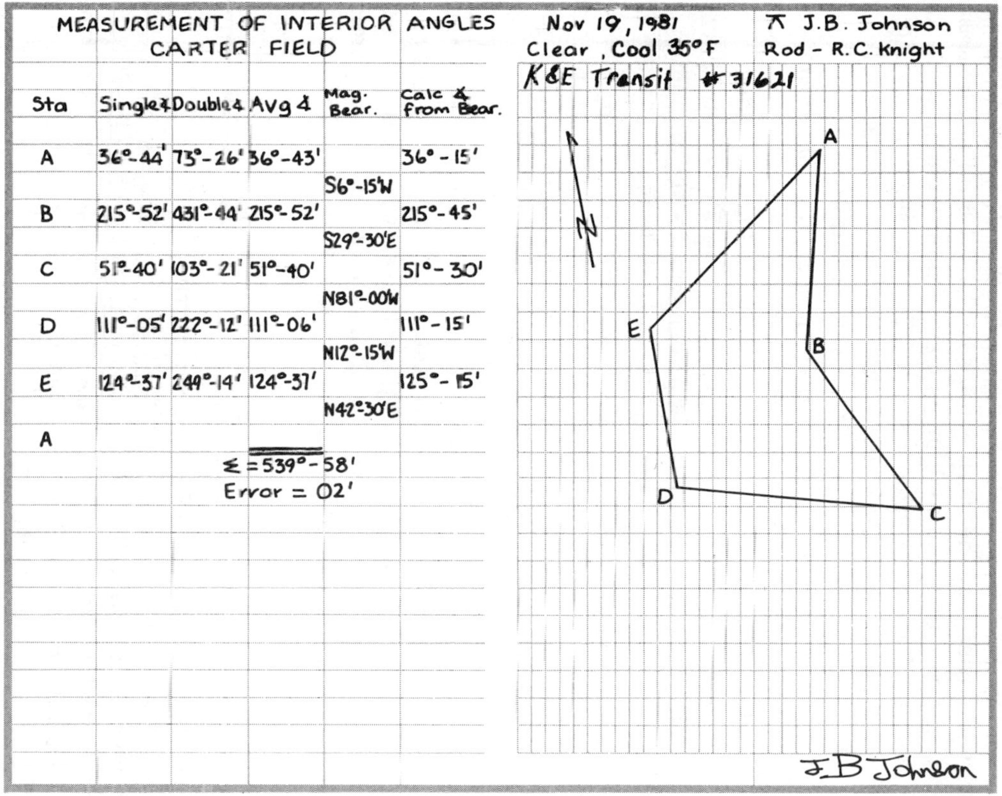

MEASUREMENT OF INTERIOR ANGLES
CARTER FIELD

Nov 19, 1981
Clear, Cool 35°F
K&E Transit #31621

☋ J.B. Johnson
Rod - R.C. Knight

Sta	Single & Double & Avg ⦞	Mag. Bear.	Calc ⦞ from Bear.
A	36°-44' 73°-26' 36°-43'		36°-15'
		S6°-15W	
B	215°-52' 431°-44' 215°-52'		215°-45'
		S29°-30'E	
C	51°-40' 103°-21' 51°-40'		51°-30'
		N81°-00W	
D	111°-05' 222°-12' 111°-06'		111°-15'
		N12°-15W	
E	124°-57' 249°-14' 124°-57'		125°-15'
		N42°-30'E	
A	⦞ =539°-58'		
	Error = 02'		

J.B. Johnson

Figure 12-8

motion tangent screw are used to sight the telescope on the initial point again. This means that the reading at the vernier should still be the same as when the angle was initially measured.

With the aid of the upper motion clamp and tangent screw, the telescope is sighted on the second point and the reading is taken at the vernier. Obviously, the measured value should equal twice the value obtained initially for the angle. If not, a mistake has been made. The student will notice that the sample notes in Fig. 12-8 for angle measurement provide a column for doubling angles. The instrumentman should do his best to forget the value recorded for the single angle and not compare the values until the double angle has been read and recorded. Otherwise, he will have lost much of the value of this check.

For more precise surveying, horizontal angles may be repeated six or eight times. The value of the angle is added each time to the previous value on the vernier. The resulting value is divided by the number of measurements in order to obtain the angle. Actually, an angle can be measured with a 1-minute transit to approximately the nearest ±30 seconds. If an angle is turned six times and if the final total is read, it should contain roughly the same error. Dividing it by 6 should then give a reading within ±05 seconds. Actually, repeating an angle more than six or eight times does not appreciably improve the precision of its measurement because of accidental errors in sighting and instrumental errors of the transit.

When sights are relatively short, say 300 ft or less, there is actually little advantage in repetition because the errors made in pointing the telescope and in setting up over the points take away all or almost all of the increased precision supposedly obtained by repetition. Similarly the precision of a 1-second theodolite is probably wasted if the sight distances are short.

12-10 LAYING OFF AN ANGLE BY REPETITION

In the preceding section the measurement of an angle by repetition to a precision greater than the least reading obtainable with the vernier was described. In this section a related procedure is described for laying off an angle to an equivalent precision.

For this discussion, it is assumed that with a 1-minute transit it is desired to lay off an angle equal to $42°16'10''$ from an existing line AB. The transit is used to turn an angle α of $42°16'$ with the existing line AB. Point C is set several hundred feet (say 500 ft here) from the transit in the new direction, as shown in Fig. 12-9.

The newly established angle is measured by repetition six times. It is assumed here that the result is $42°15'50''$. Thus, the established

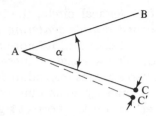

Figure 12-9

angle is in error by 20″. A new point C' (see Fig. 12-8) is set a distance from C equal to (500) (tan $0°0'20''$) = 0.048 ft. The new angle is measured again by repetition in order to check the work.

12-11 MEASURING VERTICAL ANGLES

Vertical angles may be measured with the transit in much the same way that horizontal angles are measured. The horizontal plate levels are carefully leveled, the horizontal cross hair is sighted on the point to which the vertical angle is to be measured, and the vertical scale and its vernier are used to read the angle.

As will be described in Chapter 16, the usual procedure involves (1) measuring the HI of the transit telescope by holding a level rod on the ground by the instrument, (2) sighting the telescope on the level rod at the other point with the center horizontal cross hair adjusted on the rod to the HI of the instrument, and (3) reading the vertical angle. This procedure is illustrated in Fig. 12-10.

The transit is not constructed to permit the measurement of vertical angles by repetition. Vertical angles are referred to as either plus or minus (the sign must be carefully recorded) depending on whether the telescope is sighted above or below the horizontal.

A transit has a bubble tube attached to its telescope. The bubble tube is as sensitive as the plate bubble tube. When the telescope bubble is centered, the vernier on the vertical scale should read $0°00'$. Actually, it is easy for these scales to get out of adjustment. The discrepancy is called the *index error* of the vertical circle.

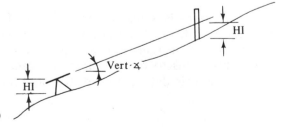

Figure 12-10

For an instrument that has a full vertical circle, the index error problem is easily handled by the so-called *double-centering* procedure. The angle is measured once with the telescope in its normal position and once with the telescope inverted. The average of these two readings is taken. For transits that have vertical arcs instead of full vertical scales, it is necessary to correct the readings by an amount equal to the index error. Great care must be taken to use the correct sign for such corrections.

12-12 THEODOLITES

The vernier transit which has been discussed in the preceding sections of this chapter and which has been used in the U.S. for many decades is gradually being replaced with optically read theodolites. Theodolites are manufactured to accomplish the same purposes as transits, that is to determine horizontal and vertical angles and to prolong straight lines. Optically read theodolites, however, usually enable the user to make single observations of angles as precisely or more precisely and in less time than those which can be made by several repetitions with a transit.

Theodolites have horizontal and vertical circles for angle measurements, as do transits, but the circles are made of glass instead of metal. Light passes through the glass circles, and with the aid of glass prisms the readings from the circles are reflected into the eyepiece. The values are greatly magnified, enabling the user to make readings without eyestrain.

Angle-measuring instruments fall into two general classes: the *repeating instruments* and the *direction instruments*. Transits which have been discussed in the last few sections of this chapter are repeating instruments. They have upper and lower motions and can be used to add or accumulate angles on the horizontal circles as was described in the repetition procedure of Section 12-9.

Theodolites may be either repeating or directional. As with transits the user of a repeating theodolite can measure angles as many times as he likes by adding them successively on the instrument circle. (See Fig. 12-11.)

The directional theodolite is a nonrepeating type of instrument and does not have a lower motion. The horizontal circle remains fixed during a series of readings. The telescope is sighted on each of the points in question and directions to those points are read. Horizontal angles are determined by calculating the differences between the directions.

Figure 12-11 Zeiss Th43 repeating theodolite. (*Courtesy Keuffel & Esser Co.*)

Since directional theodolites have only one vertical axis, they cannot be used to measure angles by repetition. However the horizontal circle is approximately set to a desired reading and a number of positions observed to make a set as described in Section 12-16.

12-13 SETTING UP THE THEODOLITE

The purpose of this section is to provide a general discussion applicable to the setting up of all theodolites. What follows, however, is probably more oriented to the Wild T-2 instrument than to others. More specific information for each type of theodolite can be found in its operation manual.

Before the theodolite is unpacked, the tripod should be carefully set over the point to be occupied with the plate approximately level. The instrument is set up on the tripod and the centering screw which is located under the tripod head is screwed up into the base of the instrument—otherwise the instrument will fall off the tripod when picked up. The instrument can be shifted laterally when the centering screw is loose. It is centered over the tripod and the screw is tightened.

The bases of theodolites, called *tribrachs*, are usually designed so as to permit the interchange of various instruments and accessories such as theodolites, EDMs, targets, subtense bars, and so on without disturbing the centering of the instrument.

The tripod is located over the point to be occupied as closely as possible, the same as previously described for the transit. Once the instrument is approximately centered over the point with the tripod legs it can be centered more exactly with a plumb bob or preferably with an optical plummet. If the plumb bob is not over the point or if the cross hairs of the optical plummet are not centered on the point, the centering screw can be loosened and the instrument shifted over the point.

Next the theodolite is approximately leveled with the bull's eye or circular level and then it is very carefully leveled with the plate level. The alidade is turned so that the level vial is parallel to two of the three leveling screws and the bubble is brought to the center with these two screws. The instrument is rotated 90° and the bubble is centered using the third screw. Then the instrument is rotated back 90° to its position parallel to the two leveling screws to see if the bubble is still centered. After a few trials the bubble is centered and the alidade is rotated through 180°. If the bubble remains centered, the instrument is level. If not, it is out of adjustment.

Unless the bubble moves a large distance, the instrument probably should not be adjusted because it takes considerable time and repeated adjustments tend to cause significant wear on the adjusting screws. Instead a reversing procedure is recommended. After the normal leveling procedure is completed, the alidade is rotated so the bubble axis is in line with one of the foot screws. The position of the lower end of the bubble (that is, the end of the bubble toward the vertical circle) is noted. The alidade is rotated through 180° and the position of the lower end of the bubble is again noted. The mean of the two positions is the *reversing position* of the bubble.

The alidade is now turned 90° so that the axis of the bubble tube is parallel to the other foot screws. By turning the two screws in opposite directions, the bubble is brought to its reversing position. To level the instrument once the reversing position is known, the user needs only to bring the instrument to its reversing position for each of the two directions of the alidade 90° apart.

Once the reversing position is determined and the theodolite leveled by this method, the vertical axis will be truly vertical and the bubble will remain in that same position whichever way the telescope is pointed.   The reversing position may gradually change and thus its position should be checked before and after each series of observations.[2]

12-14 CROSS HAIRS

In older theodolites the cross hairs were made from very fine spider webs.   One problem with spider web cross hairs is that they tend to become slack when moist.   For some instruments, cross hairs have been made from fine tungsten or platinum wires.   In modern theodolites, however, the cross hairs consist of very fine lines etched into thin glass plates.

12-15 READING THE THEODOLITE

Sample readings for a directional theodolite are shown in Fig. 12-12. With this type of instrument the observer actually views both sides of the circle simultaneously by means of internal instrument optics.   Each of these readings therefore represents the mean of two opposed sides of the circle.   This is in effect equivalent to averaging the readings of the A and B verniers of the transit.

The upper scale seen in the eyepiece is the vertical circle while the lower scale is the horizontal circle.   With some instruments, however, only one circle can be seen at a time and an inverter knob is available to select one circle or the other.

There are quite a few optically read direction theodolites on the market today but they all have approximately the same basic characteristics.   The readings shown in Fig. 12-12 are for a Wild T-2 theodolite. The student can easily verify the readings given in the figure.   For the several other types of optical theodolite scales in use, the surveyor can study the manufacturer's service manual before making readings.

For the sample readings shown, the vertical scale is based on the 360° or sexagesimal system which is almost exclusively used in the U.S.   The horizontal circle shown in the figure is based on the 400 grad or centesimal system which is widely used in Europe.   These systems were discussed in Section 11-2.

[2] J. F. Dracup et al., *Surveying Instrumentation And Coordinate Computation Workshop Lecture Notes* (Falls Church, Va.: American Congress on Surveying and Mapping, 1973), pp. 1–5.

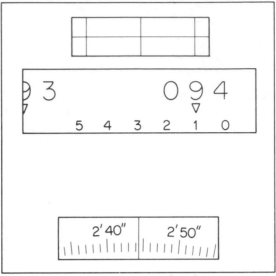

Vertical circle reading (360°) 94° 12′ 44″

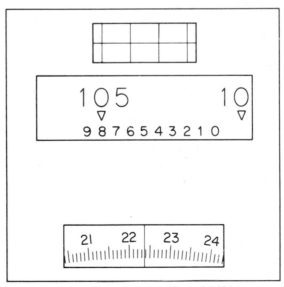

Horizontal circle reading (400 g) 105.8224 g

Figure 12-12 Wild T-2 theodolite. (*Courtesy Wild Heerbrugg Instruments, Inc.*)

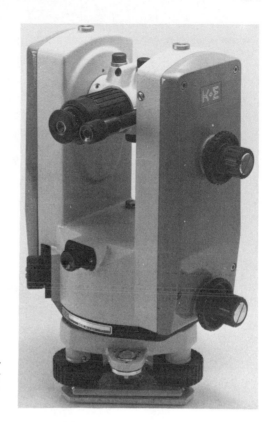

Figure 12-13 Eagle 60 optical reading theodolite.   Manufactured by Keuffel & Esser in U.S.A. (*Courtesy of Keuffel & Esser Co.*)

12-16 MEASURING WITH A DIRECTION THEODOLITE

An example set of readings taken with a direction theodolite is presented in Fig. 12-14.   The instrument was located at station A and directions were read from that point to stations B, C, and D.

These notes include directions from station A to each of the other three stations with four different positions.   Two readings were taken on each position (one with the telescope normal or direct and one with the instrument reversed or plunged).   Thus a total of eight readings were made to each station.

Though directional theodolites do not have a lower motion, they can be approximately set to certain desired values.   For position 1 the direction to station B was set near $0°$.   (It is impossible to set the circle very close to zero but it's not really necessary to do so anyway.)   The observer made a pointing on station B and read the direction and then proceeded to make pointings in a clockwise order to stations C and D. He reversed the telescope and sighted again on station D and then in a

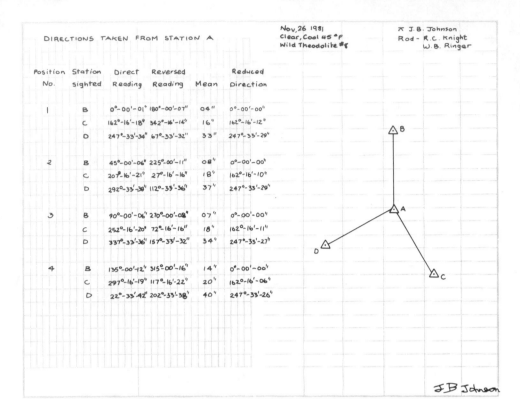

Position No.	Station sighted	Direct Reading	Reversed Reading	Mean	Reduced Direction
				DIRECTIONS TAKEN FROM STATION A	
1	B	0°-00'-01"	180°-00'-07"	04"	0°-00'-00"
	C	162°-16'-18"	342°-16'-14"	16"	162°-16'-12"
	D	247°-33'-34"	67°-33'-32"	33"	247°-33'-29"
2	B	45°-00'-06"	225°-00'-11"	08"	0°-00'-00"
	C	207°-16'-21"	27°-16'-16"	18"	162°-16'-10"
	D	292°-33'-38"	112°-33'-36"	37"	247°-33'-29"
3	B	90°-00'-06"	270°-00'-08"	07"	0°-00'-00"
	C	252°-16'-20"	72°-16'-16"	18"	162°-16'-11"
	D	337°-33'-36"	157°-33'-32"	34"	247°-33'-27"
4	B	135°-00'-12"	315°-00'-16"	14"	0°-00'-00"
	C	297°-16'-19"	117°-16'-22"	20"	162°-16'-06"
	D	22°-33'-42"	202°-33'-38"	40"	247°-33'-26"

Nov, 26 1981
Clear, Cool 45°F
Wild Theodolite #5

π J.B. Johnson
Rod - R.C. Knight
W.B. Ringer

J.B. Johnson

Figure 12-14

counterclockwise order sighted on stations C and B. These observations make up one position.

Additional positions are set so they are spread around the circle at intervals approximately equal to $180°/n$ where n is the number of positions taken. For this illustration four positions were planned to make up the "set"; therefore, the circle settings were varied by approximately $180°/4 = 45°$. The purpose of spreading out the positions around the circle is to minimize possible errors in the circle graduations and any eccentricity of the plate. The number of positions to be used for a particular survey depends on the required accuracy for the work. The number is usually specified for high-order government surveys, as described in Chapter 23.

For position 1 with the telescope normal the instrument was pointed at station B and the direction was read as $0°00'01''$. Then it was pointed at stations C and D and the readings were respectively $162°16'18''$ and $247°33'34''$.

The telescope was reversed and again sighted on station D. The

Figure 12-15 RED-1A (rapid electronic distance measurement). Infrared EDM mounted on one second theodolite TM-1A. Measures distance up to 2000 m (1.25 mi). Manufactured by Sokkisha Instruments of Japan. (*Courtesy of the Lietz Company.*)

circle reading should have changed by 180° but as shown in the notes the difference was not quite 180° due to sighting and instrumental errors. In this case the difference was 6″. The other pointings were made in a counterclockwise order.

The average or mean value (04″ for position 1) is shown in the "Mean" column. The mean direction reading to station B is actually 0°00′04″ but this was reduced to 0°00′00″ in the "Reduced Direction" column of the notes. Similarly the mean directions to stations C and D for position 1 were reduced by 04″ in the last column. A similar procedure was followed for the reversed readings.

For position number 2 the mean direct reading to station B was 45°00′06″. This was reduced to 0°00′00″ and the directions to stations

C and D were similarly adjusted by the same amount to $162°16'10''$ and $247°33'29''$, respectively.[3]

PROBLEMS

12-1. The least divisions on a transit scale are one fifth of a degree. If 24 divisions on the vernier cover 23 divisions on the main scale, what is the least count of the vernier?   (*Ans.:* $0.5'$)

12-2. The smallest divisions on a transit scale are one third of a degree. How many divisions should a vernier have in order to enable its user to make a reading to the nearest 30 seconds?

12-3. A horizontal angle was measured by repetition six times with a transit. If the reading on the scale was $21°33'$ after the angle was read once and if the final reading was $129°17'$, determine the value of the angle measured.

(*Ans.:* $21°32'50''$)

12-4. A horizontal angle was measured by repetition eight times with a transit. The initial reading was $76°17'$ and the final value was $250°14'$. What is the value of the angle measured?

12-5. A 30-second transit was used to measure an angle by repetition ten times. If the first reading was $71°36'30''$ and the final value was $356°7'30''$, what is the value of the angle?   (*Ans.:* $71°36'45''$)

[3] R. C. Brinker and P. R. Wolf, *Elementary Surveying*, 6th ed. (New York, N.Y.: Intext Educational Publishers, 1977), pp. 197–198.

Miscellaneous
Angle Discussion

13-1 COMMON ERRORS IN ANGLE MEASUREMENT
WITH TRANSITS

Most of the errors commonly made in angle measurement with transits are probably obvious enough, but they are nevertheless listed here together with comments on their magnitudes and methods of reducing them.

Imperfect vernier readings. Accidental errors are made when vernier readings are taken because it is impossible to read them perfectly. For 1-minute transits, these errors range in magnitude up to ±30 seconds, with average values in the range of ±20 seconds. Vernier errors may be significantly reduced if angles are measured by repetition. In addition, it is wise to take readings with magnifying glasses.

Instrument not centered over point. If the instrument is not centered exactly over a point, an error will be introduced into an angle measured from that position. Here it is necessary for a person to use his sense of proportion. If the points to be sighted are distant, errors caused by imperfect centering will be small. If, however, sight distances are very short, centering errors may be very serious.

Pointing errors. If the vertical cross hair in the telescope is not perfectly centered on observed points, errors similar to those described for imperfect centering of the instrument will occur. The most important method of reducing errors in pointing is to keep the sight distances

as long as possible. In fact this is a basic principle of good surveying—
avoid short sight distances if at all possible.

If the points sighted are close to the instrument, the width of a
range pole is an appreciable factor. Either a plumb bob may be held
over the point with observations being made on the plumb bob string
or it may be possible actually to sight on tacks in the stakes and so on
as described in Section 12-6. A good rule to follow in this regard is to
sight only on vertical targets (range poles, taping pins, plumb bob
strings, etc.) which appear to be only a little wider than the cross-hair
thickness when looking through the telescope.

In using the telescope, the longer a person stares at the point the
more difficult it will be to obtain a good reading because after a while
the point will seem to move. As previously described the surveyor
should use the first clear sighting of the target, since that will in all
probability be the most accurate.

Instrument inadjustment.   Since no instrument is perfect, there
are instrumental errors. If the instruments are out of adjustment, the
magnitudes of these errors will be increased, but they are greatly reduced by the process of double sighting in which the readings are taken
once with the telescope in its normal position and once in its inverted
position. Then the results are averaged. The operation of rotating the
telescope about its horizontal axis is called *plunging* or *inverting* the
telescope. If the telescope level tube is below the telescope, the telescope is in its normal position. If the telescope level tube is above the
telescope, the telescope is plunged.

Unequal settling of tripod.   Tripod legs should be firmly pushed
into the ground in order to provide solid support for the transit. The
instrumentman must be careful not to brush against the instrument and
not to step too closely to the tripod legs if the ground is soft. A good
practice, as in leveling, is to check the bubble tubes before and after
readings to make sure that they are still centered. In very soft or
swampy ground it may actually be necessary to provide special supports
for the tripod legs, for example, stakes driven into the ground.

Improper focusing of telescope (parallax).   In order to minimize
errors caused by improper focusing, the instrumentman should carefully focus the eyepiece until parallax disappears.

Miscellaneous errors.   Several miscellaneous errors that are generally not too serious except in extremely precise work include the
following: temperature changes (use an umbrella over the instrument),
heat waves affecting sights (come back later); horizontal refraction (try
to keep transit lines away from items that radiate a lot of heat, for
example, buildings, pipes, and so on); vertical refraction (read a vertical

angle from both ends of line and average—uphill angle is too large by amount of refraction error; downhill angle is too small by amount of refraction error); and wind (shield instrument as much as possible and use an optical plummet, if available, for centering instrument over point).

13-2 COMMON MISTAKES IN MEASURING ANGLES

Listed in this section are the most common mistakes made in angle measurement. If angles are measured two or more times by the repetition method, mistakes caused by any of the first four items will be discovered and can be eliminated by repeating the measurement.

Reading the wrong vernier.   The correct vernier to read is the one whose numbers are increasing in the same direction as the numbers are increasing on the scale being used.

Misreading the vernier.   The average transit has a horizontal scale divided with one-half degree divisions.   To read an angle the vernier reading should be added to the last main scale division.   If the vernier reading is 12 and the last division on the main scale is $36°30'$, the total reading should be $36°42'$, but occasionally the instrumentman will read this as $36°12'$.   A good habit to form is to estimate the value of an angle from the scale by eye before the vernier is read.   This practice will help to eliminate these one-half degree mistakes.   In addition, the measurement of angles by repetition will reveal mistakes that have been made so the work can be repeated.

Reading the wrong circle.   Sometimes the surveyor may read the angle on the wrong circle, particularly if the angle is near $180°$.   For angles in this range, the instrumentman should carefully check the adjacent numbers to be sure that he is reading the scale whose numbers are increasing in the proper direction.

Using wrong tangent screw.   This is probably the most common mistake made by the beginner.   He learns by experience the uses of the clamp and tangent screws.   The measurement of angles by repetition will reveal if mistakes have been made because of this or one of the preceding three items.   The angles may be redone until the repetition process checks.

Recording wrong values.   If the recorder calls out the angles aloud as he writes them down, he should be able to eliminate this kind of mistake.   Of course, writing down the initial value of the angle and the value after repetition will also reveal mistakes of this kind.

Vertical angles not recorded as plus or minus. Obviously, surveyors must record this information so that anyone using the notes will know whether the sights were taken uphill or downhill. Oral checks by the instrumentman with the rodman should prevent this mistake.

13-3 ANGLE-DISTANCE RELATIONSHIPS

For a particular survey it is logical for the angles and distances to be measured with comparable degrees of precision. It is not sensible to go to a great deal of effort to obtain a high degree of precision in distance measurements and not do the same with the angle measurements, or vice versa. If the distances are measured with a high degree of precision, the time and money spent have been partially wasted unless the angles are measured with a corresponding precision.

If an angle is in error by 1 minute, it will cause the line of sight to be out of position by 1 ft at a distance approximately 3440 ft (that is, 3440 times the tangent of 1 minute = 1 ft). Therefore, an angle that is in error by 1 minute is said to correspond to a precision in taping of 1/3440. It should be noted that angles measured with the 1-minute transit are usually measured a little closer than 1 minute, so that reading them to the nearest 1 minute (without repetition) probably corresponds to a precision of approximately 1/5000 in taping.

A similar discussion may be made for the relative precision obtained for angles measured to the nearest 30 seconds, 20 seconds, and so on, or for angles measured with 1-minute transits by repetition. Table 13-1 presents the angular errors that correspond to the degrees of precision described in this section. It will be remembered that the precision obtained with stadia measurement of distance varies between 1/250 and 1/1000. The table shows that in order to obtain comparable

TABLE 13-1

Angular Error	Angular Precision
5 min	$\dfrac{1}{688}$
1 min	$\dfrac{1}{3440}$
30 sec	$\dfrac{1}{6880}$
10 sec	$\dfrac{1}{20,600}$

precision in the angle measurements, the angles should be measured to approximately the nearest 5 minutes.

It should be realized that this is not the whole story on angle precisions. For instance, the trigonometric functions do not vary directly with angle sizes. In other words, the tangent of an angle of $1°11'$ which is 1 minute in error does not miss its correct value by the same amount as the tangent of an angle of $43°46'$ which is 1 minute in error. Nevertheless, the approximate relations given in Table 13-1 are a satisfactory guide for almost all surveying work.

13-4 TRAVERSING

As described in Section 11-10, a traverse consists of a series of successive straight lines that are connected together. The process of measuring the lengths and directions of the sides of a traverse is referred to as *traversing*. Its purposes are to find the positions of certain points.

Open traverses, which are normally used for exploratory purposes, have the disadvantage that arithmetic checks are not available. For this reason, extra care should be used in making their measurements. The angles should be measured by repetition and should be roughly checked by using magnetic bearings. Distances should be taped forward and back and roughly checked by taking stadia readings. The use of true directions with occasional checks made by observation of the sun or other stars is another wise procedure.

Closed traverses are much to be desired whenever feasible because they offer checks for angles and distances as will be described in Chapter 14.

Traverse directions may be determined in the following different ways: deflection angles, angles to the right, interior angles, or azimuths. These are briefly described in the following paragraphs. For each of these methods, it is wise to at least double the angles as a check. For more precise work, the angles may be measured several times by repetition.

A *deflection angle* is the angle between the extension of the preceding line and the present one. In order to measure a deflection angle, the telescope is inverted and sighted on the preceding point with the *A* vernier set to zero and then the telescope is reinverted and turned to the left or to the right as required to sight on the next point. The angle is read and a magnetic bearing may be taken as a check. This method permits easy visualization of traverses, facilitates their representation on paper, and, in addition, makes the calculations of successive bearings or azimuths very simple. Deflection angles are sometimes used for route surveys, for example, for highways, railroads, or transmission lines. Overall, their use has greatly decreased because of the frequency of mistakes in reading and recording angles as being right or left, especially when small angles are involved. The algebraic sum of all the deflection angles for a closed traverse (with no lines crossing) equals 360°. Note that right and left deflection angles must be given opposite signs in the summation.

Perhaps the most common method of measuring the angles for a traverse is the *angle to the right* method. With this method, which has partly supplanted the deflection angle method, the *A* vernier is set to zero and the telescope is sighted on the preceding point. The upper motion is released and the telescope is turned in a clockwise direction until the next point is sighted and the angle is read.

In *interior angle traverses* (used for closed traverses) the telescope is sighted on the preceding point with the *A* vernier set to zero. Then the telescope is turned either in a clockwise or counterclockwise direction so that the interior angle may be measured. It will be noted that if the surveyor proceeds in a counterclockwise direction while traversing, he will always turn clockwise angles. For closed traverses, the sum of the interior angles should equal $(n - 2)(180°)$, where n is the number of sides of the traverse.

Azimuth traverses are frequently used for surveys in which a large number of details are to be located, for example, for topographic work. If the transit is set up so that azimuths may be directly read to each point, the work of plotting is appreciably simplified. For an azimuth traverse, the *A* vernier is set to read the back azimuth of the preceding line when the telescope is sighted to the preceding point. With the upper motion the telescope is sighted on as many points as desired from that transit position and the *A* vernier is used to read the forward azimuth to each of those points. Because it is impossible to double azimuths, it is wise to read both the *A* and *B* verniers for very important lines.

13-5 INTERSECTION OF TWO LINES

A frequent surveying problem involves the intersection of two lines. For this discussion, reference is made to Fig. 13-1 in which points *A*, *B*, *C*, and *D* are established on the ground. It is desired to find the intersection of lines *AB* and *CD* (shown as point *x*).

If two transits are available, the problem can be handled easily. One transit can be set up at *A* and sighted toward *B*, and the other transit can be set up at *C* and sighted toward *D*. A rodman is sent to the vicinity of the intersection point and is waved back and forth until his range pole (or plumb bob string) is lined up with both instruments.

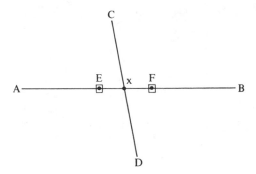

Figure 13-1

When only one transit is available, the problem can still be handled with little difficulty with the aid of some stakes and tacks and a piece of string. It is assumed that the transit is set up at A and sighted on B. Two stakes E and F are set on the line so that line EF straddles line CD. A tack is set sticking out of the top of each of these stakes (properly on the line) and a string is tied between them. After the position of the string is carefully checked, the instrument is moved to C, lined up with D, and sighted on the string. This is the desired point x.

13-6 MEASURING AN ANGLE WHERE THE INSTRUMENT CANNOT BE SET UP

Another common problem faced by the surveyor is the one of measuring an angle at a point where the transit cannot be set up. Such a situation occurs at the intersection of fences or between the walls of a building, as illustrated in Fig. 13-2.

In order to handle this problem, line AB is established parallel to the upper fence, and line CD is established parallel to the lower fence. The lines are extended and intersected at E, as described in Section 13-5, and the desired angle α is measured with the instrument. In establishing line AB, point A is located at a convenient distance from the fence. The shortest distance to the fence is measured by swinging the

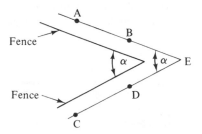

Figure 13-2

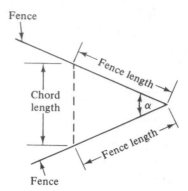

Fence

Chord
length

Fence length

α

Fence length

Fence   **Figure 13-3**

tape in an arc with *A* as the center. The desired distance is the perpendicular distance. Similarly, point *B* is located the same distance
from the fence. The same process is used to locate points *C* and *D* near
the other fence in order to establish line *CD*.

A quicker and more precise method that is frequently used (described in Section 6-5) involves the measurement of convenient distances up each fence line and the chord distance across from one fence
to the other, as shown in Fig. 13-3. The desired angle α may then be
computed by trigonometry. If the distances measured up the fence
lines are equal, the angle may be obtained from the following expression:

$$\sin \frac{1}{2}\alpha = \frac{\text{chord length}}{2 \text{ fence lengths}}$$

13-7 PROLONGING A STRAIGHT LINE
BY DOUBLE CENTERING

An everyday problem of the surveyor is that of prolonging straight
lines. In Fig. 13-4 it is assumed that line *AB* is the line to be extended
beyond point *B*. The transit is set up at *B* and backsighted on *A*. Then
the telescope is plunged to set point *C'*. If the instrument is not in
proper adjustment (if the line of sight of the instrument is not perpendicular to its horizontal axis), point *C'* will not fall on the desired
straight line. For this reason *double centering* is the method chosen
for this problem.

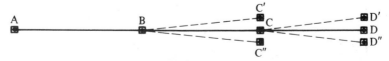

A   B   C'   C   D'
                                   C   D
                      C"   D"

Figure 13-4

With the double-centering method the telescope in its normal position (bubble tube down) is sighted on A and plunged to set point C'. The telescope is then rotated horizontally about its vertical axis until point A is sighted. The telescope is now inverted or upside down (bubble tube up). It is once more plunged and point C'' is set. The correct point C is halfway in between the two points C' and C''. Of course, if the instrument is properly adjusted, points C' and C'' should coincide. This procedure substantially reduces errors caused by instrument inadjustment and gives the instrumentman a check against the presence of other errors and mistakes. From point C the straight line is continued to point D, and so on.

The technique of double centering is a very important one for the surveyor. If he can learn to measure all of his angles (angles to the right, deflection angles, and so on) by double centering, the precision of his work will be appreciably improved and he will eliminate many blunders. When he is measuring angles by repetition, he should make half of the measurements with the telescope in the normal position and half with the telescope inverted.

13-8 ESTABLISHING POINTS ON A STRAIGHT LINE BETWEEN TWO GIVEN POINTS

Points intervisible. If the entire line is visible between the two points, the surveyor has no problems. He can set up the instrument at one end, sight on the other end, and then establish any desired points in between. If a large vertical angle is involved in setting any of the points, the careful surveyor may very well set the points with the telescope in its normal position and then check them with the telescope inverted.

Balancing in. If the end-points are not intervisible from each other but if there is an area in between from which both the points may be seen (a surprisingly common situation), the process called *balancing in* or *wiggling in* may prove beneficial. In this procedure the surveyor sets up the transit at a point he thinks is on a straight line from A to C (see Fig. 13-5). He sights on one point and then plunges the telescope to

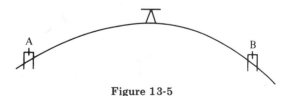

Figure 13-5

see if he hits the other point. If he doesn't, he moves the instrument to another position and tries again. Since it is very difficult to estimate closely the first time, the instrument may have to be moved several times, with the final adjustment probably made by loosening the leveling screws and shifting the head of the instrument. Once the instrument is in line, any desired intermediate points from A to C may be set. When balancing in is possible, it may very well save the time involved in running trial lines, as described in the following paragraph, and the delay required to make the necessary calculations to finish the problem.

Two points not intervisible.   When it is desired to establish intermediate points between two known points that cannot be seen from each other and cannot be seen from any point in between, the surveyor may very well use a random line procedure. This is very common in surveying whenever hills, many trees, and great distances are involved. The surveyor can run a trial straight line (measuring the distances involved) in the general direction from A toward E (Fig. 13-6). He sets points B, C, and D (perhaps by double centering, and when he is near E, he measures the distance DE and the interior angle at D. From this information he computes the desired interior angle at A and runs the straight line from A toward E, establishing points in between. If he misses E slightly, he can readjust the intermediate points proportionately.

Another random line method which may be a little more practical will become obvious after study of Chapter 14 is completed. The surveyor runs a random set of straight lines from A to E, as shown in Fig. 13-7, after which he computes the desired length and direction of line AE. Then he can either return to A and run the straight line AE or he can compute the distances required to set points on the desired line by measuring over from the random trial lines.

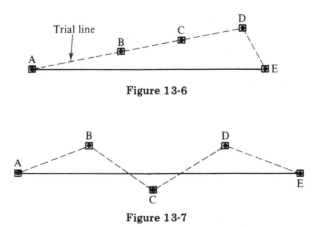

Figure 13-6

Figure 13-7

13-9 ADJUSTMENTS OF TRANSITS

The importance of proper instrument adjustments was discussed in Section 9-13. In this section six common relations that should exist in the transit are presented together with the required adjustments.

Plate bubble tubes. The axes of the plate bubble tubes should be perpendicular to the vertical axis of the instrument.

This relation and adjustment are identical with those previously described for the dumpy level. The instrument is carefully leveled with each bubble tube being parallel to a pair of leveling screws. The telescope is rotated about its vertical axis by $180°$. If either bubble moves, the bubble is moved halfway back to center with the adjusting screw on the tube in question. Then the test is rerun.

Cross-hair ring. The vertical cross hair should lie in a plane perpendicular to the horizontal axis of the instrument.

The telescope is sighted so that either the bottom or top of the vertical cross hair is focused on a well-defined point, for example, on a building. The telescope is moved up or down (rotated about its horizontal axis) to see if the point remains on the vertical cross hair. If it doesn't, the four cross-hair ring capstan screws are loosened and the ring is rotated by light tapping. Then the screws are tightened and the test is rerun.

Line of sight. The telescope line of sight should be perpendicular to the horizontal axis.

After the instrument is set up and carefully leveled, it is sighted on point A which has been set in the ground (perhaps a chaining pin) at least 250 or 300 ft from the instrument. The telescope is plunged and point B is set a similar distance from the instrument. Then the telescope is turned horizontally $180°$ about its vertical axis and sighted on point A again. The telescope is once more plunged. If it is lined up with point B, it is properly adjusted. If not, point C is set in line with the new sighting and opposite point B and another point called D is set one-fourth of the distance from C to B. In order to adjust the instrument, the cross-hair ring must be moved laterally until the vertical hair is lined up on point D. This is done by using the horizontal capstan screws on the cross-hair ring. One is loosened and the other tightened until the necessary adjustment is made.

The horizontal and vertical axes. The horizontal axis of the instrument should be perpendicular to its vertical axis.

The transit is set up near a building or other object where a clearly defined point involving a large vertical angle can be seen. After the instrument is carefully leveled, the vertical hair is sighted on the point. Then the telescope is depressed and point A is set on the ground. The telescope is plunged and rotated $180°$ about the vertical axis and sighted up to the original point. Once again the telescope is plunged. If the vertical hair hits points A, no adjustment is required. Otherwise, point B is set next to A and point C is established halfway in between. Point C will lie in the same vertical plane with the original point up on the building.   The horizontal axis can be adjusted with the capstan screws at one end of the axis. The vertical hair is sighted on point C and the telescope raised and sighted on a point opposite the original point.   The axis is adjusted until the vertical hair hits the original point. Then the test is rerun in order to check the correctness of the work.

Telescope bubble tube.   The line of sight of the telescope should be parallel to the axis of the bubble tube. Since this test (the "two-peg test") and adjustment are handled exactly as described for the dumpy level in Section 9-13, it is not repeated here.

The vertical scale.   When the plate bubble tubes and the telescope bubble tube are centered, the vertical scale should read zero.

If this is not the case, the surveyor can skip the adjustment and read the angle as previously described in Section 12-11 or he can adjust the scale to zero with the capstan screws that hold the vernier plate. The screws are loosened and the plate is moved until the zero marks are at the proper position.   Then the capstan screws are retightened. One wishes for three hands to make this adjustment.

13-10 ADJUSTMENT OF OPTICAL THEODOLITES

For the past century, surveying textbooks have included a discussion of transit adjustments similar to the ones described in the preceding section.   The reader may wonder why we haven't included a similar section for the adjustment of optical theodolites.   These modern theodolites are manufactured so carefully and their parts fit together so well that they very seldom get out of adjustment unless they are severely abused. Should one of the instruments be damaged or need cleaning or reconditioning it should be sent to one of the service departments of the manufacturer for adjustment and repair.   Actually the manuals which are furnished with optical theodolites explain a few simple adjustments which can be made by the user.   In general these adjustments are those which pertain to the level tubes and they are therefore similar to those used for transits.

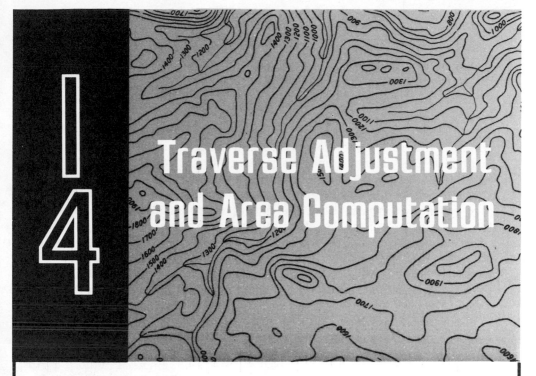

14-1 COMPUTATIONS

Almost all surveying measurements require some calculations in order to reduce them to a more useful form for determining distance, earthwork volumes, land areas, and so on. This chapter is devoted to the calculation of land areas.

Perhaps the most common need for area calculations arises in connection with the transfer of land titles, but areas are also needed for the planning and design of construction projects. Some obvious examples are the laying out of subdivisions, the construction of dams, and the consideration of watershed areas for designing culverts and bridges.

14-2 METHODS OF CALCULATING AREAS

Land areas may be calculated by several different methods. A very crude approach that should only be used for rough estimating purposes is a graphical method in which the traverse is plotted to scale on a sheet of graph paper and the number of squares inside the traverse are counted. The area of each square can be determined from the scale used in drawing the figure, and thus the area is roughly estimated.

A similar method that yields appreciably better results, but is again only satisfactory for estimating purposes, involves the use of the planimeter (see Chapter 19). The traverse is carefully drawn to scale and a planimeter is used to measure the traverse area on the paper. From this value the land area can be computed from the scale of the drawing. It is probable that with careful work, areas may be estimated by this method within a range of from $\frac{1}{2}$ to 1% of the correct values.

A very useful and accurate method of computing the area of traverses that have only a few sides is the triangle method. The traverse is divided into triangles and the areas of the triangles are computed separately. The necessary formulas appear in Fig. 14-1 in which several traverses are shown. If the traverse has more than four sides, it is necessary [as shown in part (c) of the figure] to obtain the values of additional angles and distances either by field measurements or by lengthy office computations. For this five-sided figure, the triangle areas *ABE* and *CDE* can be obtained as before, but additional information is necessary in order to determine the area *BCE*. The problem is amplified for traverses that have more than five sides. For such cases as

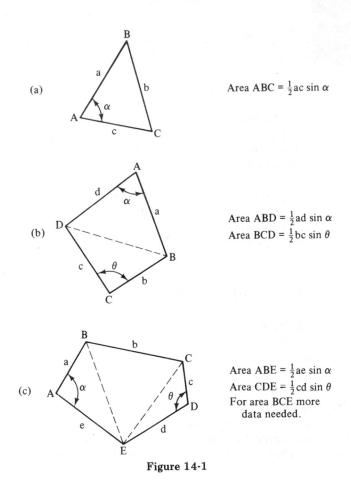

Area ABC = $\frac{1}{2}$ac sin α

Area ABD = $\frac{1}{2}$ad sin α
Area BCD = $\frac{1}{2}$bc sin θ

Area ABE = $\frac{1}{2}$ae sin α
Area CDE = $\frac{1}{2}$cd sin θ
For area BCE more
data needed.

Figure 14-1

these, the surveyor is probably better advised to use one of the methods described later in this chapter.

Other methods of computing areas within closed traverses are the double meridian distance, double parallel distance, and coordinate methods discussed in Sections 14-7 through 14-9. In addition, several methods are presented in Section 14-10 for computing the areas of land within irregular boundaries. Finally, computer methods of handling the same problems are presented in Sections 14-11 through 14-15.

14-3 BALANCING ANGLES

Before the area of a piece of land can be computed, it is necessary to have a closed traverse. The first step in obtaining a closed figure is to balance the angles. The interior angles of a closed traverse should total $(n - 2) (180°)$, where n is the number of sides of the traverse. It is unlikely that the angles will add up perfectly to this value, but they should be very close. The usual rule for average work is that the total should not vary from the correct value by more than approximately the square root of the number of angles measured times the least division readable with the vernier (see Section 2-11). For an eight-sided traverse and a 1-minute transit, the maximum error should not exceed

$$\sqrt{8} \times 1' = 2.83' \quad (\text{say } 3')$$

(For very good work with the 1-minute transit, the maximum error should probably not exceed $\pm 30'' \sqrt{n}$).

It is customary for the instrumentman to check the sum of the angles for his traverse before he leaves the field. If the discrepancies are unreasonable, he must remeasure the angles one by one until he finds the source of the trouble.

If the angles do not close by a reasonable amount, a mistake in the data has been made. If a mistake has been made in only one angle, that angle can often be identified by plotting the lengths and directions of the sides of the traverse to scale. Then if a line is drawn perpendicular to the error of closure, it will often point to the angle where the mistake was made. It can be seen in Fig. 14-2 that if the angle containing the mistake is reduced it will tend to cause the error of closure to be reduced.

When the angular errors for a traverse have been reduced to reasonable values, they are distributed among the angles so that their sum will be exactly $(n - 2) (180°)$. If the sum of the angles for an eight-sided traverse is in error by minus 3 minutes, the surveyor will probably just arbitrarily add 1 minute to each of three angles. Other alternatives are to make the adjustments to angles where the adjacent sides are

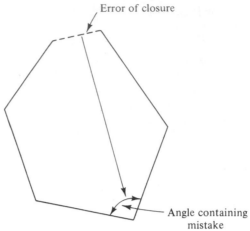

Error of closure

Angle containing
mistake   **Figure 14-2**

short, or where difficulties were encountered in measuring the angles, or where the greatest differences occurred in the doubling procedure.

It is probably not advisable to make the adjustments in units smaller than the least values readable with the instrument. If $+\frac{3}{8}'$ or 22.5″ is added to each of the eight angles of the problem just described, the math will become a little more complicated, and the precision will

TABLE 14-1

Point	Measured ∡	Corrected ∡	Calculated Bearings from Corrected ∡s
A	36°43′	36°44′	
			S6°15′W
B	215°52′	215°53′	
			S29°38′E
C	51°40′.	51°40′	
			N81°18′W
D	111°06′	111°06′	
			N12°24′W
E	124°37′	124°37′	
			N42°59′E
A			

$\Sigma = 539°58'$ $\Sigma = 540°00'$

probably be unimproved. After the angles are balanced, the bearings of the sides of the traverse are computed. The initial bearing is preferably a true one, but a magnetic or assumed bearing may be used and the other bearings computed from the balanced traverse angles.

The interior angles measured for the traverse in Fig. 12-8 are added together in Table 14-1 and total $539°58'$, which is 2 minutes less than the correct value of $540°00'$ for the interior angles of a closed five-sided figure. These are balanced in the table by merely adding 1 minute to angles A and B. If the magnetic bearing of side AB is used as a reference, the bearings of the other sides are calculated in the table from the balanced angles.

14-4 LATITUDES AND DEPARTURES

The closure of a traverse is checked by computing the latitudes and departures of each of its sides. The *latitude* of a line is its projection on the north–south meridian and equals its length times the cosine of its bearing. In the same manner, the *departure* of a line is its projection on the east–west line (sometimes called the *reference parallel*) and equals its length times the sine of its bearing. These terms (illustrated in Fig. 14-3) merely describe the x and y components of the lines.

For the calculations used in this chapter, the latitudes of lines with northerly bearings are designated as being north or plus. Those in a southerly direction are designated as south or negative. Departures are east or positive for lines having easterly bearings and west or negative for lines having westerly bearings. For example, line AB in Fig. 14-3(a) has a northeasterly bearing and thus a + latitude and a + departure. Line CD in Fig. 14-3(b) has a southeasterly bearing and thus a – lati-

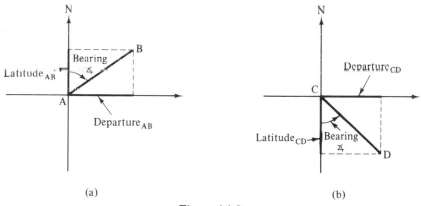

(a)   (b)

Figure 14-3

tude and a + departure. Calculations of latitudes and departures are
illustrated in Section 14-5.

14-5 ERROR OF CLOSURE

If one starts at one corner of a closed traverse and walks along its lines
until he returns to his starting point, he will have walked as far north
as he has walked south and as far east as he has walked west. This is the
same thing as saying that for a closed traverse the sum of the latitudes
should equal zero and the sum of the departures should equal zero.
When the latitudes and departures are calculated and summed, they
will never be exactly equal (except by accidental results of canceling
errors).

When the latitudes are added together, the resulting error is referred to as the *error in latitude* (E_L); the error that occurs when the
departures are added is referred to as the *error in departure* (E_D). If
the bearings and distances of the traverse of Fig. 12-4 are exactly
plotted on a sheet of paper, the figure will not close (see Fig. 14-4)
because of E_L and E_D. The magnitude of these errors is greatly exaggerated in this figure.

The error of closure can be easily calculated as follows:

$$E_{\text{closure}} = \sqrt{(E_L)^2 + (E_D)^2}$$

And the precision of the measurements can be obtained by the expression:

$$\text{Precision} = \frac{E_{\text{closure}}}{\text{Perimeter}}$$

Table 14-2 presents the calculations necessary for obtaining the
latitudes, departures, E_L, E_D, E_{closure}, and the precision for the previously used traverse.

Should the precision obtained be unsatisfactory for the purposes

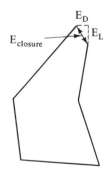

Figure 14-4

TABLE 14-2

Side	Bearing	Length (ft)	Cosine	Sine	Latitude		Departure	
					+N	-S	+E	-W
AB	S6°15'W	189.53	.99406	.10887	—	188.40	—	20.63
BC	S29°38'E	175.18	.86921	.49445	—	152.27	86.62	—
CD	N81°18'W	197.78	.15126	.98849	29.92	—	—	195.50
DE	N12°24'W	142.39	.97667	.21474	139.07	—	—	30.58
EA	N42°59'E	234.58	.73155	.68179	171.61	—	159.93	—
		939.46			+340.60	-340.67	+246.55	-246.71

$E_L = -0.07$ $\quad$ $E_D = -0.16$

$E_{\text{closure}} = \sqrt{(.07)^2 + (.16)^2} = 0.17$ ft

$\text{Precision} = \dfrac{0.17}{939.46} = \dfrac{1}{5526}$

of the survey, it will be necessary to recheck the work. Of course, the math should be checked carefully for mistakes as a first step and if mistakes cannot be found, the field measurements must be checked.

A great deal of field time will usually be saved by the surveyor if he will carefully study his calculations before he starts rechecking the field measurements. If the angles balance, a mistake in distance is to be suspected. Usually it will have occurred in a side roughly parallel to the direction of the error of closure line. For instance if a traverse has a large error in latitude but a small error in departure, one might look for a side running primarily in the north–south direction. If such a side exists it would be logical to remeasure the distance for that side first. In a similar manner if the error in departure is twice the error in latitude, the surveyor would want to check the distances for any sides whose calculated departures and latitudes were approximately in that proportion.

Despite these ideas small mistakes may not be located so easily and the surveyor may have to carefully remeasure all or most all of the sides of his traverse. Once the errors of closure are reduced to reasonable values they will be adjusted so that the traverse will close perfectly (to the number of places being used), as described in Section 14-6.

14-6 BALANCING LATITUDES AND DEPARTURES

The purpose of balancing the latitudes and departures of a traverse is to attempt to obtain more probable values for the locations of the corners of the traverse. If, as described in the preceding section, a reasonable precision is obtained for the type of work being done, the errors are balanced out in order to close the traverse. This usually is accomplished by making slight changes in the latitudes and departures of each side so that their respective algebraic sums total zero.

Theoretically, it is desirable to distribute the errors in some systematic fashion to the various sides, but practically, the surveyor may often use a simpler procedure. He may decide to make large corrections to one or two sides where the most difficulties were encountered in making the measurements. He may look at the magnitude of E_L and E_D and decide, after studying the lengths of the various sides and the sines and cosines of their bearings, that by changing the length of such and such a side or sides he can just about balance the traverse.

The practical balancing methods just described may not seem to give very good results to the reader. They, however, may be as satisfactory as the results obtained with the more theoretical rules to be described in the following paragraphs since these latter methods are based on assumptions that are not altogether true and may, in fact, be far from being true.

Very often the surveyor may have no idea which side should get the correction and, in fact, the person performing the calculations may be someone other than the person who made the measurements. For such cases as these a systematic balancing method such as the compass rule, the transit rule, or one of several others is desirable.

A very popular rule for balancing the errors is the compass or Bowditch rule named after the distinguished American navigator Nathaniel Bowditch (1773-1838). It is based on the assumption that the quality of distance and angular measurements is approximately the same. It is further assumed that the errors in the work are accidental and thus that the total error in a particular side is directly proportional to its length. The rule states that *the error in latitude (departure) in a particular side is to the total error in latitude (departure) as the length of that side is to the perimeter of the traverse.* For the example traverse, the correction for the latitude of side AB is calculated below.

$$\frac{\text{Correction in lat.}_{AB}}{E_L} = \frac{l_{AB}}{\text{Perimeter}}$$

$$\text{Correction in lat.}_{AB} = E_L \frac{l_{AB}}{\text{Perimeter}} = \frac{(+0.07)(189.53)}{939.46} = +0.01 \text{ ft}$$

If the sign of the error is +, then the correction will be minus. Actually, the sign of the corrections can be determined by observing what is needed to balance the numbers to zero. In Table 14-3 the latitudes and departures for the traverse are balanced by the compass rule.

A rule that is occasionally used is the transit rule. It is based on the assumption that errors are accidental and that the angle measurements are more precise than the length measurements. In this method *the correction in the latitude (departure) for a particular side is to the*

TABLE 14-3

Side	Latitude Correction		Departure Correction		Balanced Latitudes and Departures			
	N	S	E	W	N	S	E	W
AB	—	+0.01	—	+0.03	—	188.39	—	20.60
BC	—	+0.01	+0.03	—	—	152.26	86.65	—
CD	+0.02	—	—	+0.03	29.94	—	—	195.47
DE	+0.01	—	—	+0.03	139.08	—	—	30.55
EA	+0.02	—	+0.04	—	171.63	—	159.97	—
					340.65	340.65	246.62	246.62

total correction in latitude (departure) as the latitude (departure) of that side is to the sum of all the latitudes (departures).

Computer programs are readily available for solving all of the problems (from balancing to area calculations) described in this chapter. These programs use one of the systematic methods of balancing, such as the compass rule or the transit rule described here, or other methods such as the *Crandall method* (particularly applicable to traverses where the angles have been measured more precisely than the distances) or the *least squares method.* The least squares method is the best way of adjusting survey data, but it is seldom used except by geodetic organizations. The method was developed from the laws of probability and the measured values are adjusted so that the sum of the squares of the errors (residuals) is a minimum.[1] These latter two methods involve more lengthy computations, but they are easily handled by electronic computers. For very precise surveys adjustments are best made with the least squares method.

14-7 DOUBLE MERIDIAN DISTANCES (DMDs)

The best-known procedure for calculating land areas is the *double meridian distance* method. The *meridian distance* of a line is the distance (parallel to the east–west direction) from the midpoint of the line to the reference meridian. Obviously, the *double meridian distance* (DMD) of a line equals twice its meridian distance. In this section it will be proved that if the DMD of each side of a closed traverse is multiplied by its balanced latitude and if the algebraic sum of these values is determined, the result will equal two times the area enclosed by the traverse.

Meridian distances are considered positive if the midpoint of the line is east of the reference meridian and negative if it is to the west. In Fig. 14-5, the plus meridian distance of side *EA* is shown by the dotted horizontal line. For sign convenience, the reference meridian is usually assumed to pass through the most westerly or the most easterly corner of the traverse. If the surveyor has difficulty in determining the most westerly or the most easterly corners, he can quickly solve the problem by making a freehand sketch. He starts at any corner of his traverse and plots the departures successively to the east or west for each of the lines until he returns to the starting corner. The location of the desired points will be obvious from the sketch.

In Fig. 14-5 it can be seen that the DMD of side *EA* equals twice its meridian distance or equals its departure. The DMD of side *AB*

[1] R. C. Brinker and P. R. Wolf, *Elementary Surveying*, 6th ed. (New York: Thomas Y. Crowell Company, Inc., 1977), pp. 228–229.

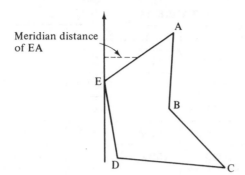

Figure 14-5

equals two times the departure of *EA* plus two times one half the departure of *AB*. In this manner the DMD of any side may be determined. By studying this process, however, the reader will develop the following rule for DMDs that will simplify the calculations: *The DMD of any side equals the DMD of the last side plus the departure of the last side plus the departure of the present side.* The signs of the departures must be used and it will be noticed that the DMD of the last side (*DE* in Fig. 14-5) must equal the departure of that side, but it will of necessity be of opposite sign.

In order to see why the DMD method for area calculation works, we refer to Fig. 14-6. In this discussion, north latitudes are considered plus and south latitudes are considered minus. If the DMD of side *AB*, which equals *B'B*, is multiplied by its latitude, the result will be plus two times the area of the triangle *B'BA* which is outside the traverse. If the DMD of *BC* is multiplied by its latitude *B'C'*, which is minus, the result will be minus twice the trapezoidal area *B'BCC'*, which is inside and outside the traverse. Finally the DMD of *CA* times its latitude *C'A* equals plus two times the area *ACC'*, which is outside the traverse. If these three values are added together, the total will equal two times the area inside the traverse because the areas outside the traverse will be canceled.

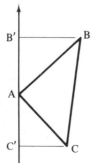

Figure 14-6

TABLE 14-4

Side	Departures			Latitudes		Double Areas	
	E	W	DMD	N	S	+N	−S
AB	—	20.60	299.34	—	188.39	—	56,393
BC	86.65	—	365.39	—	152.26	—	55,634
CD	—	195.47	256.57	29.94	—	7,682	—
DE	—	30.55	30.55	139.08	—	4,249	—
EA	159.97	—	159.97	171.63	—	27,456	—
						+39,387	−112,027

$$2A = -72,640$$
$$A = -36,320 \text{ sq ft} = 0.834 \text{ acres}$$

Table 14-4 shows the area calculations by the DMD method for the example traverse used in this chapter. Because the signs of the latitudes must be used in the multiplications, the table provides one side for positive values under the north column for north latitudes and one for negative values under the south column for south latitudes. The traverse area equals one half the algebraic sum of the two columns. It does not matter that the final value is either positive or negative. The resulting area may be quickly checked by moving the reference meridian to another corner and repeating the calculations. If it has been assumed at the most westerly corner, it will probably be moved to the most easterly corner. This same problem is solved with a programmable desk calculator in Section 14-15, with a great savings in time and effort.

When lengths are in meters the area is expressed in square meters (m^2). The SI system does not specify a particular unit for land areas, but the *hectare* which equals 10 000 m^2 is used by many countries. One hectare equals 2.47104 acres.

14-8 DOUBLE PARALLEL DISTANCES (DPDs)

The same procedure used for DMDs may be used if double parallel distances are multiplied by the balanced departures for each side. The final areas will be the same. The *parallel distance* of a line is the distance (parallel to the north–south direction) from the midpoint of the line to the reference parallel or east–west line. The parallel is probably drawn through the most northerly or the most southerly corner of the traverse.

14-9 AREAS COMPUTED BY COORDINATES

Another useful method for computing land areas is the method of coordinates. Some surveyors like this method better than the DMD method because they feel that there is less chance of making math mistakes. The amount of work is approximately the same for both methods. Instead of computing the DMDs for each side, the coordinates of each traverse corner are computed and then the coordinate rule is applied. This rule is derived in Fig. 14-7.

From this figure it can be seen that to determine the area of a traverse, each y coordinate is multiplied by the difference in the two adjacent x coordinates (using a consistent sign convention such as minus the following plus the preceding). The sum of these values is taken and the result equals two times the area. The math can quickly be checked by taking each x coordinate times the difference in the two adjacent y coordinates. The coordinates of the corners of the example traverse are given in Fig. 14-8 as determined from the previously balanced latitudes and departures, and the area is computed.

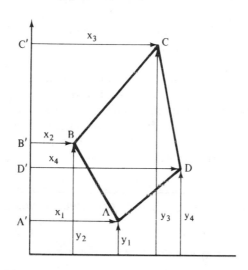

Area ABCD = area C'CDD' + area D'DAA' − area C'CBB' − area B'BAA'

Area ABCD = $(\frac{1}{2})(x_3 + x_4)(y_3 - y_4) + (\frac{1}{2})(x_4 + x_1)(y_4 - y_1) - (\frac{1}{2})(x_3 + x_2)$
$(y_3 - y_2) - (\frac{1}{2})(x_2 + x_1)(y_2 - y_1)$

Multiplying these values and rearranging the results yields
2 area = $y_1(-x_2 + x_4) + y_2(-x_3 + x_1) + y_3(-x_4 + x_2) + y_4(-x_1 + x_3)$

Figure 14-7

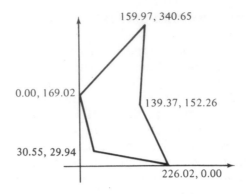

$$2A = (0.00)(-340.65 + 29.94) + (159.97)(-152.26 + 169.02) +$$
$$(139.37)(-0.00 + 340.65) + (226.02)(-29.94 + 152.26) +$$
$$(30.55)(-169.02 + 0.00)$$
$$2A = 72,640$$
$$A = 36,320 \text{ sq ft} = 0.834 \text{ acres}$$

Figure 14-8

14-10 AREAS WITHIN IRREGULAR BOUNDARIES

Very often property boundaries are represented by irregular lines, for example, the center line of a creek or the edge or center line of a curving road. For such cases as these, it is often not feasible to run the traverse along the exact boundary line. Instead, it may be practical to run it a convenient distance from the boundary and locate the position of the boundary by measuring offset distances from the traverse line, as shown in Fig. 14-9. The offsets may be taken at regular intervals if the boundary doesn't change suddenly, but when it does change suddenly, offsets are taken at irregular intervals, as shown by *ab* and *cd* in the figure.

The area inside the closed traverse may be computed by one of the methods previously described, and the area between the traverse line

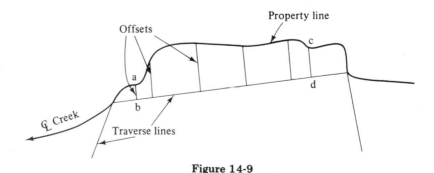

Figure 14-9

and the irregular boundary may be determined separately and added to the other value. If the land in question between the traverse line and the irregular boundary is carefully plotted to scale, the area may be determined satisfactorily with a planimeter (see Chapter 19). Other methods commonly used are the trapezoidal rule, Simpson's one-third rule, and another coordinate method involving offsets. All of these methods are described in this section, but it should be recognized that they probably do not yield results that are any more satisfactory than those obtained with a planimeter because of the irregular nature of the border between the measured offsets.

(a) Trapezoidal Rule

When the offsets are fairly close together, an assumption that the boundary is straight between those offsets is satisfactory and the trapezoidal rule may be applied. With reference to Fig. 14-10, the offsets are assumed to be located at regular intervals and the area inside the figure equals the areas of the enclosed trapezoids, or

$$A = d\left(\frac{h_1 + h_2}{2}\right) + d\left(\frac{h_2 + h_3}{2}\right) + \cdots d\left(\frac{h_{n-1} + h_n}{2}\right)$$

from which

$$A = d\left(\frac{h_1 + h_n}{2} + h_2 + h_3 + \cdots h_{n-1}\right)$$

(b) Simpson's One-Third Rule

If the boundaries are found to be curved, Simpson's one-third rule is considered better to use than the trapezoidal rule. Again, it is assumed that the offsets are evenly spaced. The rule is applicable to areas that have an odd number of offsets. If there is an even number of offsets, the area of all but the part between the last two offsets (or the first two) may be determined with the rule. That remaining area is determined separately, ordinarily assuming it to be a trapezoid.

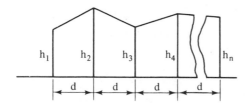

Figure 14-10

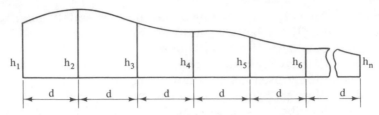

Figure 14-11

With reference to Fig. 14-11, Simpson's rule is written as follows:

$$A = (2d)\left(\frac{h_1 + h_3}{2}\right) + \left(\frac{2}{3}\right)(2d)\left(h_2 - \frac{h_1 + h_3}{2}\right)$$

$$+ (2d)\left(\frac{h_3 + h_5}{2}\right) + \left(\frac{2}{3}\right)(2d)\left(h_4 - \frac{h_3 + h_5}{2}\right), \text{etc.}$$

This reduces to

$$A = \frac{d}{3}\left[(h_1 + h_n) + 2(h_3 + h_5 + \cdots h_{n-2})\right.$$

$$\left. + 4(h_2 + h_4 + \cdots h_{n-1})\right]$$

(c) Coordinate Rule for Irregular Areas

When offsets are taken at irregular intervals, the area of each figure between pairs of adjacent offsets may be computed and the values totaled. In addition, the planimeter method is particularly satisfactory here. There are other methods of handling the problem, for example, the coordinate rule for irregular spacing of offsets which says that twice the area is obtained if each offset is multiplied by the distance to the

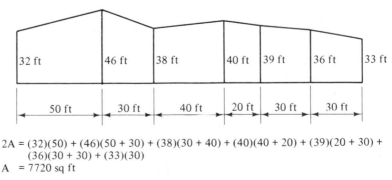

$2A = (32)(50) + (46)(50 + 30) + (38)(30 + 40) + (40)(40 + 20) + (39)(20 + 30) +$
$\quad (36)(30 + 30) + (33)(30)$
$A \;\; = 7720 \text{ sq ft}$

Figure 14-12

preceding offset plus the distance to the following offset. It will be noted that for the end offsets the same rule is followed, but there will be only one distance between offsets because the outside one does not exist. An application of this coordinate rule is shown in Fig. 14-12.

14-11 ELECTRONIC COMPUTERS

The electronic computer has now been around for some time and is widely used for lengthy computation in all phases of industry and commerce. A glance through the yellow pages of the telephone directory for any large American city will reveal hundreds of companies advertising various types of computer services. The age of these electronic marvels has arrived and their application to everyday surveying problems is a reality—so much of a reality that surveyors who are not taking advantage of electronic computers or programmable desk or pocket calculators are actually endangering their professional futures.

14-12 ACCESS TO COMPUTERS

The average surveyor cannot justify a large "in-house" computer because its cost would run several thousand dollars per month. He instead will make use of one of the following alternatives:

1. With *time-sharing* he can have a terminal installed in his office, permitting access to a remotely located computer at a fraction of the cost of having his own computer. The terminal is in actuality a typewriter that is connected by phone to the computer. The computer number is dialed on the phone and the information typed. Usually this data is pretyped on a tape that can be quickly run after the computer is dialed. He is charged a standard monthly rate for the terminal plus a fee for the time of actual usage.

2. *Computer service bureaus* do not require the surveyor to be really familiar with computers or computer programs. He need only furnish the data to the bureaus, which select an appropriate program and provide the solution. (Many companies act both as service bureaus and as time-sharing companies.)

3. Small *desk-type* or *pocket-size programmable calculators* allow the surveyor to solve problems in between those that are handled with regular desk calculators and the very large programs that require the use of the electronic computers. They are programmed to

compute precision, land areas, earthwork volumes, missing lengths, curves, coordinates, stadia reductions, and so on.

14-13 PROGRAMS

In order to apply the computer to a particular problem, it is necessary to have a program. This means that someone must translate the language of the problem into the language of the computer. Preparing such a program even for the simplest problem is a formidable task which the average practicing surveyor has neither the time nor the background to perform.

Fortunately, surveying problems for which the computer is commonly used fall into a few general classes and many organizations have successfully prepared programs for them and have made them available throughout the country. The result is that today the average surveyor uses "canned programs" that have been prepared by someone else.

14-14 AVAILABILITY OF PROGRAMS

There are today an increasing number of programs that the surveyor may use. To prepare a list of all of the companies having such programs available is an almost impossible task and the list would be obsolete before this chapter is printed. It is sufficient to say that a surveyor in any part of the United States can find suitable programs available from one organization or another for almost any problem that he faces. Many of the programs are supplied by the manufacturers of the various computers; they can often be obtained through user groups (organizations of users of a particular computer model); the libraries of service bureaus are quite extensive; and they can sometimes be obtained from other surveying firms.

The time-sharing companies briefly mentioned in Section 14-12 usually have program libraries which may include the types of programs required or they may be able to obtain desired programs from other organizations. There are some nonprofit companies which make available to users at cost programs that have been contributed by other users. A user may contribute any program he wishes to the library.

Regardless of where a program is obtained, almost any surveyor would want to take a short problem of two which he has previously solved and check out the correctness of the program. There are few surveyors who would be willing to blindly stake their professional reputations on the claims made by the representative of a service bureau, time-sharing company, or other organization without making a preliminary check.

Of particular interest to the surveyor is the fact that some computer programs are available that will not only make the computations necessary for precision, balancing, areas, and so on, but will also print out a drawing or plat of the traverse and show the bearings and distances of its sides. Other programs are available that subdivide property into lots according to given limitations and print out a drawing of the results.

One very well-known program used by surveyors, called COGO (Coordinate Geometry), is briefly discussed here. The civil engineering department of M.I.T., under the leadership of Professor C. L. Miller, embarked on a major project in programming in the early 1960s. They prepared a very broad programming system covering all phases of civil engineering and called it ICES (Integrated Civil Engineering Systems).

ICES is divided into numerous subsystems, such as COGO (which deals with the geometric problems involved in surveying, mapping, and construction), ROADS (design of highways, dams, and railways), PROJECT (for the planning and control of construction projects), and several others dealing with soils, water, bridges, buildings. Today, ICES is handled by McDonnell Automation Company of St. Louis.

With the subsystem COGO, which is available throughout the United States, the surveyor has a program with which he can handle almost any computation that he may need, such as closure, precision, area, missing lengths, curves, alignment, and subdivisions. The use of a program like this permits the calculations to be made in a very small fraction of the time that would be required with an ordinary desk calculator and, in addition, the chance of mistakes is greatly reduced. If the reader investigates further, he will find many other commercially available programs that will perform the work he needs just as well as COGO does.

14-15 PROGRAMMABLE CALCULATORS

Perhaps for the average surveying firm the programmable calculator provides the most practical method for handling the repetitious work involved in the numerical calculations for precision, traverse balancing, areas, stadia, earthwork volumes, and so on.

For a while there was a gap between the large electronic computer on the one hand and the ordinary desk calculator on the other. Some problems were too small to require the use of the large computer and yet they were too long and tedious to handle with the ordinary desk

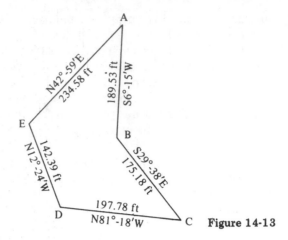

Figure 14-13

calculator. This gap has been clearly bridged with the programmable electronic desk calculators and programmable pocket calculators which provide a great deal of the computer's capability at a much lower cost and at the same time have the simplicity, availability, and "approachability" of an ordinary desk calculator. The programmable pocket calculators are now on the market at prices of only a few hundred dollars.

Programmable calculators have both keyboard programming and card reader programs in which cards can be punched and read into the calculator in seconds. Furthermore, libraries of card programs for different problems are available.

To illustrate the speed and simplicity with which the surveyor may handle numerical calculations, the traverse used earlier in this text is repeated in Fig. 14-13. Then, with the permission of Monroe, the Calculator Company, the latitudes, departures, closure error, and so on are computed as was done earlier in this chapter, with Monroe's programmable printing calculator called the "Surveyor."

It is possible to learn to use this calculator in a matter of minutes. The author spent approximately one minute putting the data into the calculator for this traverse (Fig. 14-13). The calculator took approximately one minute printing out the results shown on the following pages. Remarks are made at the side of these numbers in order to show the reader what the numbers represent.

. . . .
1 .
    6   15   0 . 0   3

      189 . 530

. . . .
2 .
  29   38   0 . 0   2

      175 . 180

. . . .
3 .
  81   18   0 . 0   4

      197 . 780

. . . .
4 .
  12   24   0 . 0   4

      142 . 390

. . . .
5 .
  42   59   0 . 0   1

      234 . 580

. . . .

. . . .
1 .
    6   15   0 . 0   3

      189 . 530

    -188 . 403
    -20 . 633

      811 . 596   n
      979 . 366   e

. . . .
2 .
  29   38   0 . 0   2

      175 . 180

    -152 . 267
      86 . 617

      659 . 328   n
    1,065 . 983   e

This is data typed into the calculator. Starting with side AB (1-2 in their notation), the bearings and lengths of each side were typed. The number to the far right indicates the quadrant of the bearing (1 for NE, 2 for SE, 3 for SW, and 4 for NW).

Here the calculator has computed the latitudes and departures of each side and then has printed the coordinates of each traverse corner.

. . . .
3 .
  81 18   0 . 0   ⌐

     197 . 780

      29 . 916
    −195 . 504

     689 . 245   n
     870 . 479   e
. . . .
4 .
  12 24   0 . 0   ⌐

     142 . 390

     139 . 068
     −30 . 576

     828 . 313   n
     839 . 903   e
. . . .
5 .
  42 59   0 . 0   /

     234 . 580

     171 . 607
     159 . 933

     999 . 920   n
     999 . 836   e
. . . .

(See remarks on preceding page)

. . . .
0 .
     −0 . 079   ⟵ $E_L.$
     −0 . 163   ⟵ $E_D.$

     0 . 0001   ⟵   Precision (see note on p. 239).

  64 11 19 . 4   ƒ   } Bearing and length of line re-

     0 . 181   quired to close traverse.

     939 . 460   ⟵   Perimeter.
36,348 . 588   ⟵   Area in square feet.
     0 . 834   ⟵   Area in acres.
. . . .
1 .
   6 14 26 . 2   ƒ

     189 . 510

   −188 . 387
    −20 . 600

(See remarks on following page.)

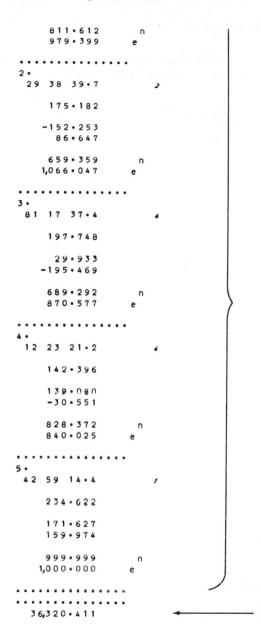

```
    811•612        n
    979•399        e
• • • • • • • • • • • • • •
2•
 29  38  39•7            2

    175•182

   -152•253
     86•647

    659•359        n
  1,066•047        e
• • • • • • • • • • • • • •
3•
 81  17  37•4            4

    197•748

     29•933
   -195•469

    689•292        n
    870•577        e
• • • • • • • • • • • • • •
4•
 12  23  21•2            4

    142•396

    139•080
    -30•551

    828•372        n
    840•025        e
• • • • • • • • • • • • • •
5•
 42  59  14•4            /

    234•622

    171•627
    159•974

    999•999        n
  1,000•000        e
• • • • • • • • • • • • • •
• • • • • • • • • • • • •
  36,320•411
```

The traverse is balanced for the computed error of closure (both bearings and lengths being adjusted), and new values of latitudes, departures, coordinates, and area are given.

← Final area in square feet.

Note regarding precision: The value 0.0001 seems to indicate a precision of 1 part in 10,000. If carried one more decimal, it would agree more closely with the value in Table 14-2.

PROBLEMS

In Problems 14-1 to 14-4, compute the latitudes and departures for the sides of the traverses shown in the accompanying figures. Determine the error of closure and precision for each of the traverses.

14-1.

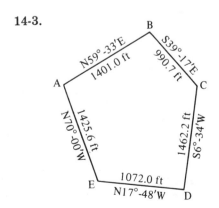

(*Ans.*: E_C = 0.14, Precision = 1/6110)

14-2.

14-3.

(*Ans.*: E_C = 1.26, Precision = 1/5040)

14-4.

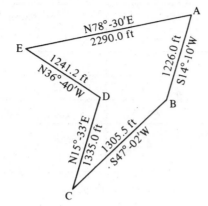

In Problems 14-5 to 14-7, balance each of the sets of latitudes and departures given by the compass rule and give the results to the nearest 0.01 ft.

14-5.

		Latitudes		Departures	
Side	Length	N	S	E	W
AB	400.00	320.00		245.00	
BC	300.00		180.00	235.36	
CA	500.00		140.24		480.00

(*Ans.:* 320.08; 179.94; 140.14; 244.88; 235.27; 480.15)

14-6.

		Latitudes		Departures	
Side	Length	N	S	E	W
AB	600.00	450.00		339.00	
BC	450.00		285.00	259.50	
CA	750.00		164.46		599.22

14-7.

		Latitudes		Departures	
Side	Length	N	S	E	W
AB	220.40	185.99		118.26	
BC	287.10		234.94	165.02	
CD	277.20		181.25		209.73
DE	200.10	187.99		68.55	
EA	147.90	42.01			141.81

(*Ans.: AB* 186.03, 118.20; *EA* 42.04, 141.85)

In Problems **14-8** to **14-10**, balance by the compass rule the latitudes and departures computed for each of the traverses of Problems 14-1 to 14-3.

(*Ans.:* Problem 14-9 for *AB* 144.79, 635.20; for *CD* 213.97, 441.66)

In Problems 14-11 to 14-13, from the given sets of balanced latitudes and departures calculate the areas of the traverses in acres using DMDs with the meridians through the most westerly points.

14-11.

	Balanced	Latitudes	Balanced	Departures
Side	N	S	E	W
AB	600		200	
BC	100		400	
CD	0	0	100	
DE		400		300
EA		300		400

(*Ans.:* 4.13 acres)

14-12.

	Balanced	Latitudes	Balanced	Departures
Side	N	S	E	W
AB	100			200
BC	100		200	
CD	200		100	
DE		700	100	
EA	300			200

14-13.

	Balanced	Latitudes	Balanced	Departures
Side	N	S	E	W
AB	200		100	
BC	100		200	
CD		150	150	
DE		50		200
EA		100		250

(*Ans.:* 1.29 acres)

14-14. Repeat Problem 14-11 with the meridian passing through the most easterly point.

14-15. Repeat Problem 14-12 using double parallel distances (DPDs) with the reference parallel passing through the most northerly point.

(*Ans.:* 1.95 acres)

14-16. Repeat Problem 14-13 using DPDs with the reference parallel passing through the most southerly point.

In Problems 14-17 and 14-18, for the given coordinates compute the length and bearing of each side.

14-17.

Point	x (ft)	y (ft)
A	0	0
B	+200	+150
C	+500	−100
D	+300	−200

(*Ans.:* AB = N53°08′E, 250.00 ft; BC = S50°12′E, 390.51 ft)

14-18.

Point	x (m)	y (m)
A	+300	+200
B	+400	−300
C	+100	0

In Problems 14-19 to 14-21, compute the area in acres by the method of coordinates for each of the traverses whose corners have the coordinates given.

14-19.

Point	x (ft)	y (ft)
A	+200	+400
B	+300	-100
C	+100	-300
D	-150	-250
E	-100	+150

(*Ans.:* 4.68 acres)

14-20.

Point	x (ft)	y (ft)
A	0	0
B	+400	+150
C	+500	-200
D	+200	-150
E	+100	-300

14-21.

Point	x (ft)	y (ft)
A	+100	+250
B	+350	+300
C	+400	-250
D	-100	-300
E	-300	-200
F	+100	-100

(*Ans.:* 4.53 acres)

14-22. Repeat Problem 14-11 using the coordinate method.

14-23. Repeat Problem 14-12 using the coordinate method.   (*Ans.:* 1.95 acres)

14-24. Repeat Problem 14-13 using the coordinate method.

14-25. Determine the area of the traverse of Problem 14-17 in acres using the coordinate method.   (*Ans.:* 1.89 acres)

14-26. Using the coordinate method compute the area in hectares (1 ha = 10 000 m^2) for the traverse of Problem 14-18.

In Problems 14-27 and 14-28, compute the area (in square feet) of the irregular tracts of land shown using the trapezoidal rule.

14-27.

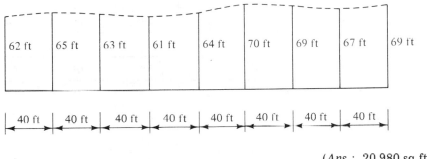

(*Ans.:* 20,980 sq ft)

14-28.

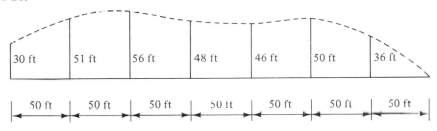

14-29. Repeat Problem 14-27 using Simpson's one-third rule.   (*Ans.* 21,000 sq ft)

14-30. Repeat Problem 14-28 using Simpson's one-third rule.

14-31. In order to determine the area between a base line *AB* and the edge of a lake the following offset distances were measured at 50-ft intervals. Compute the area (in square feet) of the tract by using Simpson's one-third rule. Offset distances: 16 ft, 35 ft, 52 ft, 63 ft, 71 ft, 68 ft, 60 ft, 55 ft, 41 ft, 37 ft.   (*Ans.:* 23,733 sq ft)

For Problems 14-32 to 14-34, the offsets are taken at irregular intervals. Determine the area (in square feet) between the traverse line and the boundary for each case. Use the coordinate method. The distances given are measured from the origin in each case.

Problem 14-32		Problem 14-33		Problem 14-34	
Distance	Offset	Distance	Offset	Distance	Offset
0	30	0	18	0	17
30	50	20	22	10	30
50	35	40	28	20	45
80	45	60	34	30	56
90	58	70	46	50	50
100	50	75	58	100	62
120	68	80	64	150	71
150	60	85	66	200	78
		100	60	210	63
		120	55	230	52
				250	50

(*Ans.:* 4905 sq ft)

Miscellaneous
Traverse
Computations

15-1 OMITTED MEASUREMENTS

Occasionally one or more angles and/or lengths are not measured in the field and their values are computed later in the office. There are several reasons for not completing the field measurements such as difficult terrain, obstacles, hostile landowners, lack of time, and so on.

Should all of the measurements for a closed traverse be completed and an acceptable precision obtained, it is perfectly permissible to compute bearings and distances within that traverse. For instance in Fig. 15-1 it is assumed that a group of lots (1–4) are being layed out in the field. The perimeter of the lots, represented by the solid lines, has been run with an acceptable precision and the errors of closure balanced. For such a case it is perfectly permissible to compute the lengths and bearings of the missing interior lot lines which are shown dotted in the figure. Furthermore a great deal of time may be saved where there is an appreciable amount of underbrush along the lines and/or where the ends of the lines are not visible from each other due to hills, trees, etc.

The omission of some of the measurements for one or more sides of a closed traverse is a very undesirable situation. Even though it is possible to calculate the values of up to 2 missing lengths or 2 missing bearings (that is the same as 3 angles missing) or 1 length and 1 bearing (2 angles) the situation is much to be avoided. The trouble with such calculations is that there is no way to calculate the precision of the field

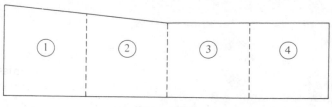

Figure 15-1

measurements that were made, since the figure was not closed. The measurements which were taken are assumed to be "perfect" in order that the missing quantities may be calculated. As a result, severe blunders can be made in the field causing the computed values to be meaningless.

From this discussion it is evident that if the measurements of any lengths or angles which are part of a closed traverse are omitted in the field, it is advisable and almost essential to use some kinds of approximate checks on the computed values. Such things as stadia distance readings, angle measurements or compass bearings, and even eyeball estimates can be critical to the success of a survey.

The next several sections of this chapter describe the calculations for the following types of omitted measurement problems: length and bearing of one side missing; length of one side and bearing of another missing; lengths of two sides missing; and bearings of two sides missing.

15-2 LENGTH AND BEARING OF ONE SIDE MISSING

For the traverse of Fig. 15-2 the lengths and bearings of sides AB, BC, and CD are known but the length and bearing of side DA are unknown. This is the same thing as saying that the angles at D and A are missing as well as the length of DA. From the assumption that the measurements for the three known sides are perfect, it is easily possible to compute the missing information for side DA. The problem is exactly the

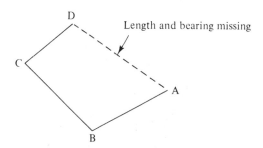

Length and bearing missing

Figure 15-2

same as the one previously handled in Table 14-2, where the error of closure of a traverse was determined.

The latitudes and departures of the three known sides are computed, summed, and the "error of closure," represented by the dotted line DA in the figure, is derived. This is the length of the missing side; the tangent of its bearing angle equals the horizontal component of its distance divided by its vertical component, and may be written as follows:

$$\text{tan of bearing } \angle = \frac{\text{Error of departure}}{\text{Error of latitude}}$$

From this expression the bearing can be obtained as illustrated in Example 15-1. A simple way to determine the direction of the missing side is to examine the latitude and departure calculations. To close the figure of this example the sum of the south latitudes needs to be increased as does the sum of the east departures. Thus the bearing of the line is southeast.

Since everyone is subject to making errors in math, it is well to compute the latitude and departure for this newly calculated line to determine if the traverse closes. Such a mathematical check is desirable for each of the omitted measurement problems described in this chapter.

Example 15-1

Determine the length and bearing of side BC of the traverse shown in Fig. 15-3. The lengths and bearings of the other sides are known.

Solution   Computing the latitudes and departures of the known sides.

					Latitudes		Departures	
Side	Bearing	Length	Cosine	Sine	N	S	E	W
AB	N57°10′E	310.20	0.54220	0.84025	168.19	×	260.65	×
BC	—	—	—	—	—	—		—
CD	S43°18′W	234.32	0.72777	0.68582	×	170.53	×	160.70
DA	N82°36′W	406.90	0.12880	0.99167	52.41	×	×	403.51
					220.60	170.53	260.65	564.21

$$E_L = 50.07 \qquad E_D = 303.56$$

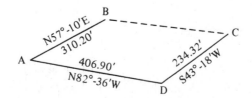

Figure 15-3

Computing the length and bearing of the missing side.

$$l_{BC} = \sqrt{(50.07)^2 + (303.56)^2} = 307.66'$$

$$\text{tan of bearing } \angle = \frac{E_D}{E_L} = \frac{303.56}{50.07} = 6.6027122$$

$$\text{bearing} = S80°38'E$$

15-3 LENGTH OF ONE SIDE AND BEARING OF ADJOINING SIDE UNKNOWN

Figure 15-4 shows a closed traverse for which the bearing of side AB and the length of side BC are unknown. (This is the same as saying that the angles at A and B are missing as well as the length of side BC.) The lengths and bearings of the other sides are known. The latitudes and departures of these known sides may be calculated and from them the length and bearing of the dotted line AC determined.

The missing information for triangle ABC which is redrawn in Fig. 15-5 may easily be determined. First of all the angle at C can be computed from the known bearings of sides CA and CB.

When one angle and the lengths of two sides of a triangle are known the sine law may be used to find the missing information. For instance the following equation may be written for the triangle ABC of Fig. 15-5:

$$\frac{\sin \angle C}{l_{AB}} = \frac{\sin \angle B}{l_{AC}}$$

With this equation the angle at B can be determined, thus giving two of the three interior angles of a triangle. Angle A equals $180° - \angle B - \angle C$ and the length of side BC can be determined from the equation:

$$\frac{\sin \angle C}{l_{AB}} = \frac{\sin \angle A}{l_{BC}}$$

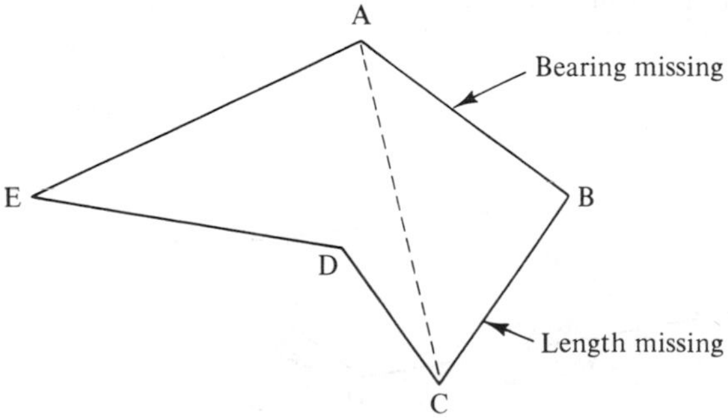

Figure 15-4

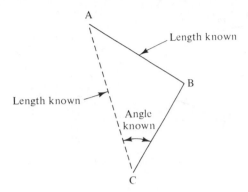

Figure 15-5

It is theoretically possible to obtain two sets of answers which will satisfy the problem (see Fig. 15-6). The lines *AB* and *BC* shown in Fig. 15-6(a) represent one possible solution while the lines *AB'* and *B'C* shown in part (b) of the figure provide another solution. If, however, the approximate bearing of side *AB* had been observed in the field the confusion would have been avoided since the correct set of answers would be obvious.

Two further comments need to be made concerning the accuracy of this method. If the angle between the side of unknown bearing and the side of unknown length is close to $90°$, the solution may be quite poor. In using the sine law it may be impossible to tell on which side of $90°$ an angle is located. (The sines of $89°$ and $91°$ are the same.) Secondly the value of the sine near $90°$ changes very slowly and thus a small variation in the calculated values can cause a relatively large error in the angle calculated by the sine law. For these reasons the answers may be unsatisfactory.

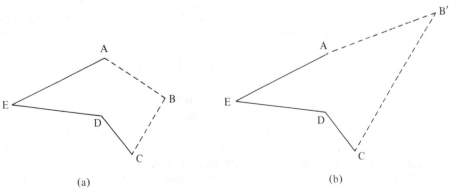

(a)   (b)

Figure 15-6

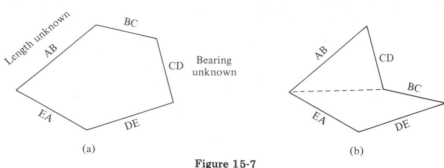

Figure 15-7

15-4 LENGTH OF ONE SIDE AND BEARING OF ANOTHER UNKNOWN—SIDES NOT ADJOINING

Should the length of one side and the bearing of another side be missing, the two sides not adjoining, the unknowns can still be determined as they were when the sides were adjoining. The latitude and departure of a line of fixed length and direction is obviously the same no matter where the line is placed. Thus it is possible to rearrange the sides of a traverse (their lengths and directions remaining the same) so that the unknown sides adjoin each other. For instance in Fig. 15-7(a) the length of side AB is unknown as is the bearing of side CD. In part (b) the figure is redrawn with sides AB and CD adjoining. The length and bearing of the dotted line is determined by summing up the latitudes and departures of the known sides BC, DE, and EA. The sine law is then applied to the triangle on top of the figure to calculate the missing information.

15-5 BEARINGS OF TWO SIDES UNKNOWN

Should the bearings of two sides of a closed traverse be missing (the same thing as saying 3 angles are unknown), they can easily be calculated with the use of a common triangle formula. The two sides in question may or may not be adjoining. If they do not adjoin, the figure can be redrawn so they are adjoining. In Fig. 15-8 it is assumed that the bearings of sides AB and BC are unknown.

The latitudes and departures of sides CD, DE, and EA are computed and summed; from the results the length and bearing of the dotted line AC is determined. Then the lengths of all sides of the oblique triangle ABC are known as well as the bearing of side AC. In Fig. 15-9 the letters a, b, and c represent the lengths of the sides of the

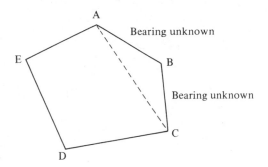

Figure 15-8

triangle and s is a constant for the triangle which simplifies the use of the angle equation.

Letting $s = \frac{1}{2}(a + b + c)$

$$\sin \tfrac{1}{2} \angle A = \sqrt{\frac{(s - b)(s - c)}{bc}}$$

After the angle at A is determined the sine law may be applied to obtain the other missing angles in the triangle.

It is necessary to know roughly the direction of at least one of the sides, since the solution of the trigonometric equation determines the shape of the triangle but not necessarily its position. In Fig. 15-10 it is observed that sides AB' and $B'C$ form another possible solution for the problem of Fig. 15-8.

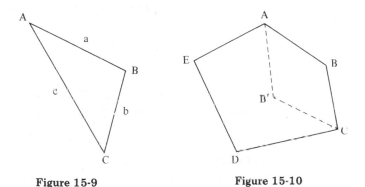

Figure 15-9   **Figure 15-10**

15-6 LENGTHS OF TWO SIDES UNKNOWN

Sometimes a surveyor may read angles to a point which he does not intend to occupy. He may read an angle to a water tank, church steeple, tower, etc., from two or more points. In other words he does not want to take the time to measure distances to the point directly,

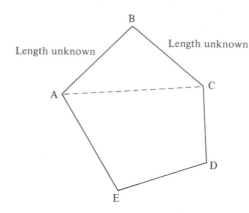

Figure 15-11

either by taping or with an EDM device. Should this situation occur the missing lengths can be computed with the sine law. In Fig. 15-11 the lengths of sides AB and BC are assumed to be unknown. The latitudes and departures of sides CD, DE, and EA are computed and summed and the length and bearing of line AC determined. For triangle ABC the bearings of all three sides are known as is the length of line AC. With this information it is possible to write

$$\frac{\sin \angle B}{l_{AC}} = \frac{\sin \angle C}{l_{AB}}$$

In this expression the only unknown is l_{AB}. After its value is determined, the following expression may be written to determine the length of side BC.

$$\frac{\sin \angle C}{l_{AB}} = \frac{\sin \angle A}{l_{BC}}$$

15-7 CUTTING OFF REQUIRED AREAS FROM GIVEN TRAVERSES

A frequent problem faced by the surveyor is that of cutting off required areas from given traverses. For instance a buyer may purchase 40 acres from a larger tract with two sides of his land made parallel to a given direction as perhaps a highway as shown in Fig. 15-12, or he may buy the land on one side of a line drawn through two points (Fig. 15-13) or he may obtain a certain number of acres below a line drawn through a certain point (Fig. 15-14) and so on.

We now consider four kinds of area cut-off problems. There are various types in addition to the ones described, but if the reader understands the cases presented here he will be able to develop a reasonable

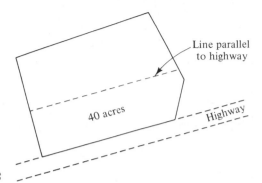

Figure 15-12

approach to the other situations. In each of the cases that follow it is assumed that:

1. A boundary survey has been made for the whole tract of land and a satisfactory degree of precision obtained.
2. The errors have been distributed throughout the traverse by one of the balancing methods such as the compass or transit rules.
3. The lengths and directions of the sides of the balanced traverse have been computed from the adjusted values of step 2 and are used for area cut-off calculations.

(a) Cutting Off an Area with a Line Drawn Between Two Given Points

For the traverse shown in Fig. 15-13 it is desired to determine the area below a line drawn from F to C.

The latitudes and departures of sides CD, DE, and EF may be computed and used to determine the length and bearing of line FC. (The same value would be obtained if sides FA, AB, and BC were used.) Once this information is obtained the area of $CDEF$ can be determined with DMDs, coordinates, or some other method.

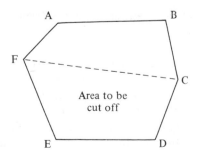

Figure 15-13

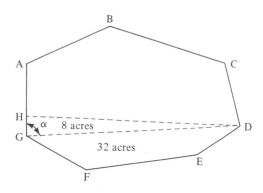

Figure 15-14

(b) Cutting Off a Required Area with a Line Drawn Through a Given Point

For the traverse of Fig. 15-14 it is desired to draw a line through point D such that it cuts off 40 acres below.

As a first step in solving this problem the surveyor might draw a line from point D to another corner such that he thinks it cuts off an acreage somewhere in the neighborhood of the desired value. In this case corner G is selected and the length and bearing of line GD is computed with latitudes and departures after which the area $DEFG$ is computed. It is assumed that the resulting area is 32 acres. It is next necessary to draw a line DH such that the triangle DGH contains eight acres. Since the bearings of DG and GH are known, the angle between them (α in Fig. 15-14) may be computed and the following equation written for the triangle area:

$$\text{Area } DGH = (8)(43{,}560) = \tfrac{1}{2} \sin \alpha \, l_{DG} \, l_{GH}$$

In this equation the only unknown is the length of side GH. Solving the equation will yield its value and then the length and bearing of DH may be determined by the sine law or by latitudes and departures. Finally the area of figure $DEFGH$ should be calculated to be sure that it equals 40 acres.

(c) To Cut Off an Area with a Line Passing Through a Given Point in a Given Direction

For the traverse shown in Fig. 15-15 it is desired to cut off the area below a line drawn through point D parallel to side EF.

The length and bearing of a line from point D to a convenient corner such as F can be computed by latitudes and departures. Then with the bearings of lines DF, FH, and HD known as well as the length of line DF it is possible with the sine law to determine the lengths of

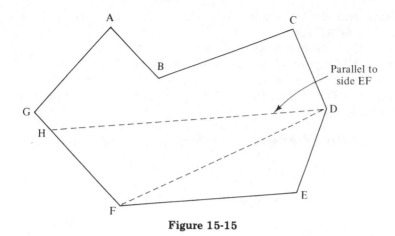

Figure 15-15

lines *FH* and *HD*. Once these values are determined the area *DEFH* can be calculated.

(d) Cutting Off a Required Area with a Line in a Given Direction

For the traverse of Fig. 15-16 it is desired to draw a line parallel to side *EF* so that it will cut off 60 acres below.

A line is drawn through a convenient corner *D* parallel to side *EF* until it intersects the line *GA* at point *H*. The corner *D* is selected by the surveyor so that a line through that point will cut off an area below it somewhere in the neighborhood of the desired value. The lengths of lines *GH* and *HD* are computed as previously described in paragraph

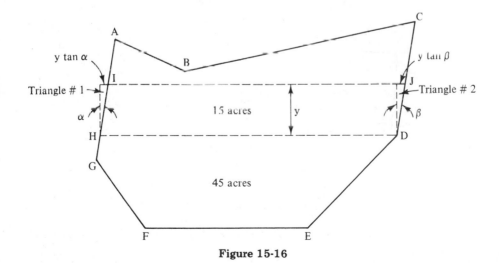

Figure 15-16

(c) above and then the area $DEFGH$ is calculated. Here the result is assumed to be 45 acres.

It is then necessary to draw a line IJ parallel to line HD such that the area $DHIJ$ equals 15 acres. It will be noted in Fig. 15-16 that since all of the bearings are known, the angles α and β shown may be easily determined. The area of the parallelogram $DHIJ$ can then be calculated as follows:

Area $DHIJ$ = Area of rectangle of y height and length l_{DH}

$\quad$ – Area of triangle #1 shown in Fig. 15-16

$\quad$ + Area of triangle #2

$$(15)(43,560) = (y)(l_{DH}) - (\tfrac{1}{2})(y)(y \tan \alpha) + (\tfrac{1}{2})(y)(y \tan \beta)$$

By solving this equation the value of y can be determined and then by trigonometry the lengths HI and JD can be calculated. The length IJ equals the length of $DH - y \tan \alpha + y \tan \beta$. Having these values the area $JDEFGI$ can be calculated to check its intended value of 60 acres.

PROBLEMS

15-1. Compute the lengths and bearings of sides BG and CF for the accompanying closed traverse. The latitudes and departures for the outer traverse (in solid lines) have been balanced.

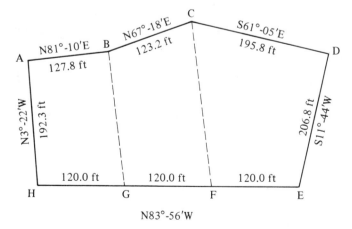

(*Ans.:* BG = S1°07′E, 224.3 ft; FC = N2°01′W, 284.7 ft)

15-2. A random traverse is run from points A to F, as shown in the accompanying figure. Compute the length and bearing of a straight line from A to F.

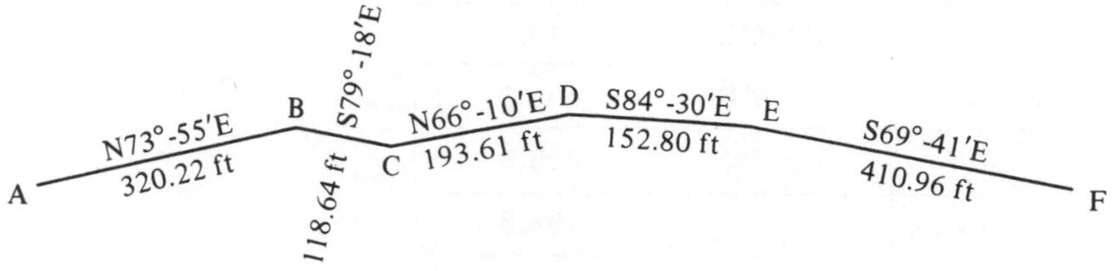

In Problems 15-3 to 15-5, for each of the closed traverses given in the tables, the length and bearing of one side are missing. Compute the latitudes and departures of the given sides and determine the length and bearing of the missing side.

15-3.

Side	Bearing	Length (ft)
AB	N46°10′E	250.00
BC	S31°18′E	300.00
CD	S40°37′W	200.00
DE		

(*Ans.:* N41°14′W, 312.52 ft)

15-4.

Side	Bearing	Length (ft)
AB	N61°17′E	334.21
BC	S17°42′W	296.32
CA		

15-5.

Side	Bearing	Length (ft)
AB	N49°10′E	210.00
BC	S41°18′E	242.00
CD	S16°46′W	233.00
DE		
EA	N21°02′W	314.00

(*Ans.:* S79°35′W, 141.02 ft)

In Problems 15-6 to 15-8, compute the missing data in each table.

15-6.

Side	Bearing	Length (m)
AB	S61°56′E	346.9
BC	S40°27′E	401.3
CD		503.6
DA		498.3

15-7.

Side	Bearing	Length (ft)
AB	N50°18′E	
BC	S17°39′W	1095.1
CD		863.2
DA	N45°45′E	742.9

(*Ans.:* length AB = 861.3 ft, bearing CD = S88°20′W)

15-8.

Side	Bearing	Length (ft)
AB	N70°16′E	642.9
BC	S37°18′E	
CD	S58°35′W	
DA	N15°10′W	811.2

15-9. For the accompanying sketch of the balanced traverse *ABCDE* it is desired to cut off the tract of land *ABFG* so that *FG* is parallel to *AB* and the distance *GA* is 150.00 ft. Determine the lengths *BF* and *FG*.

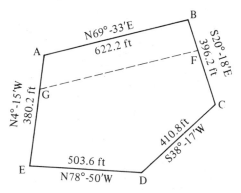

(*Ans.:* BF = 146.2 ft, FG = 689.1 ft)

15-10. For the balanced traverse shown in the accompanying illustration compute the length *DA* and the area *DEFA*.

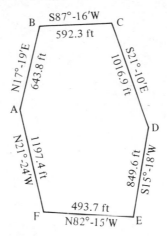

15-11. For the traverse of Problem 15-10 it is desired to cut off 6 acres below a line drawn through point *D*. Determine the lengths and bearings of the necessary sides of the figure.

(*Ans.:* Bearing and length of *FX* = N21°24′W and 113.9 ft, and for *XD* = N49°25′E and 994.2 ft)

15-12. For the traverse of Problem 15-10 it is desired to cut off 12 acres below a line drawn through point *D*. Determine the lengths and bearings of the necessary sides of the figure.

15-13. For the traverse of Problem 15-10 it is desired to cut off the land below a line drawn through point *D* parallel to line *EF*. Determine the length of the necessary sides of the figure and the area of the figure.

(*Ans.:* Length of side *FA* = 964.4 ft, area = 15.17 acres)

15-14. For the traverse of Problem 15-10 it is desired to cut off 20 acres below a line drawn parallel to line *AD*. Determine the length of the necessary sides of the figure.

Topographic Surveying

16-1 INTRODUCTION

Topography can be variously defined as the shape or configuration or relief or roughness or three-dimensional quality of the earth's surface. Topographic maps are made to show this information, together with the location of artificial and natural features of the earth including buildings, highways, streams, lakes, forests, and so on. Obviously, the topography of a particular area is of the greatest importance in planning large projects such as buildings, highways, dams, or pipelines. In addition, the reader would probably, unless he lives in fairly flat country, want to have a topographic map prepared for his own land before he located and planned the building of a house. Topography is also important for soils conservation projects, forestry plans, geological maps, and so on.

The ability to use maps is very important to people in many professions other than surveying, for example, engineering, forestry, geology, agriculture, climatology, and military science. In October 1793, Napoleon Bonaparte at the age of 24 received his first promotion because of his ability to make and use maps when he was placed in command of the artillery at the siege of Toulon.

Since the preparation of maps is quite expensive the surveyor should learn what maps have been previously made for the area in question before he begins a new one. For instance, the U.S. Geological Survey has prepared topographic maps for a large part of the U.S. These

262

maps are readily obtainable at very reasonable prices. They are drawn to scales of approximately 1 in. = 1 mile (probably 1:62,500) or smaller and with contour intervals (a term defined in Section 16-2) of 10 ft in relatively flat country and up to 100 ft in mountainous territory.

Many other government agencies, state and federal, have prepared maps of parts of our country and copies of them are easily obtained. In fact, more than 30 different federal agencies are engaged in some phase of surveying and mapping. In order to obtain information about the maps available for a particular area, one may write to the Branch of Distribution of the U.S. Geological Survey in Arlington, Va. 22202 for locations east of the Mississippi River and in Denver, Colorado 80225 for those locations to the west of the Mississippi. This organization serves as a clearing house for all other federal agencies.

If the surveyor cannot find a map from these sources that has a sufficiently large scale for his purposes, he should nevertheless be able to find one that will give general information and serve as a guide for his work.

16-2 CONTOURS

The most common method of representing the topography of a particular area is to use contour lines. A *contour* is an imaginary level line that connects points of equal elevation (see Fig. 16-1). If it were

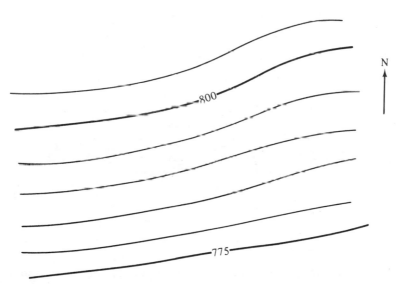

Figure 16-1 Typical contours (5-ft contour interval). Notice that the slope is fairly uniform for the lines are almost equally spaced.

possible to take a large knife and slice off the top of a hill with level slices at uniform elevation intervals, the edges of the cut lines around the hill would be contour lines (see Fig. 16-2). Similarly, the edge of a still lake is a line of equal elevation or a contour line. If the water in that lake is lowered or raised, the edge of its new position would represent another contour line.

The contour interval of a map is the vertical distance between contour lines. The interval is determined by the purpose of the map and by the terrain being mapped (hilly or level). For normal maps, the interval varies from 2 to 20 ft, but it may be as small as one-half ft for flat country and as large as 50 to 100 ft for mountainous country.

The selection of the contour interval is a very important topic. The interval must be sufficiently small so the map will serve its desired purpose while at the same time being as large as possible to keep costs at a minimum. When the maps are to be used for earthwork estimates, a 5-ft interval is usually satisfactory unless very shallow cuts and fills are to be made. For such cases a 2-ft contour interval is probably needed. When the maps are to be used for planning water storage projects, it is usually necessary to use 1- or 2-ft contours.

Figures 16-1, 16-2, and 16-3 present introductory examples of

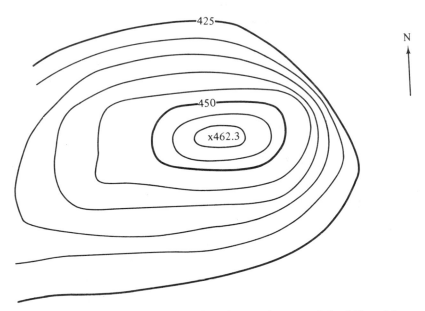

Figure 16-2 Notice that x462.3 ft indicates the top of the hill and its highest point. The close spacing of the contours on the upper right indicates a steep slope.

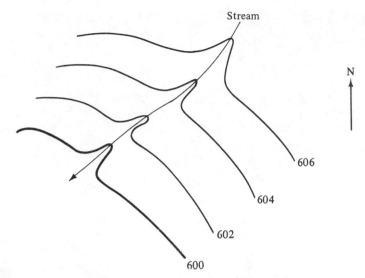

Figure 16-3 Notice how the contour lines bend upstream in crossing creeks or rivers.

contour lines together with descriptive notes.   Figure 16-4 shows a portion of a topographic map prepared by the U.S. Geological Survey.

Other methods that may be used to show elevation differences are relief models, shading, and hachures.   Relief models are a most effective means of showing topography.   They are made from wax, clay, plastic, or other materials and are shaped to agree with the actual terrain.   The Army Map Service once produced and distributed beautifully colored molded plastic relief maps for those parts of the country which have significant elevation differences.   These maps are now sold by Hubbard Scientific Co. of Northbrook, Illinois.

An old method used to show relative elevations on maps was shading.   In this method an attempt was made to shade in the various areas as they might appear to a person in an airplane.   For instance, the steeper surfaces might be in shadows and would be darkly shaded on the drawings.   Less steep slopes would be indicated with lighter shading, and so on.

A few decades ago it was common practice to represent topography by means of hachures, but their use today is more infrequent, and they have been replaced by contour lines.   Hachures are short lines of varying widths drawn in the direction of the steepest slopes. They give only an indication of actual elevations and do not provide numerical values.   By their spacing and widths, these lines produce an effect similar to shading, but they are perhaps a little more effective.

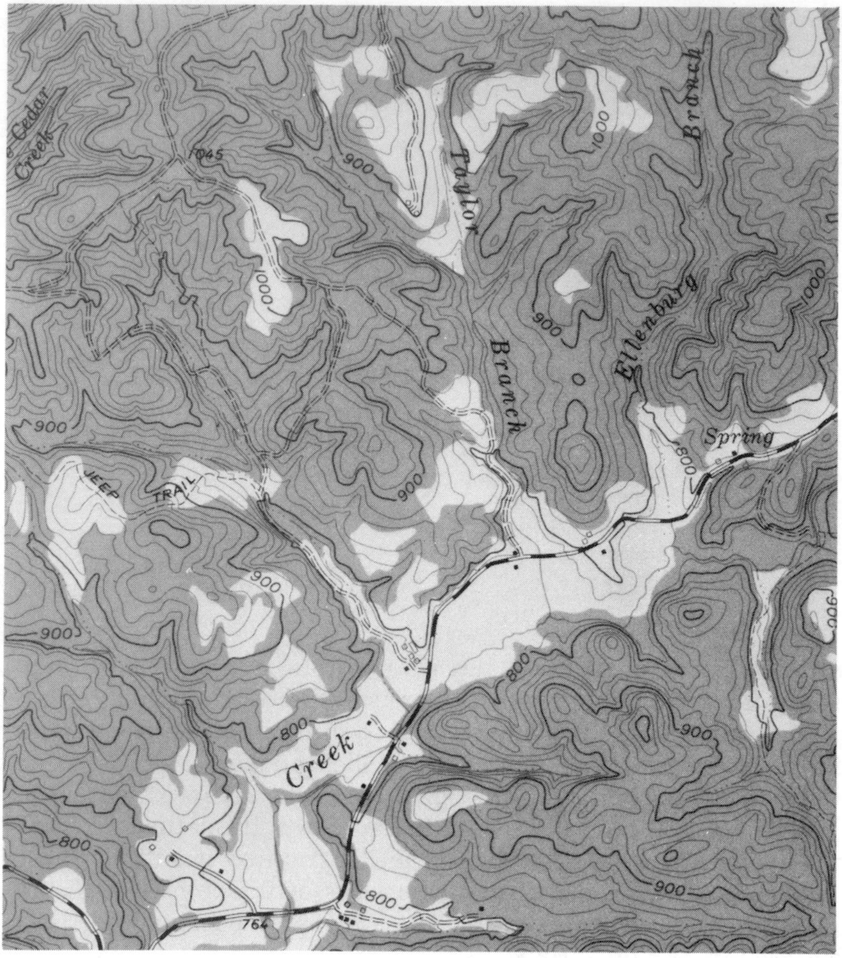

Figure 16-4 Sample contour map. (*Courtesy U.S. Geological Survey.*)

16-3 METHODS OF OBTAINING TOPOGRAPHY

There are three general methods used to obtain topography. These are the transit-stadia, plane table, and photogrammetric methods. In the transit-stadia method the necessary measurements are made in the field, recorded in the field book, and then plotted on paper in the office. In the plane table method the measurements are made in the same way as those in the previous method, but the data are plotted in the field on a paper that is attached to a drawing board mounted on a tripod. This device is called a plane table. Both of these two methods are explained in this chapter, while the subject of photogrammetry is presented in Chapter 22. Most topographic surveys for areas larger than approximately 50 to 100 acres are now handled by photogrammetry for reasons of economics.

16-4 STADIA

Stadia is commonly used to obtain data needed for preparing topographic, hydrographic, and other maps.   It is also useful for making rough surveys and for checking more precise work, but it is not sufficiently precise for most property surveying.

With stadia the surveyor can with one instrument setup make readings that will provide him with the distances, directions, and elevation differences to many surrounding points.   The usefulness of this procedure is gradually declining, however, in the face of advances made in aerial mapping.

16-5 STADIA EQUIPMENT

The equipment used for stadia surveying consists of a transit whose telescope is provided with stadia hairs and a regular level rod or a stadia

Figure 16-5 Stadia rods. (*Courtesy Keuffel & Esser Co.*)

rod. The stadia hairs are mounted on the cross-hair ring, one a distance above the center horizontal hair and one an equal distance below. A regular level rod is satisfactory for stadia readings of up to 200 to 300 ft, but for greater distances stadia rods similar to the ones shown in Fig. 16-5 are used. A study of this figure will show how easily good stadia rods can be made, and for this reason there are numerous types of rods in use, many of which are homemade. Because stadia rods are large, unwieldy, and inconvenient to transport in vehicles, a stadia design that is printed on a strip of coated woven fabric is sometimes used. The fabric can be tacked onto a board for a particular job and then it can be removed, the board can be thrown away, and the fabric can be folded up and carried to the next job.

16-6 STADIA THEORY: HORIZONTAL SIGHTS

In this section an optics discussion of transit telescopes is neglected. The result is that the theory is very simple. A detailed theoretical discussion of this matter is presented in *Elementary Surveying*, 6th ed., by R. C. Brinker and P. R. Wolf.[1] It is assumed initially that externally focusing telescopes are used. These instruments have a prominent focusing screw that causes the objective lens to move. Reference is made to Fig. 16-6 in which it is desired to determine the distance D from the center of the instrument to the rod. This distance equals c (the distance from the center of the instrument to the center of the objective lens) plus f (the focal distance) plus d (the distance from the focal point to the rod).

If we let the distance from the top stadia hair to the bottom stadia hair be i, it is possible to write the following expression and from it determine the distance D:

$$\frac{i}{f} = \frac{s}{d}$$

$$d = \frac{f}{i} s$$

$$D = \frac{f}{i} s + (c + f)$$

In nearly all transits the value of f/i or K is 100 and the value of $c + f$, called the *stadia interval factor*, varies from approximately 0.8 ft to 1.2 ft with an average value of 1 ft. Manufacturers show on their instrument boxes more precise values for the stadia constant for each instru-

[1] New York: Thomas Y. Crowell Company, Inc., 1977, pp. 257-262.

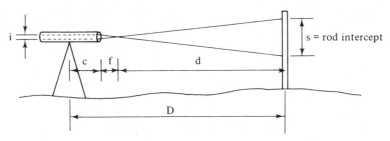

Figure 16-6

ment.   In general, when the telescope is in a horizontal position, the horizontal distance from the center of the instrument to the center of the rod is

$$H = Ks + 1$$

The stadia intervals usually measured are probably a little too large because of unequal refraction and because of unintentional inclination of the rod by the rodman.   In order to offset these systematic errors, the stadia constant is neglected in most surveys.

Furthermore, many transits of recent manufacture have internally focusing telescopes.   For these instruments, the stadia constant is probably only a few tenths of a foot and it is even more reasonable to neglect it.   The objective lens of this kind of instrument remains fixed in position while the direction of the light rays is changed by means of a movable focusing lens located between the plane of the cross hairs and the objective lens.

16-7 STADIA THEORY: INCLINED MEASUREMENTS

Since stadia readings are seldom horizontal, it is necessary to consider the theory of inclined sights.   For such situations it is necessary to read vertical angles as well as the s values or rod intercepts and from these values to compute elevation differences and horizontal components of distance.

If it were feasible to hold the rod perpendicular to the line of sight, as shown in Fig. 16-7, the horizontal and vertical components of distance would be as follows (again, the stadia constant is neglected):

$$H = D \cos \alpha = Ks \cos \alpha$$

$$V = D \sin \alpha = Ks \sin \alpha$$

Obviously, it is not practical to hold the rod perpendicular to the line of sight, that is, at an angle α from the vertical; therefore it is held

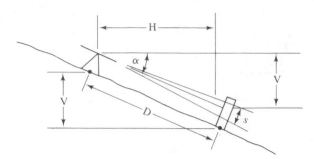

Figure 16-7

vertically. This means that the rod is held at an angle α from being perpendicular to the line of sight. As can be seen in Fig. 16-8, the rod intercept reading is too large if the rod is not perpendicular to the line of sight. If the readings are taken with the rod plumbed, they can be corrected to what they should be if the rod had been perpendicular to the line of sight by multiplying them by the cosine of α. The resulting values for H and V may then be written as follows:

$$H = (Ks \cos \alpha)(\cos \alpha)$$

$$H = Ks \cos^2 \alpha$$

$$V = (Ks \sin \alpha)(\cos \alpha)$$

$$V = \tfrac{1}{2} Ks \sin 2\alpha$$

As a practical matter, if the vertical angle α is less than $3°$, the diagonal distance may be assumed to equal the horizontal distance, but for such cases the vertical differences in elevation (the V values) definitely cannot be neglected.

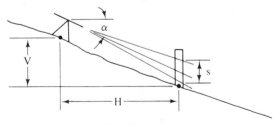

Figure 16-8

16-8 STADIA REDUCTION

The derivations of the stadia reduction formulas were presented in the last section. In this section a few remarks are made regarding their application and solution. These formulas may be used directly for calculating horizontal distances and elevation differences, but their direct application for such purposes is infrequent because more efficient methods are available. If large areas are to be surveyed by stadia, for example, for mapping purposes, there may be hundreds or even thousands of readings to be reduced. If each of these were directly solved by the formulas, the task would be very large indeed.

The most common method for making these calculations is to use the stadia reduction tables (Table A in the appendix). These tables give horizontal distances and elevation differences for different vertical angles where Ks is equal to 100, that is, where s equals 1.00. For s values other than 1.00, the needed values are simply equal to s times the numbers given in the table. These multiplications are usually rounded off to the nearest foot for horizontal distances and to the nearest 0.1 ft for elevation differences.

Other devices for stadia reduction are available, such as the stadia slide rule which includes the ordinary log scales plus two special scales, one with log $\cos^2 \alpha$ values and another with log $\frac{1}{2} \sin 2\alpha$ values. Stadia slide rules are not as accurate as the tables just mentioned, but they are sufficient for most topographical work. Furthermore, they have a slight advantage when the computations are made in the field, as, for example, in plane table mapping which will be discussed in Sections 16-11 and 16-12.

Another method of stadia reduction involves the use of the *Beaman stadia arc*, also called the *stadia circle*. This device is a specially graduated arc that either may be part of the vertical circle of the transit or alidade or may be separate. The arc consists of two scales graduated in percent: (1) one which enables the surveyor to determine the number of feet that must be subtracted from each 100 ft of stadia distance in order to obtain the horizontal distance measured and (2) one which enables him to determine the number of feet of rise or fall of the line of sight for each 100 ft of stadia distance.

There are European instruments that have curved stadia lines which permit direct readings of H and V. The lines appear to move farther apart or closer together as the telescope is raised or lowered, in other words, as the vertical angle is changed. Two lines are curved to correspond to the variations in the trigonometric function $\cos^2 \alpha$; two other lines are curved to represent the variations of $\frac{1}{2} \sin 2\alpha$.

16-9 FIELD PROCEDURE FOR STADIA MAPPING

The precision obtainable with stadia (averaging approximately 1/500) is sufficient for many preliminary or rough surveys. As such, it provides a quicker and more economical job than may be obtained with a transit-tape survey, particularly in rough country. The procedure used follows that of transit-tape surveys in that the horizontal angles or azimuths are measured with the transit. Distances are measured by stadia instead of the tape.

The most common use of stadia surveying is for obtaining the details necessary for mapping. For topography, it is necessary to measure distances, directions, and vertical angles. For most maps only the details are obtained by stadia. A control traverse is carefully made with the transit-tape method and differential levels are made to the stations of the control traverse. From these stations stadia is used in order to obtain the details needed for the map.

The transit may be set up at each of the traverse stations and a large number of stadia readings taken on points in the surrounding area in order to determine their position and elevation. If all the required area is not sufficiently covered from the traverse stations, short stadia traverses may be run out from those stations into the required areas.

The transit is set up at a particular point, its HI is measured by holding the stadia rod or a tape up to the center of the telescope, and its value is recorded in the notes. With the middle cross hair set on the HI of the rod, the line of sight is parallel to a line drawn from the point over which the transit is centered to the point under the level rod. Thus the difference in elevation between the two points on the ground is the same as the difference in elevation between the center of the telescope and the HI up on the level rod. The horizontal circle is oriented by backsighting to the last station. The scale may be set to zero in the backsight or it may be set to the back azimuth of the line. The upper motion is then used for sighting all subsequent points.

The rodman moves to the first point and the instrumentman sets the middle horizontal cross hair on the HI and reads the rod intercept s. In order to do this, he may very well move the bottom cross hair up or down to a whole foot because it simplifies the subtraction from the top cross-hair reading. The instrumentman then resets the middle cross hair on the HI and lines up the vertical hair. He waves the rodman on to the next point while he reads the horizontal and vertical angles.

A typical set of stadia notes is shown in Fig. 16-9. The first four columns identify the points by number and show the three measurements taken for each. The last three columns are used to record computed values. The remaining space on the right-hand page is used to

Sta.	Azimuth	100x Rod.Int	Vert. ∠	Elev. Diff.	Horiz. Dist.	Elev.	
STADIA SURVEY OF WILD HOG FARM						Dec. 3, 1981 Clear, mild, 60°F K&E Transit #31621	⫟ J.B.Johnson Rod – N.C.Hanson Rod – A.N Nelson
⫟ @ C, Elev = 207.30, HI = 4.7, BS on D							
1	27°-08'	145	-11°-05'	-27.4	140	179.9	₵ of creek
2	51°-56	260	-3°-06'	-14.0	259	193.3	Keller's front corner far side of road
3	77°-05'	130	-14°-20'	-31.2	122	176.1	Pt under power line
4	95°-00	170	-10°-50'	-31.4	164	175.9	" " " "
5	127°-53'	297	-3°-41'	-19.0	296	188.3	Keller's back corner
6	161°-20'	75	-4°-50'	-6.3	74	201.0	Edge of cleared field
7	270°-00'	120	+4°-40'	+9.7	119	217.0	P.O.G. (pt. on ground).
8	347°-20'	80	-4°-05'	-5.7	80	201.6	P.O.G.
⫟ @ B, Elev.=213.60, HI = 4.6, BS on C							
1	8°-20'	148	+4°-08'	+10.6	147	224.2	Top of creek bank
2	13°-50'	96	0°-00'	0.0	96	213.6	" " " "
3	56°-35'	93	+4°-48'	+7.8	92	221.4	" " " "
4	177°00'	200	-8°-34'	-29.5	196	184.1	₵ of creek
5	196°-25'	96	-17°-38'	-27.7	87	185.9	" " "
6	266°-15'	195	-8°-15'	-27.7	191	185.9	" " "

J.B.Johnson

Figure 16-9

carefully describe the points. Obviously, these descriptions are very important to the map plotter back in the office. In addition, the recorder should make any necessary explanatory sketches in the fieldbook.

As an example, suppose that for a particular point with the telescope sighted on the HI, a vertical angle of $-6°18'$ and a rod intercept of 3.20 are read. Using the stadia reduction tables of Appendix A, the horizontal distance and the difference in elevation can be determined as follows:

$$H = (98.80)(3.20) = 316.2 \text{ ft} \quad \textbf{say 316 ft}$$

$$V = -(10.91)(3.20) = -34.91 \text{ ft} \quad \textbf{say -34.9 ft}$$

Should it be assumed that a level with an externally focusing telescope and a stadia constant of 1.25 were used for the readings, the value can be corrected as described below. As previously mentioned in Section 16-6, it is quite common to neglect the stadia constant and the

resulting corrections as we did in the preceding calculations. If, however, it is desired to make a correction for the constant, it can be
handled by using the values listed opposite C (the stadia constant) given
at the bottom of the stadia reduction tables. For the above example,

$$H = (98.80)(3.20) + 1.24 = 317.4 \text{ ft} \quad \textbf{say 317 ft}$$

$$V = -(10.91)(3.20) - 0.14 = -35.05 \text{ ft} \quad \textbf{say –35.1 ft}$$

For the stadia notes shown in Fig. 16-9 the stadia constant is
neglected.

For mapping purposes, it is necessary to determine the elevations
of a sufficient number of points in order to picture the contour or
relief of the area accurately. Several approaches are possible in selecting these points. One is the so-called *checkerboard method* in which
the desired area is divided into squares or rectangles and the elevation
determined at the corners of each of these figures. In addition, elevations are determined at points where slopes suddenly change, as, for
example, at valley or ridge lines. This method is best suited to small
areas which have little relief or roughness.

The most common method of selecting the points is the *controlling point method*. Elevations are determined for key or controlling
points and the contour lines are interpolated between them. The controlling points are usually thought of as those points between which
the slope of the ground is approximately uniform. They are such
points as the tops of hills, the bottoms of valleys, and tops and bottoms
of the sides of ditches, and other points where important changes in
slope occur. If there is a uniform slope for a long distance, it is theoretically necessary to obtain only one elevation at the top of the slope
and one at the bottom, for the map plotter will merely interpolate
between those points in order to obtain his contour lines. Practically
speaking, however, since the eye is easily fooled by slopes, it is well to
take an occasional intermediate point, even though the slope seems
constant.

A third method which makes the plotting of contours quite easy
but which is not so simple to apply is the *tracing contour method*. A
number of points are located at the elevation of a desired contour and
their locations are plotted on the map and connected by the contour
line. In order to do this, it is usually easier to level the telescope and
calculate from the HI what the FS rod reading should be when the
rodman is on a particular contour line. For instance, if the HI is
457.2 ft, the rodman will be on the 450-ft contour line if the FS reading is 7.2. He can be waved up or down the slope until the reading is
7.2. After a number of points are located for the 450-ft contour line,

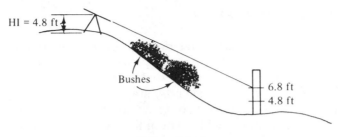

Figure 16-10

the surveying party can work on, say the 448-ft contour line, by positioning the rodman at points where the FS readings are 9.2 ft, and so on.

Sometimes in sighting to a point the instrumentman is unable to set the center cross hair on the HI because there is an obstruction, for example, the bushes shown in Fig. 16-10. If it is not feasible to cut the bushes down, it will be necessary to sight on some point on the rod other than the HI. It is usually convenient to select a value that is an exact number of feet above or below the HI. The difference involved is carefully recorded in the notes and must be added (or subtracted) to the calculated vertical distance between the instrument and the point. In Fig. 16-10 the correction is 2.0 ft and the calculated elevation difference V must be increased by that amount.

A similar problem may occur when in sighting to a point the instrumentman is unable to make the readings at both the top and bottom stadia hairs. If, however, he is able to make the readings at the middle hair and one of the other two, he may take this value (which is one-half the desired rod intercept) and double it.

Some stadia rods may be read for distances considerably greater than 1000 ft. It will be noted, however, that because of the inconvenience of handling the bulky stadia rods, they are usually limited to maximum lengths of 10 or 12 ft. If the full intercept between the top and bottom hairs is read, the maximum sight distance then is theoretically limited to 1000 or 1200 ft. Actually though, the presence of grass, bushes, and so on often keeps the instrumentman from being able to see the full length of the rod. Therefore, it is usually necessary for sights in the range of 1000 ft or greater to be obtained by reading only one-half the intercept and then by doubling it. Some instruments are equipped with a quarter hair between the middle cross hair and the upper stadia hair. By using this hair and taking a one-quarter intercept reading, it is occasionally possible to take a sight for a distance as great as 4000 ft.

16-10 SAMPLE STADIA WORK

This section presents a very brief description of the steps involved in working from the stadia notes to a completed topographic map. In Fig. 16-11 the control traverse for a tract of land is plotted along with the elevation of a group of points taken by stadia.

In Fig. 16-12, 2-ft interval contours are drawn from the elevations that were plotted in Fig. 16-11. These lines were sketched in freehand by "eyeball" interpolation of the proportionate distances between elevation points. For very precise maps, the interpolation may be made with a calculator and the proportionate distances scaled between the points.

Finally, in Fig. 16-13 the control traverse is removed from the drawing, and the locations of important objects, such as roads, buildings, power lines, and so on, are plotted. On many topographic maps, for example, those prepared for house lots or building lots, the property lines and corners are so important for the project in mind that they are included on the map. Final details (the legend, title box, meridian arrow) are not shown in this figure.

Within the past few years much work has been done in the area of plotting maps with computers. There are several texts and programs

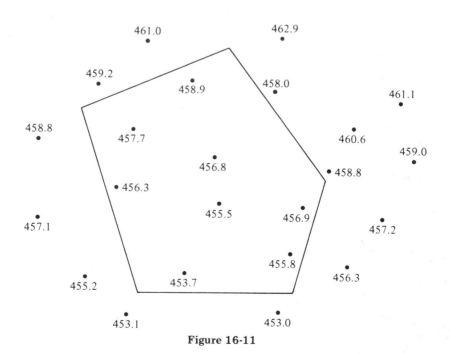

Figure 16-11

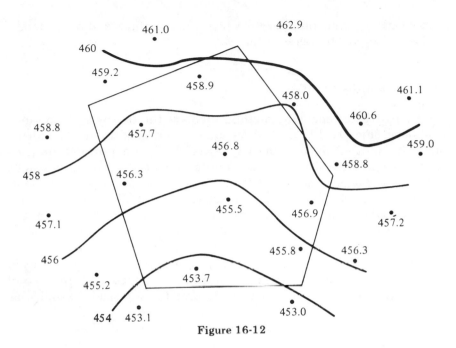

Figure 16-12

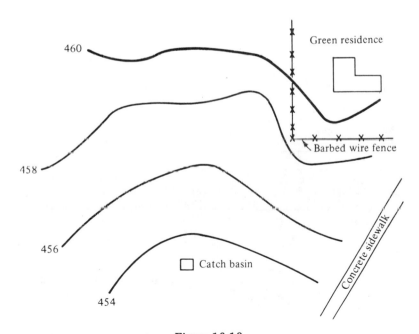

Figure 16-13

available around the county with which contour lines and other details may be plotted for topographic maps.[2]

16-11 PLANE TABLE SURVEYS

A very satisfactory mapping method involves the use of a plane table. An alidade (Fig. 16-14) is used to take stadia measurements and the results are plotted on drawing paper attached to a portable drafting board or plane table. The table is mounted on a tripod so that it can be rotated and leveled. Map details are plotted on the paper as soon as they are measured. The contour lines are drawn as the work proceeds. The great advantage of plane table mapping is that the plotter can compare or correlate his drawing with the actual terrain around him as he works, thus greatly reducing the chance of making mistakes.

The alidade consists of a telescope mounted on a brass ruler of approximately $2\frac{1}{2}$ in. by 15 in. The ruler is aligned with the line of sight of the telescope which enables its user to take readings and scale them off in the proper direction along the ruler's edge. The telescope is similar to those used on transits and has the usual stadia hairs. In addition, the alidade has a trough compass mounted in a small metal box, a bubble tube mounted on the straight edge for leveling purposes, and a

Figure 16-14 Plane table and alidade. (*Courtesy Wild Heerbrugg, Inc.*)

[2]D. L. Ryan, *Computer-aided Graphics and Design* (New York: Marcel Dekker, Inc., 1979) pp. 113–114.

vertical arc that permits the measurement of vertical angles needed for determining elevation differences.

If the surveyor has his position located on the drawing and if the drawing is oriented in the proper direction, he can sight on a desired point and determine the distance and elevation difference to that point by stadia and immediately plot the information on the paper.

Plane table paper is subject to rough field conditions, particularly dirt and moisture. Consequently, a special tough paper is used that minimizes expansion and contraction. This paper has been subjected to alternating drying and wetting cycles, in order to make it more resistant to moisture changes. Sharp, hard pencils (6H to 9H) are used for plotting details, and great care is needed in handling the alidade in order to reduce smearing. It is lifted and moved rather than slid on the paper.

The plane table is set up at approximately waist height so that the surveyor or topographer can bend over it without resting on it. The topographer sets up, levels, and orients the table with great care. For small-scale maps (say those having a scale smaller than 1 in. = 50 ft), it is generally satisfactory to center the table over each traverse station by eye, but for larger-scale maps, it is necessary to center the table more carefully. Each plane table is equipped with an arm under the board from which a plumb bob may be hung and used to center the table over the desired point.

16-12 PLANE TABLE DEFINITIONS

Several terms pertaining to plane table surveying that must be defined in order to enable the reader to understand how the work is accomplished are listed below.

Foresight. Sighting a point with the alidade and drawing a line from the plotted point occupied by the table to the other point is called a foresight.

Orientation. The process by which the plane table is set up and aligned so that the lines on the drawing paper are parallel to the lines on the ground which they represent is called orientation. Orientation may be accomplished by the use of the compass, by backsighting, and by resection.

Backsight. A sight that is taken from the position of the plane table at a known and plotted point to another known and plotted point is called a backsight. The edge of the alidade scale is placed along the line between the two known points and the table is turned until the telescope line of sight hits the other point. When this is done the table is oriented in relation to the points on the ground.

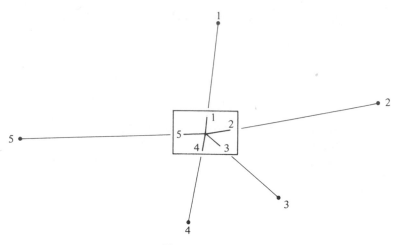

Figure 16-15

Radiation. The process by which a series of points are located in relation to the plotted position of the table is referred to as radiation. The plane table is set up and oriented, and rays are drawn in the direction of each of the points whose elevation and position is desired. Figure 16-15 illustrates this process. The orientation of the table should occasionally be checked to see if any slippage of the table has occurred. This might happen if the topographer has leaned against the table, or bumped it.

Traversing. In plane table surveying, traversing involves a procedure very close to the one followed with the transit-tape procedure. The plane table is moved successively to each of the traverse stations. At each station the table is oriented, ordinarily by backsighting, after which a foresight is taken to the next station. A line of the proper length is drawn to that station by the radiation method just described. The traversing procedure is illustrated in Fig. 16-16 in which points A, B, C, D, and E are the traverse stations on the ground and a, b, c, d, and e are the plotted points for those stations on the drawing. Notice that when the table was set up at E, the whole traverse was previously completed while the table was at D. This means that the setup at E can be used for checking.

Resection. If the table is set up at a point on the ground that has not previously been plotted on the drawing, the process of locating the point is called resection. This may be done by sighting on two or more previously plotted points whose positions are visible from the table and which have been previously plotted on the drawing. The topographer should not get himself into this situation and we do not undertake here

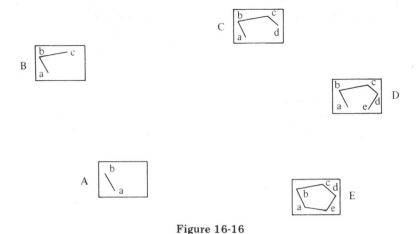

Figure 16-16

to describe its solution. Several excellent references are available which discuss the subject in detail.[3]

16-13 MAP SYMBOLS

It is quite convenient to represent various objects on a map with symbols that are easily understood by the users. Such a practice saves a great deal of space (as compared to the lengthy explanatory notes which might otherwise be necessary) and results in neater drawings. Many symbols are standard across the United States while others vary somewhat from place to place. In Fig. 16-17 a sample set of rather standard symbols are presented. Very detailed symbol lists are available from the U.S. Geological Survey, these being the ones they use for their maps.

Occasionally the surveyor encounters some feature for which he does not have a standard symbol. For such a case he may make up a symbol but he must identify it in his legend.

16-14 COMPLETING THE MAP

The appearance of the completed map is a matter of the greatest importance. It should be well-arranged and neatly drawn to a scale suiting the purpose for which it is to be used. It should include a *title*, *scale*, *legend*, and *meridian arrow*. The lettering should be neatly and

[3]R. E. Davis, F. S. Foote, and J. W. Kelly, *Surveying Theory and Practice*, 5th ed. (New York: McGraw-Hill Book Company, 1966), pp. 431–436.

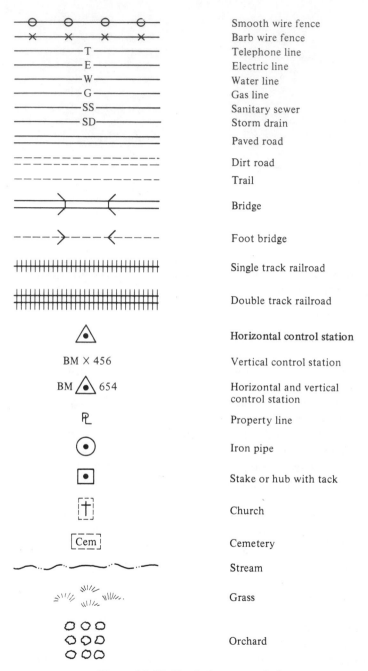

	Smooth wire fence
	Barb wire fence
—T—	Telephone line
—E—	Electric line
—W—	Water line
—G—	Gas line
—SS—	Sanitary sewer
—SD—	Storm drain
	Paved road
	Dirt road
	Trail
	Bridge
	Foot bridge
	Single track railroad
	Double track railroad
	Horizontal control station
BM × 456	Vertical control station
BM 654	Horizontal and vertical control station
PL	Property line
	Iron pipe
	Stake or hub with tack
	Church
Cem	Cemetery
	Stream
	Grass
	Orchard

Figure 16-17 Typical map symbols.

282

carefully performed or perhaps lettering sets or press-on letters should be used.

The title box may be placed where it looks best on the sheet but is frequently placed in the lower right-hand corner of the map. It should contain the title of the map, the name of the client, location, name of the project, date, scale, and the name of the person who drew the map and the name of the surveyor.

17

Land Surveying or Property Surveying

17-1 INTRODUCTION

Land surveying or property surveying is concerned with the location of property boundaries and the preparation of drawings (or plats) which show these boundaries. In addition, it involves the writing and interpretation of land descriptions involved in legal documents for land sales or leases. More precisely, land surveys are made for one or more of the following purposes:

1. To reestablish the boundaries of a section of land that has been previously surveyed.
2. To subdivide a tract of land into smaller parts.
3. To obtain the data needed for writing the legal descriptions of a piece of land.

The location of property lines began before recorded history and for all of the subsequent ages it has been necessary for the surveyor to reestablish obliterated land boundaries, establish new boundaries, and prepare boundary descriptions. The need for people to locate, divide, and measure land caused the development of the land surveying profession. Today, our expanding population, the demand for second homes, and the formation of many new industries are creating a demand for more and more property surveys.

In general, land prices have been climbing rapidly and the accurate location of property lines is becoming increasingly important. A century or two ago land was so plentiful that describing a distance such as: "as far as a fast horse can run in 15 minutes" might have been adequate, but it is certainly no longer true. A few decades ago the compass and chain were satisfactory for making property surveys, but this is no longer the case. Today, the transit and steel tape are still in common use but the situation is rapidly changing. Recent technical advancements, such as electronic distance measuring devices, are commonly applied in the field all over the U.S.

17-2 TITLE TRANSFER AND LAND RECORDS

Title to land may be transferred by deeds, wills, inheritance without a will, or by adverse possession (to be discussed in Section 17-8). A deed for a piece of land contains a description of the boundaries, the monuments, and important information concerning adjoining property.

For a deed to be legally effective, it must be recorded in the office of the proper public official. Usually, this office is located in the county court house and called the Registry of Land Deeds or the Recorder of Deeds, or some similar name. At this office the information is copied into the records and is available for anyone to see. You as an individual can go to the office and see the records that show the description of anyone's land in the county and, in addition, you can determine how much he or she paid for it.

These offices keep indexes so that it is reasonably easy to find the records desired. The indexes are kept chronologically and indexed in the name of both the grantor (the seller) and the grantee (the buyer). When the surveyor goes to this office, he is usually looking up the legal description of certain pieces of land. There he will find copies of the land description and generally copies of plats of the land which have been filed at the office.

If a person becomes involved in land purchases, he will want to make sure that he obtains full or clear title to his land. Various attorneys and private title searching companies are by far better prepared to search these indexes for information relating to a title than is the ordinary individual. They will, for a fee, search the records and provide information regarding the title. The reader will readily understand the importance of such careful checks if he will just think for a moment about the legal problems that could arise if he were to build a house on a lot for which he thought he had title but did not.

For many important property sales the purchaser may want the seller to provide title insurance guaranteeing clear title to the property before he will buy. Title companies usually combine or can arrange both services (title checks and title insurance).

17-3 COMMON LAW

Most of the law pertaining to property surveying is common law or, as it is frequently called, unwritten law (that is, it is not statute law passed by political bodies). Common law is that body of rules and principles that has been adopted by usage from time immemorial. It has been formed by transforming custom into rules of law. When disputes between two or more parties arose in early English days the courts decided what to do from established custom. If situations occurred that did not seem to be adequately covered by custom, the judges based their decisions on their own ideas of right and wrong.

As an illustration of the kind of problems settled by these courts, consider the case in which a man excavated for a building foundation on his land and a nearby building on another person's land caved in because of the subsidence of the ground. The injured party went to court to ask for compensation for his loss. From such cases as these the principle evolved that each landowner owes lateral support to adjoining land owners.

Beginning with the Year Book of 1272 A.D., the English courts began to keep records of their decisions and have kept them continuously since. Since the beginning of these records judges have searched through them to determine if their particular case had previously been decided.

English common law is the basis of jurisprudence in 49 of the states of the United States. Louisiana was originally French and its legal foundation is based on the Roman common law. Even though the Romans occupied most of the British Isles for six centuries, England was the only European country to develop an independent system of common law. (English common law, however, did draw heavily from Roman common law.)

17-4 MONUMENTS

Corners are points that are established by surveys or by agreements between adjacent property owners. The usual practice is to mark these corners with a relatively permanent object called a *monument*. Monuments may be natural features such as rocks, trees, springs, and so on, or they may be artificial objects such as iron pipes driven into the ground, blocks of concrete, mounds of stone, wooden stakes perhaps with some more permanent material buried at their bases such as charcoal, glass, etc., or (in areas of low rainfall) mounds of earth or pits. Although wooden stakes alone would seem inadequate as monuments because of their temporary nature, some courts have held that substantial wooden stakes may be so classified.

The surveyor must be reasonable in his method of marking corners. He may describe a particular corner in a manner that is perfectly clear to everyone at the time of marking, for example, the northeast corner of Joe Smith's barn. A few decades later, however, Mr. Smith's barn may be completely obliterated and there will be no one around who can prove where it was located. There are plats on record that describe a corner as being that spot where so and so shot a bear on such and such a date. Needless to say, it is quite a challenge to locate that point 5 or 50 years later. Clearly, that kind of description is a good start toward future litigation.

Just about anything may be used for monuments, but long iron pipes or concrete monuments are generally more satisfactory and may even be required by law in some areas. Whatever type of monument is used, it should be carefully described in the surveyor's notes on his plats.

The life of the land surveyor and the problems of landowners (small or large) would be much simpler if they both would understand and follow this important rule: *Establish property monuments so carefully that their obliteration is unlikely, but reference them in such a manner that they may be easily and economically replaced if they are obliterated.*

Unfortunately, monuments are frequently destroyed and resurveys are required to replace them. (Deliberate damage or destruction of monuments is prohibited by law.) In those places where monuments may be easily destroyed unintentionally, for example, near busy streets or in construction areas, it is a good practice to use *witness corners.* These might be iron pipes (or other types of monuments) placed some convenient distance back along property lines in more protected locations. These monuments are also described in the surveyor's notes and are shown on his plats. Another place where witness corners are used is where the actual corners are located in places difficult of access, for example, in streams or lakes.

If the location of a property corner can be fixed beyond any reasonable doubt, it is said to *exist.* If its position cannot be found, it is said to be *lost.* On occasions when the monument used to mark the corner cannot be found, the corner is said to be *obliterated.* This does not necessarily mean that the corner is lost because it may very well be possible to reestablish its original position.

Corners may often seem lost when they really aren't. For instance, if the corners had been marked with wooden stakes, cutting off slices of earth with a shovel might reveal a change in soil color where the stake had decayed. If an iron pipe or concrete monument had been removed, the void space filled in with the surrounding soil might show a slight change in coloration. Among several other methods that may be used to find old corners are the locating of old fence corners, statements by neighboring people, the use of metal detectors to locate iron pipes, and so on.

17-5 BLAZING TREES

Another aid in locating property lines and corner monuments in forest areas is blazing and hacking trees. This practice was very common with surveyors in the past, but it is not as common today because so many people object to having their trees marked. A *blaze* is a flat scar made with an ax (or machete) on a tree at about breast height. The bark and a small amount of the tree tissue are removed and the tree is marked so that it can be identified for several decades. A *hack* or *notch* is referred to as a V-shaped cut in the trunk of a tree.

One common method of marking trees will be described in this section, but the reader should realize that surveyors in various parts of the country mark trees in different ways. In the author's locality it was the practice along boundary lines to put two hacks on the sides of trees nearest the boundaries. Two hacks were used to distinguish them from accidental marks from other causes. The trees hacked were those which the surveyor could reach while walking along the line. If a tree was exactly on the boundary (sometimes called a *line tree*), it was marked with a blaze with two hacks underneath. These marks were placed on both sides of the tree in line with the boundary.

It was also the custom to place three hacks on trees near property corners. These hacks were placed on the trees facing the corner and at a distance above the ground equal to the distance from the particular tree to the corner. If a tree was the corner monument, it was marked with an X and three hack marks underneath it.

Blazing trees may often aid the surveyor in locating property lines and corners. It is true, however, that in many cases lumber operations, forest fires, and other causes may have destroyed the marked trees.

17-6 THE LAND SURVEYOR—A SPECIALIST

Disputes over land boundaries are an everyday occurrence, as may be seen by checking through court records. Although the land surveyor may on occasion be involved in original surveys or in subdividing tracts of land into smaller tracts, a very large proportion of his work deals with resurveys in which he tries to reestablish old boundaries that have been previously surveyed.

The surveyor is faced with many problems in relocating old land boundaries. Incomplete data on plats and deeds, missing monuments on the ground, conflicting claims by adjoining property owners, and inaccurate original measurements greatly magnify his problems. These are only a few of the problems besetting the land surveyor, but they show that in order to do his work, he must be something of a specialist. As a result, land surveying is a profession that is to a large extent learned only after considerable experience in a particular locality.

Land surveying may be learned only by field experience and by a thorough study of the laws pertaining to the subject. In order for a surveyor to become familiar with the conditions in a particular area, he must have a great deal of experience in that locality with regard to the methods of surveying used by the previous surveyors, court interpretation of land problems, and so on. As an example, the weight given to different items varies from state to state, for instance, the relative importance of monuments in place when they do not agree with the recorded values.

One frequent purpose of resurveys is to attempt to settle disputes between adjacent property owners as to where the lines should actually be. *In resurveying, the land surveyor's goal is to reestablish lines and corners in their original positions on the ground, whether or not those locations are in exact agreement with the old land descriptions. His duty is not to correct old surveys but to put the corners back in their original positions.*

Too often the inexperienced surveyor thinks that land surveying merely involves the careful measurement of angles and distances. These measurements are actually only one means of reestablishing old boundaries. This comment is not meant to imply that careful measurements are not important but rather to remind the surveyor that his duty is to find where the original corners and boundaries were regardless of the precision with which the original survey was conducted.

The location of corners and property lines is determined from the *intent* of the parties to the original establishment of the boundary. The law is not concerned with their secret intentions but with their intent as expressed by the action of their surveyors as evidenced in plats, deeds, existing monuments, and so on. As time goes by it becomes more and more difficult to determine the original intent of the parties.

The reader should understand that the surveyor has no legal authority to establish boundary locations. On many occasions, however, he is called upon to relocate lines which are so difficult to find that the results are doubtful. In such cases as these, an agreement between the parties involved is probably the most sensible and economical solution. Property owners are always free to do this as long as their agreement does not affect some other party.

In most lawsuits over land boundaries, even the winner loses unless the land is extremely valuable, because of the cost and ill-will connected with such actions. If the surveyor is able to bring about an amicable agreement between the two parties, he will probably have served them both from an economic viewpoint and perhaps even more in preserving friendship. An experienced land surveyor is probably in a better position to suggest a just settlement of this type of problem than is a court of law. If the surveyor is able to persuade the landowners to compromise in a boundary dispute, he must survey the new line and the landowners should then go through a formal and recorded acceptance of the line.[1]

17-7 MONUMENTS, BEARINGS, DISTANCES, AND AREAS

Among the factors involved in relocating boundaries are existing monuments, adjacent boundaries, bearings, distances, and areas. In considering the relative importance of these items in reestablishing property boundaries, natural monuments are given the greatest preference. Artificial monuments are given the next highest preference. It is thought that natural objects (springs, streams, ridge lines, lakes, and beaches) offer a greater degree of permanence (they are less likely to be destroyed or relocated) than do artificial monuments such as set iron pipes, concrete markers, or stones. Of course, some people will deliberately move one of the latter monuments in order to better their land position. Other people, not realizing the importance of these monuments, will pick them up and take them home for their own use. Occasionally, concrete monuments are found being used as doorsteps at nearby houses. A person who deliberately moves a property monument is breaking the law and may be fined and/or imprisoned.

The courts feel that because the original owners could *see* the monuments, they more nearly express the original intent of the parties to land agreements than do the measurements of directions and distances which are subject to so many errors and mistakes. After natural and artificial monuments come adjacent boundaries and then bearings and distances, generally in that order. It should clearly be noted that bearings and distances cannot control the reestablishment of land boundaries if the monuments that are in place actually show the original boundaries. Bearings and distances can control the outcome,

[1] A. H. Holt, "The Surveyor and His Legal Equipment," *Transactions of the ASCE*, **99**(1934), 1155–1169.

however, if there have been mistakes in the placing of the monuments or if the monuments are lost.

Land area is considered the least important of the factors listed, unless the area is the very essence of the deed as, for example, when an exact amount of land is clearly conveyed. The layman may be puzzled by the following description commonly used in deeds regarding acreage: "26 acres more or less." The purpose of the words "more or less" is to indicate that all of the land within the specified boundaries is being conveyed, even though the area stated may vary greatly from the actual area. *This doesn't mean that the surveyor can do a sloppy job and get by with it by using such a term as "more or less." He is responsible for the quality of his work and if it is not up to the standard which is expected of a member of his profession, he is liable for any damages which result.*

17-8 MISCELLANEOUS TERMS RELATING
TO LAND SURVEYING

In this section are defined several terms that frequently occur in dealing with the transfer of land title.

Adverse possession.   If a person occupies and openly uses land not belonging to him for a specific length of time and under the conditions described by the laws of his state, he may acquire title to the land under the U.S. doctrine of adverse possession. The possession must be open and hostile and is usually for a period of 20 years but perhaps less under certain conditions as, for example, when the original title to the land was not clear.

Even though a surveyor is able to reestablish without question the original boundaries of a tract of land, it is possible that because of adverse possession the original lines are no longer applicable. If the owner fails to act during the designated time, he loses his right to act. It is generally held that a private citizen does not have the right to acquire government land by adverse possession. If the owner of the land gives permission to another person to occupy the land, that person can never acquire title to the land by the adverse possession doctrine no matter how long he uses the land after the permission is given.

Riparian rights.   The rights of persons who own property along bodies of water are referred to as riparian rights. Following are a few comments concerning these rights as they apply to property lines. Creeks, rivers, and lakes are very natural and convenient boundaries between various tracts of land because they are easy to describe and are easy to locate. For small streams that are not navigable, property lines usually go to the center line of the stream or to the "thread of the stream." The thread of the stream is generally thought of as being the center line of the main channel.

The U.S. Supreme Court has decided that it is up to the individual states to decide where private property lines run along navigable streams. Some states have decided on the center lines of streams, some on the threads of the streams, and others have selected the high water marks (those marks to which the water has risen so commonly that the soil is marked with a definite character change as to vegetation, and so on).

Alluvium and avulsion. Although a stream or river is an excellent boundary, it can't always be depended upon to stay in the same place. Its position may shift so very slowly and imperceptibly as a result of erosion, current, or the force of waves that the property owners cannot recognize the short-term change in position. When bodies of water change by this method, called *alluvium*, property lines are held to move with the change. If, however, the change is very sudden and perceptible, as, for example, when a large amount of soil is suddenly moved from one landowner to another or when a stream changes its bed completely, it is called *avulsion*. When avulsion occurs property lines are held not to change from their original positions. This description seems simple enough, but unfortunately, many years after the fact it is difficult to tell whether a change was caused by avulsion or alluvium.

Accretion and reliction. When avulsion occurs and the water deposits materials on the bank of a stream or other body of water it's called *accretion*. Should a body of water recede as when it partly or wholly dries up the land area adjoining it is increased and this is called *reliction*.

When land areas change by accretion or reliction the courts will attempt to establish the new boundaries of the adjoining tracts of land in an equitable manner. If water frontage is involved it may be of the greatest importance in the decisions made. They may divide the water frontage in some proportion to the frontage of the old boundaries.

Further comments on riparian rights. There are so many complicated problems which can arise regarding riparian rights that a special body of law has evolved on the subject. Some of the numerous problems which may be encountered are related to property lines, navigation, docks, water supply, fishing rights, oyster beds, artificial fills or cuts, erosion, accretion, and many others. The laws pertaining to these situations vary quite a bit from state to state and court decisions in different states have been entirely different for similar cases.

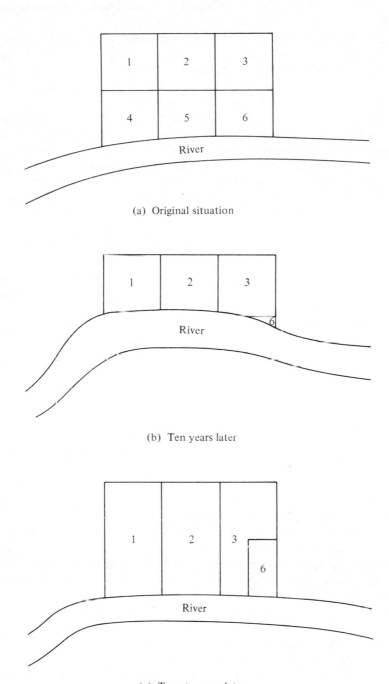

(a) Original situation

(b) Ten years later

(c) Twenty years later

Figure 17-1 Possible ownership changes as water boundaries change (depending on state).

One illustration is presented here to show how involved the legal situation can become concerning riparian rights. For this discussion reference is made to Fig. 17-1, where both erosion and accretion are involved. The discussion presented here can apply to the legal situation in some states while it may not apply to others.

The purpose of this discussion is to show how complicated the legal problem may be. In part (a) of Fig. 17-1 six lots are shown. The lots numbered 4 through 6 bordered the river and had riparian rights. During the next ten years lots 4, 5, and most of 6 eroded away leaving the situation shown in part (b) of the figure. At this time lots 1, 2, and the parts of 3 and 6 bordering the river have water rights. In the next ten years it is assumed that the river gradually moved back to its original position and the replaced land now could belong to lots 1, 2, 3, and 6 as shown in part (c) of the figure. A similar discussion applies to beachfront property in some states.[2]

17-9 RESURVEYS

The average client of a surveyor has no idea what a time-comsuming and hair-pulling job a resurvey can be. Ideally, the original surveyor painstakingly measured distances and directions and carefully set monuments and witnessed them with equal care. Sadly, the truth is that often the monuments were wooden stakes placed at points that were located by haphazardly determined distances and bearings. The result is that there is no problem in surveying so difficult and requiring so much patience, skill, and persistence as that of relocating old property lines.

If, as previously stated in Section 17-6, the old lines and monuments can be reestablished, they still govern the boundaries no matter how poorly the old survey was performed. For a surveyor to reestablish old property lines, he must follow in the footsteps of the original surveyor. In order to be able to do this, he must have a good idea of how the original survey was performed. Was a compass or transit used to measure the angles? Were distances measured on slopes or horizontally?

When the surveyor cannot definitely relocate boundaries, he has no power to establish them. If there are disputes between adjacent landowners over boundaries and if he cannot persuade the owners to compromise, the courts will have to settle the matter. In court the surveyor can only serve as an expert witness and present the evidence

[2]P. Kissam, *Surveying For Civil Engineers*, 2nd ed. (New York: McGraw-Hill, 1981), pp. 327–328.

that he has found. Once a dispute is settled, whether by mutual agreement or by court action, the surveyor should do all he can to see that a precise survey is made of the settled boundaries and good monuments established.

To begin a resurvey the surveyor carefully studies available plats and deeds of the property as well as those of adjacent tracts. As a part of this study he typically calculates by latitudes and departures the precision obtained in the original survey of the tract in question. Surprisingly enough, the surveyor may frequently spend more than half of his time studying and planning, while the actual field work may not take very long at all.

Next, the surveyor begins a careful examination of the tract of land in the field. If the original survey was precisely done and if one or more of the original corners can be found, the survey can be handled just as was the five-sided traverse used for an example in Chapters 5 and 12 in which the distance and angle measurements were made.

The trouble, however, is that so many original surveys contain major errors and mistakes and often several or all of the monuments are gone. One frequent cause of poor measurements was the equipment used, often the compass and link chain. If the original survey was made with a surveyor's compass, the surveyor may attempt to rerun the lines by using a compass and by making proper allowance for the estimated change in magnetic declination since the time of the original survey.

If a surveyor can definitely establish the corners at the ends of at least one line, he is off to a good start in the whole relocation job. He can measure that line and compare its distance with the original measurement and thus obtain an idea as to how the original value was obtained. Was it measured on the slope or horizontally? By writing the proper proportions he can compute proportional lengths of the other sides. He can determine the true bearing of the known side (as described in Chapter 18) and from that value compute the estimated magnetic declination at the time of the original survey. With this magnetic declination he can compute the estimated true bearings of the other sides.

With these computed distances and bearings he can start running the sides. At each estimated corner location he will carefully search for evidence of the original monuments. If he finds one of the old monuments, he sets a new one (if necessary) and carefully references it. If he is unable to find a monument, he sets a temporary point and moves on to the next line. If he finds monuments farther along, he tries to work back to those that he couldn't find by using new proportions based on the found monuments. He continues in this fashion until he locates all of the old monuments or sets temporary monuments at all of the various corners. Several ideas will be helpful to him in looking for old corners: hacked trees, old fence lines, road beds, places where vegetation is different, and so on. After he carefully studies the gathered information, he may very well return to the field for more measurements.

In a resurvey the surveyor may initially be able to find only one corner or perhaps, as is often the case, none at all. If only one corner is evident and the original survey was run with a compass, he can estimate the magnetic declination at the time of the survey, convert the bearings to estimated true bearings, and try to run the lines as described in the last paragraph.

If no corners are evident, the surveyor will probably begin his work by studying the records of the adjoining land tracts and try to run their lines in order to see if he can locate some of their monuments.

When he has finished his work for any of the cases mentioned, he will give to his client his best judgment as to the location of his lines and corners based upon his opinion as to the true intent of the parties to the original survey.

17-10 METES AND BOUNDS

The oldest method of land surveying is the metes and bounds system. (The term *metes* means to measure or to assign a measure, while the term *bounds* refers to boundaries.) Thus a metes and bounds description gives the measurements and limits of the outside boundaries of the tract of land in question. The length and direction of each of the sides of a tract of land are determined and monuments are established at each of the property corners. Creeks, lakes, or other natural landmarks are occasionally used to define property corners. Most land surveys in the original thirteen states of the U.S., in Kentucky and Tennessee, and in some surveys elsewhere were made by the metes and bounds method.

Metes and bounds surveys are not usually used in those parts of the country that have been sectionalized by the government (see Section 17-11). Some exceptions occur in parts of the country where land was originally granted by France, Spain, England, or even the U.S. government before the beginning of sectionalization.

Property descriptions are usually written by lawyers, real estate persons, or surveyors. The accuracy of their descriptions is of the utmost importance because one simple mistake in their writing may lead to property disputes between adjoining landowners for several generations. A person who writes the descriptions for land should try to put himself in the shoes of someone else 5, 10, or 100 years later who will try to interpret the description as to ground location. Following is a typical metes and bounds deed description (see also Fig. 17-2).

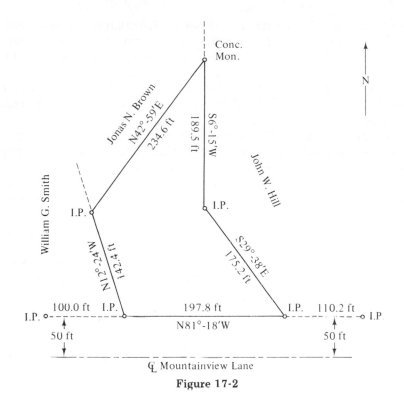

Figure 17-2

"All of that certain piece, parcel or tract of land situated, lying and being in state and country aforesaid about 1/2 mile west of the town of Tamassee, and lying on the north side of Mountainview Lane. Beginning at an iron pin at the southwesterly corner of the property of the grantor located 50 ft from the center line of Mountainview Lane and running thence along the easterly line of the William G. Smith property N12°24'W 142.4 ft to an iron pin; thence with the southerly line of the Jonas N. Brown property N42°59'E 234.6 ft to a concrete monument; thence _____ to the point of beginning. All bearings are referred to the true meridian, the tract contains 0.83 acres more or less, and is shown on a plat dated October 16, 1981, drawn by Arthur B. McCleod, registered land surveyor #1635, state of _____."

17-11 THE U.S. PUBLIC LAND SURVEY

In colonial America, the manner of obtaining land from the governments involved (Dutch, Spanish, English, French, etc.) and the methods of describing the land grants varied widely. The boundaries of these tracts of land consisted of streams, roads, fences, trees, stones and other natural features. The tracts were generally irregular in shape except for some subdivision in towns and cities. Furthermore, no overall control method for the surveys was available. The legal descriptions of the land were often vague and the measurements involved contained many mistakes. As a result of all of these factors, when boundaries were obliterated it was and is difficult if not impossible to restore their original positions.

Because of the problems involved in the early colonies the Continental Congress in 1785 established the U.S. Public Lands Survey System with the objective that the same mistakes would not be repeated for the remaining land which the Federal government possessed. This land, called the *public domain*, is the land held in trust for the people by the federal government. About 75% of the land in the U.S. was once part of the public domain. Today approximately 30% of the country's land is still in the public domain. Of the approximately 2.3 billion acres of land in the 50 states, over 1.35 billion have been surveyed by the public land survey system. There are about 350 million acres in Alaska which are unsurveyed.

The U.S. Public Land Survey, which is a rectangular system, has been used to subdivide the states of Alabama, Alaska, Florida and Mississippi as well as all of the states to the west and north of the Mississippi and Ohio Rivers except for Texas. Actually Texas has a system similar to that of the U.S. Public Land Survey but it was affected by the Spanish settlers before its annexation into the U.S. Figure 17-3 shows the parts of the country covered by the U.S. Public Land Survey.

It would certainly have been ideal if all of the land of the U.S. could have been obtained at one time and set up under one survey system. Although this was not possible the country is indeed fortunate to have so much of its land under the public land survey system. Although the system does have its problems, it has nevertheless been a tremendous asset to the country. The success of the system can be verified by the fact that the original procedures have not been greatly changed up to the present day.

With the public land survey system each parcel of land whether it be $2\frac{1}{2}$, 5, 10, 40 or 160 acres is described in such a manner that the description will not apply to any other parcel of land in the entire system. This simplicity of describing land tracts has made the system one of the most practical methods ever devised for land identification and description.

When the rectangular system was introduced it was not applied to the original thirteen states because of the enormous problems involved in changing the existing descriptions of the countless thousands of land tracts involved.

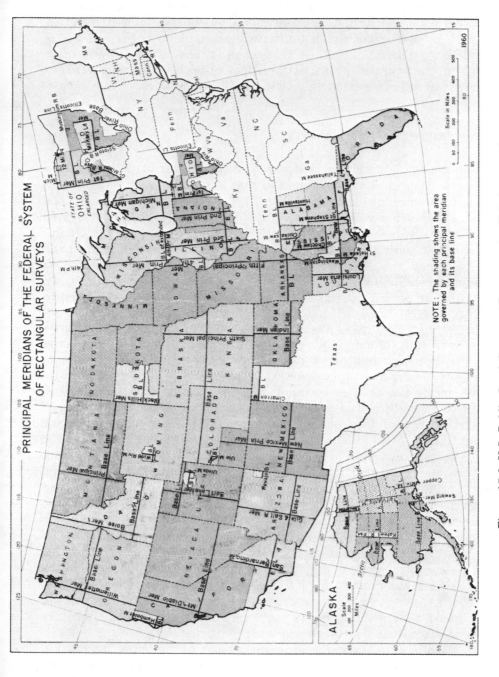

Figure 17-3 U.S. Public Land Survey. *(From map by U.S. Bureau of Land Management.)*

299

Most of the property surveys in the original thirteen states as well as Kentucky and Tennessee were made by separate closed traverses, that is, by the metes and bounds system described in Section 17-10. There is no definite overall control system in those areas and very often some of the boundaries such as the edges of lakes or streams were not even measured but were merely stated as the boundaries.

17-12 EARLY DAYS OF THE SYSTEM

The purpose of the public land survey was to devise a rectangular system and establish it on the ground so that it would provide a permanent basis for describing land parcels. In 1796 the post of Surveyor General was established at a salary of $2000/yr; the first appointment was given to General Rufus Putman, an experienced surveyor and aide to George Washington during the Revolutionary War. In 1812 the General Land Office (now the Bureau of Land Management) was established to manage, lease, and sell the vacant public lands of the United States. To identify and describe the land involved it was of course necessary first to survey it.

At that time there was a great demand for land and yet its price was very low (about $1.25 per acre for "homesteads" of 160 acres each). For such low prices it was impossible to justify very accurate surveys. The surveying instruments available were rather crude in today's terms and the survey points were marked with wooden stakes, mounds of earth placed over buried pieces of charcoal or by pieces of stone. Many of these monuments have been destroyed through the years.

The surveys were handled on a contract basis and sometimes different rates were paid depending on the relative importance of the lines and on their characteristics (as wooded or swampy or hilly). The surveyors were paid about $2 per mile until 1796 and about $3 per mile thereafter. The amount of money which a surveyor could make depended entirely on how fast he could complete the survey.

Although the maximum permissible errors were specified from the early days of the system, the standards were vague and little supervision or checking was done to see if the work met the requirements until the 1880s. Another reason for large errors and mistakes in the system was the speed with which some surveys were handled in Indian territory (where often little field work was actually carried out).

Despite all of these problems a large part of the public land survey was handled very well. Generally inaccuracies in the old surveys cannot be corrected if there is evidence of the location of the original land corners. In other words, once corners are established and used to mark property boundaries, they cannot be changed regardless of the magnitude of the mistakes made.

17-13 OUTLINE OF THE SYSTEM

This section presents a quick summary of the procedure involved in the U.S. public land survey system while subsequent sections present a more detailed description of the system. The area is divided into smaller and smaller approximate squares as follows:

1. The land is divided into *quadrangles* approximately 24 miles on each side. (See Fig. 17-7.)
2. The quadrangles are each divided into 16 *townships*, the sides of which are approximately 6 miles. (See Fig. 17-8.)
3. The townships are divided into 36 sections each approximately 1 mile square. (See Fig. 17-9.)
4. The sections are further subdivided into quarter sections and smaller sections as will be later described. (See Fig. 17-10.)

Since meridians converge to the north, it is impossible for all townships, sections, and so on to be in the shape of exact squares. In the subdivision of townships it is, however, desirable to lay out as many sections as possible in 1-mile squares. To do this, convergence errors are thrown as far west as possible by running the lines parallel to the eastern boundary. In a similar manner the errors in distance measurement are thrown as far to the north as possible by locating the monuments at 40-chain (or $\frac{1}{2}$-mile) intervals along the lines parallel to the eastern boundary so that any accumulated error will occur in the most northerly half mile.

17-14 MEANDER LINES

Navigable waterways or streams of three or more chains in width, as well as lakes covering 25 acres or more (except those formed after the state in question was admitted to the Union) are not part of the public domain and thus were not surveyed nor disposed of by the federal government. The individual states have sovereignty over such bodies of water.

The traverses of the margins of these bodies of water are called *meander lines*. These traverses consist of straight lines which conform as closely as possible to the mean high water marks along the banks or shore lines involved. Meander lines were not run as boundary lines and when the bed of the lake or stream changes the high water marks and the property lines change also.

17-15 DETAILED SUBDIVISION

From 1815 to 1855 the subdivision of the public lands were handled in accordance with instructions from the office of the Surveyor General. Since 1855, however, the detailed directions for subdivision have been given in the *Manual for Instructions for the Survey of Public Lands of the United States* published by the U.S. Government Printing Office. This manual has been revised at various intervals since its original publication. When surveying is being done in a particular area it is necessary to be familiar with the instructions which were in effect when that land was originally surveyed. The following paragraphs contain a detailed description of the procedure which was used in subdividing the public land.

Initial points. Beginning points called initial points were established in each area. Their locations were determined on the basis of astronomical observations. There are 37 initial points, 5 of which are in Alaska.

Principal meridians. True meridians called principal meridians were run through each initial point and extended as far as necessary to cover the area involved. They were identified by number or by name as say the sixth principal meridian which is in Nebraska and Kansas or the Willamette Meridian which is in Washington and Oregon. These are shown in Fig. 17-3.

As the principal meridians were run, monuments were set at intervals of 40 chains and at the intersection with meander lines. The directions of the meridians were usually determined from solar attachments with which true directions could be mechanically obtained. These devices are discussed in Section 18-21 of the next chapter. Supposedly the directions obtained for the meridians were to be within three minutes of true directions. The solar compass was developed by William A. Burt and introduced in 1836. In 1850 the Surveyor General specified that surveyors were to use Burt's improved solar compass or its equivalent for public land surveys. This was a much better compass for determining astronomical north than were the usual magnetic compasses.

With the Manual of 1890 the use of the magnetic compass was forbidden except for subdivision and meandering and in those situations it could supposedly be used only in areas which were free from local magnetic attractions. With the Manual of 1894 the use of the magnetic compass was discontinued for all new public land surveys. In attempting to rerun old lines or "follow in the footsteps of the original surveyors" the regular magnetic compass is still frequently used.

Base lines.   Through each initial point, lines are run out at 90° to the principal meridian and extended as far to the east and/or west as required to cover the area involved. These lines, which are referred to as base lines, are run as true parallels of latitude. They are, therefore, curved lines and are run by laying off 40-chain chords along the curves. Three different methods may be used for laying out the base lines. These are the solar, tangent and secant methods.

In the *solar method* true north is measured at each 40-chain interval, usually with a solar attachment; a 90° angle is measured from true north and another 40 chains is layed off and the process is repeated. This procedure is illustrated in Fig. 17-4.

With the *tangent method* a 90° angle is layed off to the east and/or west as required from the principal meridian and a straight tangent line is layed off for 480 chains or 6 miles as shown in Fig. 17-5. Corners are set at 40-chain intervals. Since this is a straight line, corrections will have to be made to correctly set the corners on the curved base lines. The corrections are made by offsets from the originally set straight line. The magnitudes of the offsets are available in *Standard Field Tables* from the U.S. Bureau of Land Management. They are given in links (1 link = 1/100 × 66 ft = 0.66 ft) and run as high as 37 links at the 480-chain corner. The offset corrections, a few of which are shown in Fig. 17-5, are measured in true north directions to the base line. The errors resulting for this correction procedure are thought to be negligible.

Finally, the *secant method* of laying out base lines is illustrated in

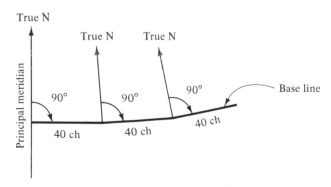

Figure 17-4 Solar method of laying off base lines.

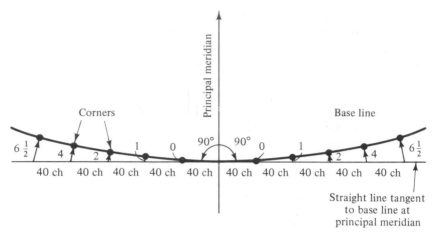

Figure 17-5 Tangent method of laying off base lines.

Fig. 17-6. This method is quite similar to the tangent method. However an angle of less than 90° is measured from the meridian and a straight line is run as shown in Fig. 17-5. Since the rate of curvature of base lines depends on the latitude, the sizes of the angles laid off from the meridian vary. Their values are available in the *Standard Field Tables.* It is necessary to set a point south of the beginning corner at a distance specified in the tables. Then the correct angle is measured from the meridian east and/or west and projected for 480 chains or 6 miles. The offset corrections are then measured to the north or south of the projected lines as required (see Fig. 17-6). These offsets are relatively small compared to some of the tangent offsets.

Standard parallels. After the principal meridians and base lines have been established the standard parallels (often called correction lines) are run in true directions at 24-mile intervals along the principal merid-

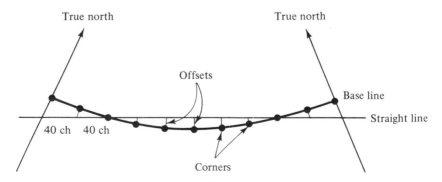

Figure 17-6 Secant method of laying off base lines.

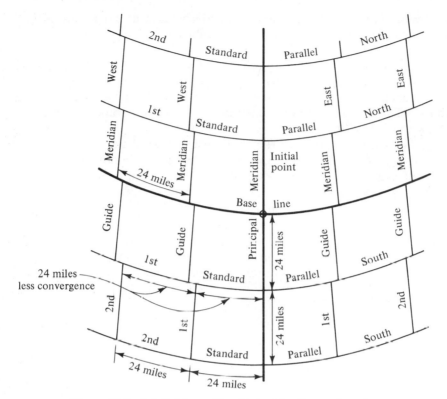

Figure 17-7 Subdivision of land into 24-mile quadrangles.

ian by the same method used for laying out the base line. (In some of the earlier public land surveys the intervals between the parallels were 30 or 36 miles.) Corners are also set at 40-chain intervals along the parallels. As shown in Fig. 17-7, the parallels are numbered north and south as the 1st Standard Parallel North, 2nd Standard Parallel North, etc.

Guide meridians. The so-called guide meridians are run as true directions at 24-mile intervals east and west of the principal meridians. There also are shown in Fig. 17-7. The guide meridians and standard parallels form approximately square quadrangles. From the 24-mile corners on the most southern standard parallel the guide meridians are run in true northerly directions until they intersect the next standard parallel. At these points corners are established as shown in Fig. 17-7 and the distances to the standard township corners already in place are measured. Then the surveyor moves to those township corners and runs true north meridians to the next standard parallel. He continues this procedure until he reaches the upper boundary of the area being surveyed. These guide meridians are designated at 1st Guide Meridian East, 2nd Guide Meridian East, etc., as shown in the figure.

Townships. Once the quadrangles are established they are divided into townships (approximately 6 miles by 6 miles). Range lines are established at 6-mile intervals along the parallels between the guide lines. These lines are run continuously between standard parallels as shown in Fig. 17-8.

The surveyor starts at the southeast corner of the southwest township (point *A* in Fig. 17-8) and runs a line in a true northerly direction for 480 chains setting corners at 40-chain intervals along the way. He then sets township corner *B* and runs a random line due west for 480 chains toward the principal meridian setting temporary corners at 40-chain intervals. If the random line hits within 3 chains in the east–west direction and within 3 chains in the north–south direction of the previously set corner *F*, then his work can be considered acceptable. Then the line and the temporary corners are adjusted proportionately

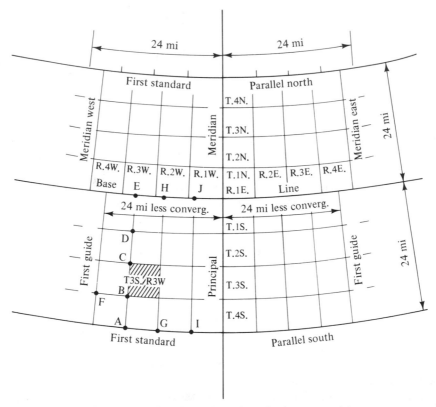

Figure 17-8 Division of quadrangles into townships.

in their positions to agree with the correct position of corner F. Any error in length from B to F is put in the most westerly half mile.

The same procedure is followed in setting points C and D in the figure. Then the range line is continued north until it intersects the standard parallel at E. If there is an error in the 24 miles from points A to E, it is put in the last or most northerly half mile of range line DE.

In the same fashion the range lines GH and IJ are run. As range line IJ is established the random lines are run to the east and corrected backwards so that any error is placed in the most westerly half mile.

A row of townships extending north and south is called a *range*, and a row extending east and west is called a *tier*. Figure 17-8 shows this arrangement in detail. Notice the numbering system for the cross-hatched township (Tier 3 south, Range 3 west).

Sections. The final work done in the rectangular survey is that of laying out the sections as shown in Fig. 17-9. The townships are divided into 36 sections each approximately 1 mile square and containing 640 acres. The sections are numbered as shown in the figure. In addition, the government surveyors mark the quarter section corners at 40-chain intervals and the local surveyors (state, county, or private) perform any further subdivision.

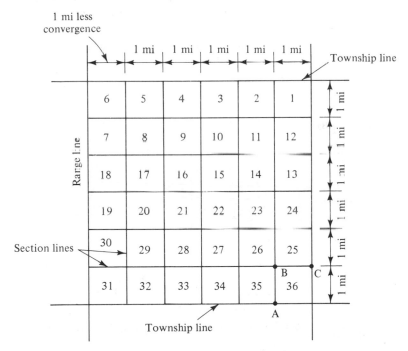

Figure 17-9 Subdivision of township into sections.

To lay out the sections in a township the surveyor starts at the southeast corner of the township (Section 36 in Fig. 17-9) and reruns the south and east boundaries for one mile to check the earlier work and to check his instruments against the ones previously used. He then starts at point A in the figure and runs a line for 80 chains or one mile to the north, parallel to the eastern side of the township, and sets point B as well as a one-quarter-section corner at the 40-chain point in the middle. He runs a random line from B to the east toward the previously set section corner at C setting a temporary half-mile corner in between. If he comes within 50 links of C he adjusts proportionately the position of the quarter section corner at the midpoint of line BC.

A similar procedure is followed for setting the remaining section and quarter section corners for the township. Space is not taken up here to describe this rather repetitious process. The laying out of the U.S. public land system throughout its existence is covered in great detail in an excellent book by McEntyre.[3]

The public system called for the disposal of the land in units equal to quarter-quarter sections of 40 acres each. This division is seen in Fig. 17-10. It will be noted that this procedure can be continued because the quarter-quarter sections can be divided into quarters, each containing 10 acres each.

In Fig. 17-10 a particular quarter or half of a quarter section is readily identified. In the full description of one of these 40-acre pieces, the quarter of the quarter section is listed, then the quarter section, then the section number, then the township, range, and principal

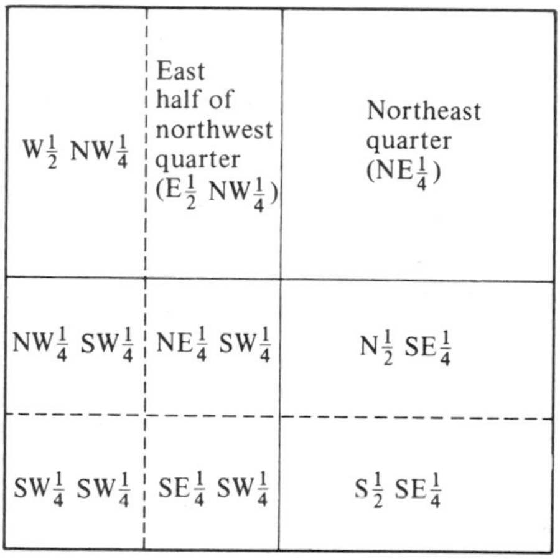

W$\frac{1}{2}$ NW$\frac{1}{4}$	East half of northwest quarter (E$\frac{1}{2}$ NW$\frac{1}{4}$)	Northeast quarter (NE$\frac{1}{4}$)
NW$\frac{1}{4}$ SW$\frac{1}{4}$	NE$\frac{1}{4}$ SW$\frac{1}{4}$	N$\frac{1}{2}$ SE$\frac{1}{4}$
SW$\frac{1}{4}$ SW$\frac{1}{4}$	SE$\frac{1}{4}$ SW$\frac{1}{4}$	S$\frac{1}{2}$ SE$\frac{1}{4}$

Figure 17-10 Subdivision of Section 22 of Fig. 17-9.

[3] J. G. McEntyre, *Land Survey Systems* (New York: John Wiley & Sons, Inc., 1978), pp. 32–199.

meridian.   In this manner a tract of land could be described as the NE 1/4SW1/4, Section 22, T2S, R2E of the third principal meridian.

17-16 CORNER MARKERS

Quite a few different materials have been used for marking the township, section, and quarter-section corners. Markers have included wood posts, stones, broken bottles, pits filled with charcoal and other types. When stones or posts were used it was common to mark them with from 1 to 6 notches to identify the particular section or township corners. Quarter-section corners usually had $\frac{1}{4}$ marked on them.

Since 1910 iron pipes with brass caps on which the designation of the particular corner is given have been in common use. It is thought that they will last and remain legible for perhaps 100 years or more. Stones are usually piled around the pipes to protect them and to make them easier to find.

17-17 WITNESS CORNERS

Should the location of a corner fall in an unmeandered lake or stream or be on a steep cliff or in a marsh or other inaccessible spot, a witness corner is established. A witness corner is usually placed on one of the regular survey lines of the property. However if a satisfactory point for such a corner cannot be occupied within 10 chains along one of the survey lines it is permissible to locate a witness corner in any direction within 5 chains of the corner position.

17-18 RESTORING LOST OR OBLITERATED CORNERS

When resurveys are made in the public land area it is sometimes necessary to restore lost corners. Restoration is made only as a last resort and then is done by a procedure that is carefully spelled out in *Restoration of Lost or Obliterated Corners and Subdivision of Sections*, 1974 edition, published by the Bureau of Land Management, United States Department of the Interior. To make the restoration it is first necessary for the surveyor to fully understand the manner in which the original survey was conducted.

The corners are restored by either single or double proportionate measurements from existing corners in the area. *Single proportionate measurement* involves the new measurement of an old line to determine one or more positions on that line. As the term implies, the direction of a line is established from two existing corners and intermediate positions on that line are established by proportionate measurements.

Double proportionate measurement involves the new measurements made between four known corners (two of which establish a north–south or meridional direction and two of which establish an east–west or latitudinal direction). The lost corner is reestablished by double proportionate measurements between the four corners.

The above-mentioned publication of the Bureau of Land Management presents rules of precedence as to whether single or double proportionate measurement should be used. For instance, single proportionate measurement may be used where one line on which the lost corner is located holds precedence over the line passing through the corner in the other direction. One such case would be a section corner on a township line where the east–west township line takes precedence over the north–south section line.

17-19 DEED DESCRIPTIONS OF LAND

To describe regular tracts of land within the U.S. public lands system for legal purposes is quite simple. An acceptable description of a 40-acre quarter section was previously given in Section 17-15 with reference to Fig. 17-10.

When an irregular tract is involved, however, or one which is not a regular part of the public land system, it is first necessary to tie the description carefully into the rectangular system. Then a length and bearing or metes and bounds description of each side of the tract is given. A description of such a tract might be as follows: "Beginning at a point, marked by an iron pin 300.00 ft North of the NE corner of the SE1/4 of the SW1/4 of Section 28, T3S, R1E, 3rd P.M.; thence North 998.00 ft to an iron pin; thence East 864.00 ft to an iron pin; and so on back to the point of beginning.

Surveying Astronomy

18-1 INTRODUCTION

There are actually three slightly varying definitions of true north: astronomic north (as determined by observations of the stars), geodetic north (based on a reference spheroid), and geographic north. The directions obtained by the three methods vary by a few seconds from each other. True north as defined in this text refers to astronomic north, and the terms *true direction* and *astronomical direction* are assumed to be identical. Furthermore this is the definition of true north used for land surveying.

To present a complete theoretical background and explanation of astronomical observations would require an entire textbook in itself. Fortunately, however, it is possible to present in a single chapter sufficient information to enable the surveyor to work effectively in this area.

Up to this point, only magnetic directions have been discussed. And yet directions read with a compass are very inaccurate. For this reason the compass is used only for rough or preliminary surveys, certain mapping activities, and for rough checking of more precise surveys. A very large proportion of old property surveys were done with a compass and on many occasions it is necessary to be able to use a compass in order to rerun or retrace old boundary lines. For such cases it is to be remembered that magnetic declinations change through the years, often by very large amounts.

18-2 ADVANTAGES OF TRUE OR ASTRONOMICAL DIRECTIONS

The use of true or astronomical directions has several advantages over the use of assumed or magnetic directions. These include the following:

(a) Permanence is given to the direction of boundaries in land surveying as compared to magnetic directions where the magnetic directions are forever changing. If only one corner of a piece of property which was surveyed on the basis of true directions can be located, all the corners can be established assuming the distances were measured with reasonable precision.

(b) As previously described, true directions are quite useful for correlating surveys and for checking the angles for long open or closed traverses.

(c) Astronomical directions are useful for orienting important maps and charts as well as orienting radio and radar antennae.

18-3 CELESTIAL BODIES TO BE CONSIDERED

True directions may be obtained by sighting on the sun or on one of several thousand stars if their positions are known. (Actually the sun is also a star but that name is not frequently applied to it.) In this text only observations on Polaris the North Star and the sun are presented because it is felt that they are the most practical for the surveyor located in the northern hemisphere. In the southern hemisphere there is not a bright star located near the South Pole, and the Southern Cross is commonly used for observations.

18-4 ASTRONOMICAL TABLES USED BY THE SURVEYOR

An *ephemeris* (plural ephemerides) is an astronomical almanac which shows the position of the sun, the planets, the moon, and various stars at certain time intervals. The position of any one of these celestial bodies at times other than those listed in the tables can be determined by interpolation. The various companies which manufacture surveying instruments provide ephemerides at very low cost. In the following discussion, reference is made to the tables of the 1981 Keuffel & Esser (K & E) solar ephemeris. Some of the appendix tables are reproduced with the permission of K & E. Included is the information necessary to solve the example and exercise problems contained in this chapter. For practical work it is necessary to use the ephemeris of the year in question although the tabular values for a particular date of the year change slowly year by year.

More extensive ephemerides are available such as the *American Ephemeris and Nautical Almanac* published annually by the U.S. Government Printing Office and the *Ephemerides of the Sun and Polaris* which is prepared annually by the U.S. Naval Observatory for the U.S. Department of the Interior's Bureau of Land Management.

Though this chapter only presents a discussion of observations of Polaris and the sun, a little additional study will enable the reader to determine true directions by making observations of other stars for which information is given in the ephemerides.

18-5 INTRODUCTORY DEFINITIONS

For the purposes of surveying astronomy, it is assumed that the stars are fixed on the inside of a gigantic sphere with an infinite radius. This tremendous sphere, called the *celestial sphere*, includes all of the heavenly bodies and is assumed to rotate around the earth (its speed of rotation being $360°59.14'$ per 24 hours). Since it is of infinite size, the earth in comparison seems a mere dot. The observer is assumed to be at the center of this sphere, the radius of the earth being negligible in comparison to the distance to the stars. When this assumption causes an appreciable error (as it may in sun shots, depending on the accuracy desired) a correction can be made as will be described in Section 18-17.

The assumption of the existence of this tremendous sphere is perfectly consistent with the view seen by an observer standing on the earth's surface. As he or she looks toward the heavens, the stars are so far away that they all appear to be equidistant and they seem to lie on the inside of a gigantic sphere.

Several definitions relating to the celestial sphere follow.

Celestial poles. If the axis of rotation through the north and south geographic poles of the earth is extended in both directions until it intersects the celestial sphere, the intersection points are referred to as the celestial poles.

Great circle. If a plane were passed through the center of the celestial sphere (that is, the center of the earth) and extended outward in all directions until it intersected the celestial sphere, its line of intersection would be a great circle.

Celestial equator. A great circle which is perpendicular to the polar axis of the celestial sphere is called the celestial equator. It is the same as a plane passing through the earth's equator extending outward until it intersects the celestial sphere.

Zenith and nadir. If a plumb line at the observer's position on earth were extended upward until it intersected the celestial sphere, that point would be the zenith. If the plumb line were extended in the other direction through the earth until it intersected the celestial sphere, that point would be the nadir.

Ecliptic. The sun goes around the celestial sphere once each year and its path across that sphere is a continuous curved line called the ecliptic. The sun moving northward on this line crosses the equator on March 21 and reaches a maximum northerly point $23\frac{1}{2}°$ above the equator on June 21. Similarly, moving southward it crosses the equator on September 22 and reaches a maximum southerly point $23\frac{1}{2}°$ below the equator on December 21.

Vernal equinox. On its northward journey each spring the sun crosses the celestial equator on March 21 and that imaginary point on the celestial sphere is called the vernal equinox. It is referred to in ephemerides as "The First Point of Aries" or just Aries and is often represented by the zodiacal symbol ♈.

Horizon. A great circle passing through the observer's position and perpendicular to a plumb line at the observer's position is called the horizon. This circle is halfway between the observer's zenith and nadir and is the plane in which azimuth is measured.

Vertical circle. A great circle passing through the observer's zenith and any celestial body is referred to as a vertical circle. Such a circle is perpendicular to the horizon.

Meridian. The celestial meridian for an observer's position on the earth's surface is a vertical great circle at that point which passes through the zenith and nadir for the point in question as well as through the celestial poles.

Reference meridian. The meridian passing through the Royal Observatory at Greenwich, England, is generally taken as the reference meridian. Longitude is measured east or west from the reference meridian from 0 to 180°.

True meridian. The true meridian at any point on the earth's surface is the great circle which passes through the point in question and the celestial poles.

18-6 LATITUDE AND LONGITUDE

Positions on the earth's surface are described in terms of latitude and longitude. A brief definition of these terms follows.

Latitude is normally defined as the angular distance (0 to 90°) that a point is above or below the equator. It is also equal to the angle between the plane of the earth's equator and a plumb line held at the point in question.

As previously defined *longitude* is the east or west angular distance (0 to 180°) measured from the Greenwich meridian. It can be expressed either as an angle or in terms of time. If the sun is considered to travel around the earth in 24 hours that would be equal to 15° per hour or 1° in 4 minutes. A longitude of 78°20' west could be expressed in time as $5^h13^m20^s$ west (where the abbreviations h, m and s represent hours, minutes and seconds of time respectively).

For astronomical observations it is possible to scale latitudes and longitudes from a good map. The topographical maps published by the U.S. Geological Survey are usually handy. If these or other good maps are not available, the Director of the National Geodetic Survey, National Oceanic and Atmospheric Administration (NOAA) at Rockville, Md. 20882, can provide values. The individual will need to describe his location as carefully as possible, such as giving the name of the nearest post office and in Public Land Survey areas giving his Section number and perhaps the name of his township.

It will be noted that 1 statute mile on the ground is equal to a little less than 1' of latitude. If in estimating latitude from a map the surveyor has an error of 1 mile, it will cause an error in the computed true direction equal to approximately 1.5" at a latitude of 45°.

18-7 APPARENT MOTION OF THE CELESTIAL SPHERE

The celestial sphere seems to move because the earth rotates from west to east about its polar axis in approximately 24 hours and makes a complete circuit around the sun in slightly more than 365 days. For convenience in making astronomical observations, the earth is assumed to be stationary while the celestial sphere is considered to rotate around the earth. Thus the celestial bodies appear to be moving from east to west.

18-8 CIRCUMPOLAR STARS

If the axis of the earth were extended upward until it intersected the celestial sphere, that point (as previously defined) would be called a celestial pole. Unfortunately that point is not marked in the sky and we therefore sight on a nearby star, usually Polaris, the north or polestar. Polaris is located approximately one degree from the celestial pole and as the earth rotates about its axis toward the east; Polaris appears to move in an orbit or small counter-clockwise circle around the celestial pole. The radius of this circle (called the polar distance) is approximately one degree.

A star which rotates around the celestial north pole and never goes below the observer's horizon is referred to as a *circumpolar star*. Polaris is such a star and it rotates very close to the celestial north pole with an angular distance of approximately 1°. When a circumpolar star is directly above the celestial pole it is said to be at *upper culmination* (UC) and when directly below is said to be at *lower culmination* (LC). When the star is at its most easterly point it's said to be at eastern elongation (EE) and when at its most westerly point at western elongation (WE).

Figure 18-1 shows the apparent motion of a circumpolar star such as Polaris as it is seen from an observer's position X on the earth's surface. It will be noted that two times per day the star will be lined up with true north or the celestial north pole shown in the figure.

If the observer is able to sight on Polaris at UC or LC, he only needs to depress his telescope and set a point on the ground and he will have a true north direction from his position to the point set on the ground. If more convenient he can turn an angle from some other line to the line to Polaris and calculate the true azimuth or bearing of the other line.

Actually, however, there are some distinct disadvantages to making sights on a circumpolar star such as Polaris at UC or LC. The star moves quite rapidly west to east or east to west at culmination and the observer will have to sight on the star at a very precise time. A little error in time can cause an appreciably large error in azimuth. Furthermore the observer will probably be able to make only one sighting on the star at each culmination. As a result he cannot check his measurement, and if the instrument is out of adjustment that fact will cause an error when the telescope is turned downward and used to set a point on the ground.

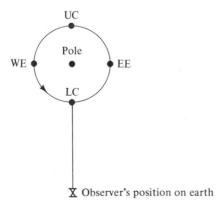

Figure 18-1 Movement of a circumpolar star.

In its rotation about the celestial pole Polaris moves on an almost vertical path for about 10 minutes before and 10 minutes after WE and EE. For all practical purposes, therefore, the star remains at elongation for about 20 minutes, enabling the observer to make numerous observations (some with the telescope in its normal position and others with it inverted). The bearing of Polaris does not change more than 5 seconds of arc during the 10 minutes before and after elongation.

An ephemeris provides the exact time of UC, WE, LC, and EE for Polaris at Greenwich, England. To determine the corresponding times at some other locations, a few simple calculations have to be made as shown in Section 18-12.

18-9 TIME

There are several systems for reckoning time with which the reader needs to be familiar in order to understand astronomical observations. These include (a) apparent time or true solar time, (b) mean solar time, and (c) sidereal time. The conversion of one kind of time to another is of the greatest significance in all phases of astronomical observations. The next several paragraphs present a brief description of these various times.

(a) Apparent Time or True Solar Time

An apparent solar day is the time required for one apparent revolution of the true sun about the earth. The earth moves along an elliptical orbit about the sun and thus the apparent movement of the sun around the earth is not constant. The sun which we see is called the *apparent sun* or the *true sun*. During four periods of the year its velocity across the sky is greater than its average velocity and during four periods its velocity is less than its average velocity. Thus the days given by the apparent sun are not of uniform length.

(b) Mean Solar Time or Civil Time

In order that our solar days will be of equal length, astronomers created the *mean sun* which is a fictitious sun that moves at a constant rate around the earth. This sun apparently makes one complete circuit from west to east among the stars and around the earth in one year. A *mean solar day* is the time required for one revolution of this fictitious or mean sun. This is the same as a *civil day*.

Civil time is the same as *mean solar time* and is the *standard time* generally used. *Local civil time* is the time based on the central meridian of an observer. Civil time based on any other meridian is designated by name, as for instance *Greenwich Civil Time*.

The various ephemerides provide the time intervals between times as determined from the apparent sun and the mean sun. This interval is referred to as the *equation of time* and can be written as follows:

Equation of time = Apparent solar time – Mean solar time

As previously described the equation can give a plus answer (if the real sun is ahead of the mean sun) or a negative answer (if the real sun is behind the mean sun) and the interval can be as large as 16 minutes.

The prime meridian for figuring longitudes anywhere in the world passes through the observatory at Greenwich, England. The standard time for Greenwich which is commonly used for astronomical observations is called *Greenwich Civil Time* or *Universal Time* or *Greenwich Mean Time*.

As we have noted, the sun makes one apparent revolution of the earth in 24 hours or that is it apparently moves 15° per hour. The longitude of the Greenwich time zone is 0° or 0^h.

The U.S. is divided into four time zones: Eastern, Central, Mountain and Pacific times. Through each of these zones there is a central meridian at 75°, 90°, 105° and 120° west of the Greenwich meridian, respectively. It will be noted that these central meridians differ from each other by one hour of time.

A person might think that noon should coincide with the appearance of the sun at its highest point over his position. If such a practice were followed in every town there would be an endless number of time zones throughout the country. Therefore within a given time zone all the watches are set to the time which applies to the central meridian in that time zone.

Since the sun's apparent daily motion is from east to west, it reaches its highest point at locations to the east of the central meridian in a particular time zone before it does at the central meridian. In a like manner it reaches noon at a later time for the western part of the time zone.

To determine GCT (Greenwich Civil Time) at any point in the U.S., the observer must add to his watch time the number of hours of longitude from Greenwich to the central meridian of his time zone.

If daylight saving time is being used in a particular zone standard time will be the same as daylight time for the time zone just to the east. For instance 8 A.M. Pacific Daylight Time is the same as 8 A.M. Mountain Standard Time. For the usual property survey watch time is satisfactory for astronomical observations but for some observations the accurate determination of time to the nearest second is quite important. The observer may need to check his watch against some standard clock or time signal. In some cities, tape-recorded time checks can be obtained by telephone; there are short wave radio time signals broadcast by the U.S. Bureau of Standards; finally, railroad and telegraph companies have standard times available.

(c) Sidereal Time

As we have seen, an apparent solar day is the amount of time required for the sun to make one complete revolution of the earth. Thus apparent solar time may simply be called sun time. Sidereal time, on the other hand, is star time. A sidereal day is the time required for the vernal equinox to make one complete revolution of the earth. A mean solar day is 3 minutes 56 seconds longer than a sidereal day which is 23 hours 56.1 minutes. Sidereal time at any point is equal to the hour angle of the vernal equinox at that point.

18-10 LOCATING POLARIS

In the Northern Hemisphere, the most commonly observed star for surveying purposes is Polaris. This star is the last star in the tail of the constellation Ursa Minor (Little Dipper). To locate Polaris correctly, it is necessary to also be familiar with the constellations Ursa Major (whose seven brightest stars are known as the Great or Big Dipper) and Cassiopeia. These are sketched roughly in Fig. 18-2.

In this figure it will be noted that a line drawn through the two end (or pointer stars) in the Big Dipper points almost exactly to Polaris. On the opposite side of Polaris from the Big Dipper is the W-shaped constellation Cassiopeia which is also shown in the figure. As a further aid in locating Polaris it will be noted that its approximate location can be determined with a compass (perhaps making a correction for magnetic declination). Its altitude will equal approximately the latitude of the point where the observation is being made.

The *declination* of a star is defined as the angular distance the star is above or below the celestial equator. The declination of the sun varies during the year from about $23\frac{1}{2}°$ South to about $23\frac{1}{2}°$ North. For Polaris the declination is in the range of $89°$ and for such a star it may be more convenient to refer to its *polar distance* instead of its declination. The polar distance of a star is the angular distance of the star from the north pole. The polar distance of Polaris would then be about $1°$. Thus the declination of a star plus its polar distance is equal to $90°$. From an ephemeris the surveyor can obtain the polar distance of Polaris for the date of his observation. If the declination is given

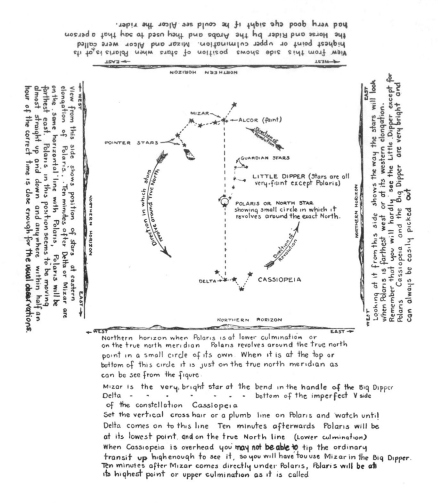

Figure 18-2 [From K. W. Leighton, *Gilbert Civil Engineering (Surveying) for Boys* (New Haven, Conn.: A. C. Gilbert Company, 1920), p. 88.]

instead of the polar distance, he can subtract it from 90° to obtain the polar distance.

In the ephemeris the civil time of elongation on the Greenwich meridian is listed for the date in question. Civil time is standard or watch time for a given zone. It is necessary to correct this value to watch time at the point of the observation. This type of calculation is explained in Section 18-12. If this calculation is made within several minutes of the correct value of elongation it will be satisfactory because Polaris will not move enough horizontally to be detected with the usual transit for 10 minutes before and 10 minutes after elongation.

About 20 or 30 minutes before the calculated time of elongation the instrument should be set up and carefully leveled. (Unless the plate bubbles are quite sensitive, leveling them may be insufficient and it may be necessary to go through a double centering process to be sure the vertical axis of the instrument is vertical.) The horizontal scale should be set on $0°$ or on some definite angle and sighted on a reference point on the ground.

A star may be sighted at dusk through the telescope without a light, but after dark it is necessary to use some kind of light to be able to see the cross hairs. A flashlight pointed diagonally into the objective lens is satisfactory. The observer has to adjust the flashlight's position so that with the indirect light he can see the cross hairs but not have so much light that the star cannot be seen. One thing that will help reflect the light into the telescope is the replacing of the sunshade with a 3- or 4-inch piece of rolled up paper held in place with a rubber band.

Polaris is located as previously described and the vertical hair is focused on it. The star is followed until a time about 10 minutes before elongation. Then the horizontal angle is read and recorded. The telescope should then be rotated through $180°$ and plunged and the horizontal angle read again and the two values averaged. More observations may be taken and averaged if desired, since there is plenty of time.

In any case, the true azimuth to the star at elongation is known from the ephemeris and may be used with the angle from the reference point to determine the true azimuth from the instrument to the reference point. The K & E ephemeris tables for elongation were obtained for latitudes of $40°$. Should the observer be at some other latitude, the azimuth will have to be corrected as described in Section 18-13.

18-11 CHOICE OF SIGHTING ON POLARIS OR THE SUN

A few remarks are presented in this section concerning the desirability of using Polaris or the sun for determining true north.

(a) Polaris

Polaris provides a small excellent slowly moving target which can be sighted clearly and directly with the telescope. Sighting on Polaris at elongation is probably the most satisfactory and surest method, as will be described in the next section of this chapter. Polaris observations do have the disadvantage that they must be made at times other than normal working hours. Also just a little bit of haze or just a few clouds may obscure the star so that it cannot be seen. The errors and mistakes which result while working in the dark can be significant. Furthermore, instrumental errors can be significant for Polaris observations particularly where the vertical angle to the star is more than 40° or 50° (at latitudes of more than 40° or 50°).

(b) The Sun

The sun presents a large target which can be observed during regular working hours. It can be seen even if there is considerable haze and/or if partly cloudy conditions exist. THE SUN SHOULD NOT BE SIGHTED DIRECTLY THROUGH A NORMAL TELESCOPE BECAUSE INJURY TO THE EYES MAY RESULT. It is necessary to use a solar attachment such as a dark filter. Alternatively, the telescope can be pointed toward the sun and a fieldbook page or white card held a few inches away from the eyepiece and the telescope moved until the sun is seen on the page or card. This topic is continued in Section 18-15. It is generally felt that Polaris observations are much superior to sun observations. Nevertheless sun shots provide very satisfactory values of true north for many purposes if several careful observations are made and if the correct standard times are used in the calculations.

18-12 OBSERVATION OF POLARIS AT CULMINATION

An ephemeris gives the times of culmination every 10 or 15 days at Greenwich, England. By interpolation between the given values it is possible to determine the time of culmination at Greenwich on the date in question. To observe the star at culmination at some other position, such as in the United States, it is necessary to correct the time of culmination at Greenwich to the standard watch time at the observer's position.

If Polaris moved around the earth in exactly 24 hours as does the imaginary or mean sun, upper culmination (UC) would occur at the same watch time at Greenwich as it would at the local watch time at the 60° meridian, or the 75° meridian, or the 90° meridian, etc. Polaris, however, as do the other stars, moves around the earth a little faster (24 hours – $3^m 56^s$ for one revolution) than the mean sun. As a result it is necessary to correct the time somewhat. For the 90° meridian UC would occur at $(\frac{90}{360})(3^m 56^s)$ earlier than at Greenwich. At 87° it would occur $(\frac{87}{360})(3^m 56^s)$ earlier.

Finally it is necessary to make a further time correction if the observer is not located exactly at one of the 15° central meridians. For instance if he is located at 87° he is using 90° watch time and UC at his position would occur earlier than at the 90° meridian. As the star moves at about 15° per hour this correction would be $(\frac{3}{15})$ (1 hour) earlier.

Example 18-1 shows the calculations necessary to determine UC at a particular position in the U.S. If the star is sighted at this time the observer can turn the telescope down and set a mark on the ground or he can turn an angle to some other point and he will have a true direction.

Example 18-1

Determine the local watch time of upper culmination of Polaris on December 10, 1981, at latitude 32°15'N and longitude 80°52'W.

Solution   Using Table D in Appendix which is from page 75 of the 1981 K & E Ephemeris.

(a) Determine time of UC at Greenwich on December 10.

$$\text{UC on December 6} = 21^\text{h}12.4^\text{m} \text{ GCT}$$

$$\text{UC on December 16} = \underline{20^\text{h}33.0^\text{m} \text{ GCT}}$$

$$\text{Difference} \qquad 39.4^\text{m}$$

UC on December 10 by interpolation $= 20^\text{h}33.0^\text{m} + (\frac{6}{10}) (39.4) = 20^\text{h}56.6^\text{m}$

(b) Correct UC to longitude 80°52'W.

The ephemeris shows culmination is changing at the rate of -3.94^m per day. Therefore, for longitude 80°52'W, UC will occur

$$\left(\frac{80.87}{360}\right) (3.94^\text{m}) = -0.9^\text{m} \text{ earlier than at Greenwich}$$

(c) Convert to the standard watch time of the observer (in this case the 75° meridian or the 5th time zone).

The star moves approximately 15° per hour and for it to be in the same position at a longitude 5°52' west of the 75° longitude would require an additional

$$\left(\frac{5.87}{15}\right) (60^\text{m}) = +23.5^\text{m}.$$

Combining the above results:

$$\text{UC of Polaris} = 20^\text{h}56.6^\text{m}$$

$$- \ 0.9^\text{m}$$

$$\underline{+ 23.5^\text{m}}$$

$$\text{Local watch time} = 21^\text{h}19.2^\text{m}$$

Note: This is Eastern Standard Time (EST).

18-13 OBSERVATIONS OF POLARIS AT ELONGATION

For an observation of Polaris at WE or EE for a latitude other than 40° the true bearing of the line of sight from the instrument to Polaris is given by the following equation:

$$\text{Bearing of Polaris (in minutes)} = \frac{\text{Polar distance (in minutes)}}{\cos \text{latitude}}$$

The polar distance of Polaris at elongation is given at 10-day intervals in Table E of the Appendix, which is Table 3 of the K & E Ephemeris (entitled "Polar Distance of Polaris"). Instead of using the preceding equation the bearing can be obtained by referring to Table F in the Appendix, which is Table 7 of the K & E Ephemeris (entitled "Bearing of Polaris at Elongation").

The example observation of WE which follows is for the same night and location as was the UC observation of Example 18-1, but WE falls a fourth of a sidereal day later on December 11. Notice that the times of elongation given in the Ephemeris are for the 40°N latitude. For other latitudes they have to be corrected slightly as provided in Table G of the Appendix (Table 9 of the K & E Ephemeris).

Example 18-2

Determine the time of WE and the true bearing to Polaris at that time for December 11, 1981, at latitude 32°15′N and longitude 80°52′W.

Solution (using tables of 1981 K & E Ephemeris):
  (a) Find the time of WE at Greenwich on December 11 (Table D of Appendix).

$$\text{WE on December 7} = 3^{\text{h}}08.7^{\text{m}} \text{ GCT}$$

$$\text{Correction to December 11} = (4)(-3.94^{\text{m}}) = -15.8^{\text{m}}$$

$$\text{WE at Greenwich December 11} = 2^{\text{h}}52.9^{\text{m}} \text{ GCT}$$

  (b) Correct WE to longitude 80°52′W.

$$\text{Correction} = \left(\frac{80.87}{360}\right)(-3.94^{\text{m}}) = -0.9^{\text{m}}$$

  (c) Correct to standard watch time at the 75° meridian.

$$\text{Correction} = \left(\frac{5.87}{15}\right)(60^{\text{m}}) = +23.5^{\text{m}}$$

  (d) Correct for latitude other than 40°N.
       From Table G in the Appendix:

$$\text{Correction} = +0.7^{\text{m}}$$

(e) Combining the above results:

WE of Polaris December 11 at longitude $80°52'W$ = $2^h 52.9^m$

$-\ 0.9^m$

$+\ 23.5^m$

$+\ 0.7^m$

Local watch time = $3^h 16.2^m$ (EST)

(f) To determine the bearing to Polaris, calculate the polar distance of Polaris on December 11 by interpolation between the December 7 and 17 values (Table E of Appendix) = $0°48.93'$.

Tne bearing of Polaris by formula is

$$Z = \frac{48.93}{\cos 32°15'} = 0°57.9'$$

or from Table F in the Appendix by extrapolation

$$Z = 0°57.9'$$

∴ Bearing to Polaris at WE = $N0°57.9'W$

18-14 OBSERVATION OF POLARIS AT ANY TIME

The times of elongation and culmination may often be inconvenient and as a result the surveyor may sight on Polaris at a more convenient time perhaps a little earlier in the evening. For such cases it is necessary to determine the difference from the time of culmination to the time of the observation. Then the bearing to Polaris at that time can be picked from Table H of the Appendix, which is Table 10 of the K & E Ephemeris. This table provides bearings to Polaris for various latitudes and hour angles.[1] To use the table it is necessary to know the latitude of the observer's position and to convert the time after culmination into an hour angle (labeled LHA for "local hour angle" in the tables). If an observation is made 1 hour and 30 minutes after upper culmination the LHA would be $(1.5/24)(360°) = 22.5°$ west of upper culmination.

The instrument is set up and a sight is taken to a reference point on the ground some distance away. The telescope is turned to Polaris and the horizontal angle and watch time are recorded. Then the true

---

[1] Instead of using the table, the bearing can be determined from the following expression, in which Z_n is the bearing, t is the hour angle to the next lower culmination, p is the polar distance and h is the latitude of the observation point plus $(\cos t)(p)$:

$$Z_n = \frac{\sin t}{\cos h} p$$

bearing to Polaris is determined from the Ephemeris as illustrated in Example 18-3 and the true bearing from the instrument position to the mark on the ground is calculated. The procedure should be repeated at least once (noting that Polaris will have moved) and the bearing to the mark again calculated to see if the results check. Example 18-3 illustrates the determination of the bearing to Polaris at some time between culmination and elongation.

Example 18-3

A sight is made on Polaris at 23^h10^m (EST) on December 10, 1981, at latitude $32°15'$N and longitude $80°52'$W. Determine the bearing to Polaris at that time.

Solution

$$\text{Watch time for UC (December 10) from Example 18-1} = 21^h19.2^m$$
$$\text{Actual time of observation} = 23^h \ \ 10^m$$
$$\text{Time difference} = \overline{\ \ 1^h50.8^m} \text{ later than UC}$$

$$\text{Correction at } 3.94^m/\text{day} = \frac{1.847 \text{ hrs}}{24 \text{ hrs}} (3.94^m) = +0.3^m$$

(The sign is plus because the star is moving a little faster than the mean sun and the LHA angle should be a little larger.)

$$\text{Local watch time or hour angle} = 1^h51.1^m = 1.852 \text{ hrs}$$

The bearings in Table H of the Appendix are given for a polar distance of $0°49.2'$. For a latitude of $32°15'$N and an LHA = $(1.852 \text{ hrs}/24 \text{ hrs})(360°) = 27.78°$ a bearing of $0°27.3'$ is determined by double interpolation.

The polar distance of Polaris on December 10 from Table E of the Appendix equals $0°48.93'$. From Table I a correction in bearing of $-0.3'$ is made (since the polar distance is slightly different from the $0°49.2'$ for which the bearings in Table H were computed).

$$\therefore \text{ Bearing to Polaris} = N0°27.0'W$$

18-15 OBSERVATIONS OF THE SUN

Observations of the sun, frequently called sun shots, are not as easily made as those on Polaris or other stars. The stars make small targets and move slowly, enabling the surveyor to sight directly on them with little difficulty. The sun's image on the other hand is large (about $32'$ diameter) and moves swiftly.

It is to be remembered that if a person sights the sun directly through the telescope his eyes may be injured perhaps permanently. He should either make the observations with a dark filter attached to the eyepiece or he should hold a white surface (such as page of his field-book or a white card) behind the eyepiece and observe the sun's image on that surface.

In the following discussion of sun shots it is assumed that the observation is made directly on the center of the sun. This may be done with some type of dark filter or with one of the special devices described later in this chapter. If a dark glass is placed in front of the objective lens, it should be made plane and perfectly parallel to the objective lens or it will introduce an error in the sighting. Another method mentioned for sighting on the center of the sun is to draw a circle on a page in the field book the size of a nickel or a dime. Dashed cross-hair lines are then drawn through the center of the circle. The image of the sun is focused until it falls into the circle and the cross hairs of the telescope coincide with and fill in the cross hairs drawn in the circle, as shown in Fig. 18-3.

The telescope is sighted on a previously set point on the ground some distance away from the instrument with the horizontal scale reading zero. The upper motion clamp and the telescope clamp for vertical motion are released and the telescope turned towards the sun. The fieldbook is held approximately 6 in. behind the eyepiece and the telescope is moved horizontally and vertically until its shadow on the field book is circular. When this is accomplished the upper motion and vertical clamps are tightened.

The field book is moved toward or away from the eyepiece and the eyepiece carefully focused until the cross hairs are as clear as possible. The observer must be careful not to mistake one of the top or bottom stadia hairs for the middle cross hair, which is rather easy to do. This mistake can be avoided by moving the telescope up or down about the horizontal axis until all three cross hairs are clearly seen.

To center the sun in the circle it would be convenient (though often not feasible) to have 3 persons available: one to hold the field book, one to track the sun with the upper motion tangent screw and one to track it with the vertical motion or telescope tangent screw. The sun's image will move much faster than the reader might expect. Once the sun's image is centered the watch time is recorded and the vertical and horizontal angles are read. For sun shots time is a very important matter. Readings should be taken with a good watch which has been

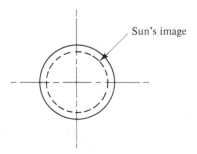

Sun's image

Figure 18-3

Figure 18-4

checked against radio signals (or other standards) before and after going into the field. Perhaps even better would be the use of a battery-powered radio in the field to check the time.

Special reticules, which can be obtained from surveying equipment manufacturers, are very useful in making solar observations. If the surveyor makes a large number of such observations he should seriously consider purchasing one of these devices. Some of the reticules contain a circle which is just a little smaller than the image of the sun, enabling the viewer to center the cross hairs accurately on the sun. This type of reticule, which is shown in Fig. 18-4, has in addition to the usual vertical and horizontal cross hairs a pair of vertical tick marks which are spaced at the same distance apart as the stadia hairs.

Another device for making observations of the sun is Roelof's solar prism. This device is fitted to the objective lens of the telescope. The viewer looking through the telescope sees four overlapping images of the sun, as shown in Fig. 18-5. If the cross hairs are centered on the little diamond-shaped area in between these images, the telescope will be pointed at the sun's center with a high degree of precision. In addition to these attachments, several other useful devices are on the market for observing the sun.

Observations of the sun are best made when its altitude or vertical angle is at least 20° above the horizon and at least 2 hours before or after noon. Thus the most desirable times are 8 to 10 A.M. and 2 to 4 P.M. (9 to 11 A.M. and 3 to 5 P.M. if daylight saving time is being used).

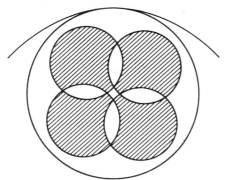

Figure 18-5 View of sun's image with Roelof's solar prism.

Should the sun's altitude be less than 20°, the effect of atmospheric refraction on the measurement of the altitude becomes significant and difficult to estimate. Observations should not be taken between 10 A.M. and 2 P.M. standard time because during those hours the rate of change of altitude of the sun with the lapse of time approaches zero. Very small errors in vertical angle measurement will cause relatively large errors in the calculated bearing to the sun. (In addition vertical angles of greater than approximately 50° cannot be read with the usual transit because the base of the instrument gets in the way of the telescope line of sight.)

When a single observation of the sun is made by projecting its rays onto a page in the fieldbook or other white surface, the surveyor will probably be able to achieve an accuracy of about ±2′. Should he make several observations and average the results he will probably obtain an accuracy of ±1′. If better accuracy is desired it is necessary to use one of the special solar prisms for observing the sun.

18-16 INDEX ERROR

A critical aspect to making observations of the sun is the measurement of the vertical angle. Very few ordinary transits have a vertical scale which reads accurately to zero when the instrument is leveled. The result is that there is an index error. If only a single observation is made, a correction will then have to be made to the vertical angle. However, if one observation is made with the telescope normal and one with the telescope inverted and the results averaged, the index error should be greatly minimized. Strictly speaking, the sun moves along a curved path; unless the readings are taken within a very small time interval (say about three minutes), it is advisable to calculate the true bearing to the point in question from each observation and average the results, rather than averaging the angle and time readings.

18-17 PARALLAX

In discussions thus far we have pictured the earth as being a mere dot in the center of the celestial sphere with a radius so small as to be insignificant with regards to the observations made. From this viewpoint a vertical angle measured from the earth to a star would be the same as if it were measured from the center of the earth. However, the actual distance to the sun is relatively small compared to the other stars and as a result it is necessary to make a correction for parallax as shown in Fig. 18-6.

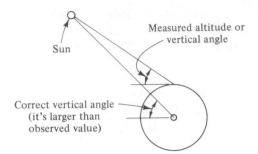

Figure 18-6

From the figure we can see that the parallax correction should be added to the measured vertical angle. The value of the correction can be computed depending on the altitude of the sun above the observer's position.

18-18 REFRACTION

The rays of light coming from the sun or stars are bent downward by the earth's atmosphere. As a result, celestial objects seem to be higher than they actually are and it is necessary to make a negative correction to the vertical angles for refraction. The magnitude of the correction is dependent on the temperature and barometric pressure of the atmosphere and also upon the altitude of the rays. The effect of refraction is much larger for lower-altitude rays. Usually parallax and refraction corrections are combined as they are in Tables K and L of the Appendix.

18-19 AZIMUTH EQUATION

There are several forms of the equation for the azimuth from the observer's position to the sun. Two convenient forms are:

$$\cos Z_n = \frac{\sin D - \sin h \sin L}{\cos h \cos L}$$

$$\cos Z_n = \frac{\sin D}{\cos h \cos L} - \tan h \tan L$$

In these expressions Z_n is the angle from the true meridian to the sun, D is the declination of the sun, L is the latitude of the observer's position and h is the altitude or the vertical angle to the center of the sun at the time of the observation.

In the morning the angle Z_n will be clockwise from true north while in the afternoon it will be counterclockwise from true north. Should the cosine of Z_n be plus, the angle will be less than $90°$ (and thus NE or NW); if minus, it will be greater than 90° (and thus SE or SW).

18-20 EXAMPLE OBSERVATIONS OF THE SUN

In Example 18-4 which follows the observer sights on a point on the ground and turns the telescope clockwise toward the center of the sun. The necessary calculations are shown to determine the true bearing from the observer's position to the point on the ground.

Example 18-4

The following measurements were taken for a sun shot. Determine the true bearing for a line from the instrument to the point on the ground.

$$Date = November 6, 1981$$
$$Latitude = 34°34'N$$
$$Longitude = 82°10'W$$
$$Vertical\ angle\ to\ sun\ (altitude) = 25°10'$$
$$Observed\ time = 3^h05^m\ P.M.\ Eastern\ Standard\ Time$$
$$Clockwise\ angle\ from\ point\ on\ ground\ to\ sun = 162°10'$$
$$Temperature = 60°F$$
$$Elevation - 800\ ft$$

Solution

1. Calculate GCT at time of observation.

$$Watch\ time = \quad 3^h05^m\ P.M.$$
$$Time\ on\ 24\ hr\ basis = \quad 15^h05^m$$
$$+\ Zone\ correction = \quad +5^h$$
$$GCT = \overline{20^h05^m}$$

2. Determine the apparent declination of the sun on November 6, 1981, at 20^h05^m GCT (using Table J in the Appendix).

$$Declination\ of\ sun\ at\ 0^h\ GCT\ November\ 6 = -S15°53.3'$$

+ Change in 20^h05^m as sun moves further south = $(20.083)(-0.75) = \quad -15.1'$

$$Declination = -S16°08.4'$$

3. Correct the measured altitude of the sun.

Refraction and parallax corrections
(obtained from Tables K and L of Appendix)

refraction by   temperature
interpolation   multiplier
↓   ↓
$$= -(2.035)\ (0.98)\ (0.98) + 0.14 = -1.81'\qquad (say\ -2')$$
↑   ↑
pressure or   parallax
elevation
multiplier

Corrected altitude = 25°10' minus 2' = 25°08'

4. Calculate the bearing to the sun.

$$\cos Z_n = \frac{\sin D}{\cos h \cos L} - \tan h \tan L$$

$$= \frac{\sin - 16°08.4'}{(\cos 25°08')(\cos 34°34')} - (\tan 25°08')(\tan 34°34')$$

$$= -0.6960463$$

Bearing to sun = $-134°07'$ counterclockwise from north or S45°53'W

Bearing to point on ground = N63°43'E

We have seen that when special prisms or reticules are not available for the observer to sight the sun directly, an indirect procedure is used. As previously mentioned, this procedure involves the observation of the sun's image on a white card, paper, or field book page held behind the eyepiece of the telescope. To get more accurate results, it is considered better practice to make such an observation with the sun's image tangent to the cross hairs instead of attempting to center it in a circle.

If the tangent method is used half of the observations should be made with the sun in one quadrant and the other half with the sun in the opposite quadrant. For this discussion it is assumed that a morning observation is to be made and further that the sun is to the south of the observer. If the observer were to look directly at the sun (he or she must not do this with an ordinary telescope), it would be moving upward and to the right. When, however, the sun's image is observed on a white card or paper held behind the eyepiece of an erecting instrument, the sun in the morning will appear to be moving downward and to the right. *The image will move in the same direction as a shadow cast by the sun.* Thus the observation would be made as shown in parts (a) and (b) of Fig. 18-7. If the two measurements made with such a procedure are averaged the results will be the same as though the observation had been made with the cross hairs centered on the sun. For the procedure described in the next few paragraphs it will be noted that the horizontal angle is the same for both the 1st and 2nd observations of parts (a)and (b) of Fig. 18-7. Should afternoon observations be made they would be as seen in parts (c) and (d) of the same figure.

There are several procedures which may be followed to observe the sun with the cross hairs tangent to its image on the paper. One such procedure is described here with respect to Fig. 18-8. It is assumed for this discussion that the observations are being made in the morning with the sun to the south of the observer's position. (The description presented is applicable to an erecting instrument in which the sun moves in the same direction as its shadow cast on the page. If Figs.

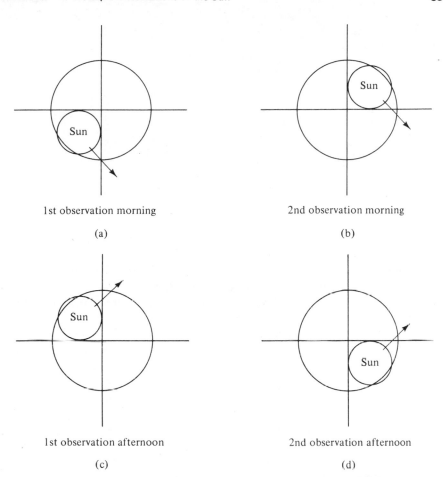

Figure 18-7 The image of the sun as seen in a fieldbook. Sun is to the south of the observer. The vertical hair is at the same azimuth for the two morning observations. Similarly the azimuth for the two afternoon observations (though different from the morning azimuth) is the same.

18-7 through 18-10 are turned upside down they will be applicable to an inverting instrument.)

The observer first moves the telescope so that the upper edge of the sun's image is above the horizontal cross hair. Then he moves the vertical hair with the upper motion until it is on the western edge of the sun. The image will be as shown in part (a) of Fig. 18-8. He then clamps the upper motion and keeps moving the vertical hair to the right with the upper motion tangent screw so that the hair remains tangent to the leading edge of the sun.

He leaves the horizontal cross hair fixed in position until the sun apparently falls far enough so that its upper edge is tangent to the hori-

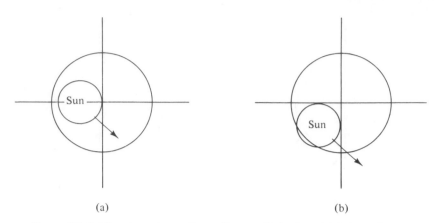

Figure 18-8 Morning observation of sun with horizontal hair stationary.

zontal cross hair as shown in part (b) of the figure. He then reads his watch time and the vertical and horizontal angles.

Next he leaves the vertical hair where it is and moves the horizontal hair until it is tangent to the bottom of the sun's image as shown in part (a) of Fig. 18-9. He keeps moving the tangent screw for vertical motion so as to keep the horizontal hair tangent to the bottom of the sun until finally the sun which is moving from left to right becomes tangent to the vertical hair as shown in part (b) of the figure. He then reads his watch and the vertical angle (the horizontal angle didn't change).

In the afternoon a similar procedure is followed except that the image of the sun in the fieldbook is apparently rising from left to right (if it's south of the observer) and it's necessary to make the observations as shown in parts (a) and (b) of Fig. 18-10.

If readings are taken by a tangent method *only* to the top *or* bot-

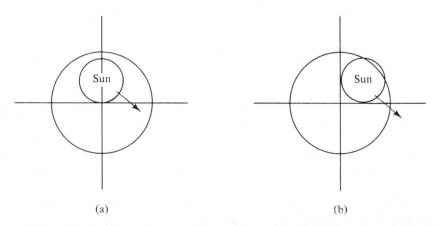

Figure 18-9 Afternoon observation of sun with vertical hair stationary.

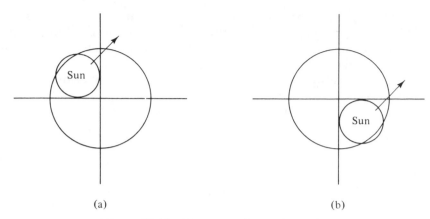

Figure 18-10 Afternoon observations of sun.

tom edges of the sun, it is necessary to make a semidiameter correction to the vertical angle or altitude read. A similar adjustment would be required if the readings were made only with the vertical hair tangent to the leading or trailing edge of the sun's image. If the vertical angle is read to the lower edge of the sun's image it is necessary to add the vertical angle. The semidiameter of the sun is given in Table M of the Appendix and varies from 16.29' to 15.76'.

When a horizontal angle is measured to the edge of the sun's image, it is necessary to correct that angle by the semidiameter times the secant of the altitude of the sun. As the sun reaches its highest point or zenith, this correction becomes very large and readings should not be taken to one edge only.

18-21 SOLAR ATTACHMENTS

There are on the market transits that are equipped with solar attachments with which the observer can make direct observations of the true north–south meridian. Depending on the particular manufacturer, they are fixed either on the sides or tops of the transits. To use these attachments the latitude of the observer, the declination of the sun and the hour angle are each set on different scales and a sighting is made on the sun. The other telescope is automatically lined up in the north–south direction, since the instrument automatically solves the azimuth equation of Section 18-19.

The three common types of solar attachments are the Burt, the Saegmuller and the Smith. They have been much used for determining true meridians for the subdivision of public lands. These devices may, however, result in directions which are appreciably in error unless they are in excellent adjustment. Even with an excellent adjustment, the results obtained are rarely as precise as those obtained with direct solar observations taken as described in the preceding sections of this chapter. Observations with solar attachments should be made at the same time of day as regular solar observations. The same comments can be made about the effect of errors in latitude and declination as they affect the directions obtained.

PROBLEMS

In Problems 18-1 to 18-3, use the appendix tables to compute for the year 1981 the time of UC of Polaris on the following dates at the locations given.

18-1. October 13, latitude $43°16'$N, longitude $87°18'$W.   (*Ans.:* $0^h 37.05^m$)

18-2. November 27, latitude $39°18'$N, longitude $103°42'$W.

18-3. December 7, latitude $41°10'$N, longitude $98°31'$W.   (*Ans.:* $20^h 41.46^m$)

18-4. Compute the time of the next WE for the data of Problem 18-1.

18-5. Compute the time of the EE immediately preceding the UC of Problem 18-3.   (*Ans.:* $14^h 45.37^m$)

18-6. An observer at point A on the ground sights on Polaris $2^h 30^m$ after UC on December 4, 1981, at latitude $34°30'$N, longitude $91°10'$W. What is the true bearing from point A to Polaris?

18-7. An observer at point A on the ground sights on point B on the ground and turns a clockwise angle of $86°10'$ to Polaris 3 hours after EE on November 16, 1981 at latitude $34°30'$N and longitude $87°30'$W. What is the true bearing of line AB?   (*Ans.:* N85°28.2'W)

In Problems 18-8 to 18-11, for each of the sun shots described a sight was taken from the instrument at point A to a point B on the ground and a clockwise angle was turned to the sun. Determine the true bearing of line AB in each case.

	Date	Time	Latitude	Longitude	Temp (°F)	Elev (ft)	Vertical Angle	Horizontal Angle
18-8.	11/20/81	$3^h 10^m$ P.M.	35°03' / Albuquerque, N.M.	106°32'	60	5500	26°13'	61°10'
18-9.	12/9/81	$10^h 16^m$ A.M.	33°45' / Atlanta, Ga (Eastern Standard Time)	84°22'	50	1200	25°08'	54°16'
18-10.	11/16/81	$10^h 33^m$ A.M.	42°25' / Boston, Mass.	71°12'	30	250	27°18'	181°53'
18-11.	10/14/81	$3^h 42^m$ P.M.	47°08' / Tacoma, Wash.	122°25'	60	350	29°10'	131°16'

(*Ans. to 18-9:* Bearing AB = S88°-18'E)

(*Ans. to 18-11:* Bearing AB = N80°56'E)

19-1 INTRODUCTION

A tremendous volume of earthwork needs to be moved for the construction of highways, railroads, canals, foundations of large buildings, pipelines, and other projects. The surveyor is often directly involved in the determination of the amounts or quantities of this earthwork. He or she is not only concerned with quantities but also with the setting of the grade stakes needed to carry out the earthwork required to bring the ground to the desired grades and elevations.

Before a construction project involving earthwork is begun, the surveyor needs to determine the shape of the ground surface. This is necessary in order that the volume of materials to be added or removed may be determined. In speaking of earthwork it is the custom to refer to excavations as *cuts* and to embankments as *fills*. The quantities of cuts and fills in the types of construction projects described herein are frequently of such magnitude as to make up appreciable percentages of the total project costs.

The principles involved in volumetric calculations are not only applicable to earthwork but also to other materials such as to volumes of reservoirs and to stockpiles of sand, gravel, and other materials. Volumes of masonry structures may be computed directly from the dimensions on plans but it is also quite common to check pay quantities by computing the volumes from the measurements of the completed structures made by the surveyor in the field.

It should also be noted that maps produced by photogrammetric methods as described in Chapter 22 enable the surveyor to estimate earthwork quantities quite well.   Either topographic maps prepared from photographs or stereoscopic models may be used.

19-2 BORROW PITS

During the construction of roads, airports, dams, and other projects involving earthwork it is often necessary to obtain or *borrow* earth from surrounding areas in order to construct embankments.   These excavations are commonly referred to as *borrow pits*.   The quantity of borrow material is very important because the contractor's pay is usually computed by taking the number of yards of borrow times the bid price per yard.   In addition, the adjacent areas where the borrow pits are located often belong to other people who are also paid on a quantity basis.

In order to determine the data needed for volume calculations, elevations may be obtained at certain points before and after the earth is removed.   Some type of grid system is normally established (say, 50-foot squares) and the elevations are determined at each of the corners.   When the excavation is completed, the levels are run again at the same points and from the differences between the original and final readings the cut made at each point is obtained.   Volumes may be approximately calculated by multiplying the average cut for a particular figure times the area of the figure.

Figure 19-1 shows a plan view of a typical borrow pit which is divided into convenient squares and rectangles and also into some triangles because of the irregular shape of the borrow pit. The numbers shown on the figure represent the dimensions of the figure and the cuts in feet at the corners.   The value of earthwork under one of the areas can be estimated as being equal to the average of its corner cuts times the area of the figure.   For instance, the approximate cut for figure *abfg* is equal to

$$(50 \times 50) \left(\frac{3.4 + 3.6 + 2.8 + 3.1}{4} \right) = 8062 \text{ cu. ft}$$

These calculations can be recorded conveniently, as shown in Table 19-1.   Instead of making the calculations for each individual figure as was done for *abfg*, it is simpler to combine the groups of figures that have the same horizontal dimensions.   For instance, *oqus* includes two rectangles.   It will be noted that in order to determine the sum of the corner cuts for both rectangles, the values at *s*, *u*, *o*, and *q*

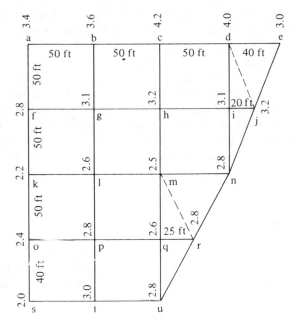

Figure 19-1

TABLE 19-1

Figure	Sum of Corner Cuts	Area of Figure in Sq. Ft	Multiplier	Volume Cu. Ft
oqus	21.4	2000	$\times \frac{1}{4}$	10,700
adnmqo	95.7	2500	$\times \frac{1}{4}$	59,800
uqr	8.2	500	$\times \frac{1}{3}$	1,370
qrm	7.9	625	$\times \frac{1}{3}$	1,650
rmn	8.1	1250	$\times \frac{1}{3}$	3,380
nij	9.1	500	$\times \frac{1}{3}$	1,520
ijd	10.3	500	$\times \frac{1}{3}$	1,720
jed	10.2	1000	$\times \frac{1}{3}$	3,400

Total volume = 83,540 cu. ft

= 3094 cu. yd

appear once while those at t and p occur twice. The remainder of the table should be self-explanatory. The numbers are rounded off a little, as is only reasonable in earthwork calculations.

A special comment should be made about the figures on the side as, for example, trapezoid *deij*. The volume of this prism may be determined by taking the sum of the four corner cuts divided by four and multiplied by the area of the trapezoid. It will be noted that this does not give the same value as that obtained by breaking it into the two triangular figures *ijd* and *jed*. If the surveyor feels that the trapezoidal formula will yield too much error, he should show the dotted lines as *dj* and *mr* in order to indicate that the triangular figures should be used.

19-3 CROSS SECTIONS

For the purposes of this discussion the construction of a highway will be considered. It is assumed that a longitudinal grade line has been selected as well as the roadway cross section. It is further assumed that cross sections of the ground surface have been taken at each station along the route by the method previously described in Section 10-6.

A cross section is a section normal to the center line of a proposed highway, canal, dam or other construction project. The cross sections of the ground surface are plotted for each station and the outline of the proposed roadway as traced from a template is superimposed on the cross sections, with the elevation of the center line of the roadway obtained from the longitudinal grade line. Examples of such cross sections are shown in Fig. 19-2.

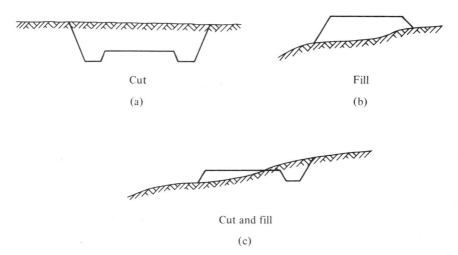

Cut   Fill

(a)   (b)

Cut and fill

(c)

Figure 19-2 Typical cross sections.

The side slopes for cuts and fills are the horizontal distances needed for a unit vertical rise. These slopes are flatter in fills than in cuts where the soil is left in its natural state. Values of $1\frac{1}{2}$ to 1 in cuts and 2 to 1 in fills are rather common but they may have to be much flatter in some soils (perhaps as much as 8 or 10 or even 12 to 1 in some extreme cases).

19-4 AREAS OF CROSS SECTIONS

To determine earthwork quantities as described later in this chapter, it is first necessary to determine the areas of the cross sections. This may be done either by means of computations or with planimeters as described in the following paragraphs.

(a) Computer Programs

The calculation of cross-sectional areas and earthwork volumes are very tedious and repetitious for large projects. As such they are ideally suited for computer applications. There are available numerous "canned programs" with which the calculations can quickly be made. Most highway departments today use computers to determine their required areas and volumes for earthwork. Nevertheless the surveyor should be familiar with the theory behind the calculations described in the remainder of this chapter.

(b) Areas of Level Cross Sections

The area of a level cross section such as the one shown in Fig. 19-3 may be calculated by multiplying the average of the top and bottom widths of the cross section by its depth as shown in the figure. Level cross sections are quite commonly encountered for highways, railroads, ditches, etc.

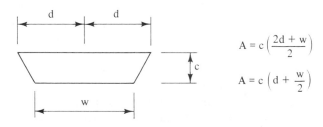

$$A = c\left(\frac{2d + w}{2}\right)$$

$$A = c\left(d + \frac{w}{2}\right)$$

Figure 19-3

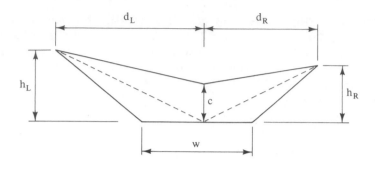

$$A = \left(\frac{1}{2}\right)\left(\frac{w}{2}\right)\left(h_L + h_R\right) + \left(\frac{1}{2}\right)(c)\left(d_L + d_R\right)$$

$$A = \frac{w}{4}\left(h_L + h_R\right) + \frac{c}{2}\left(d_L + d_R\right)$$

Figure 19-4

(c) Area of Three-Level Sections

When a three-level section is involved, such as the one shown in Fig. 19-4, its area may be determined by breaking the figure down into triangles and summing up their areas as shown in the figure.

(d) Areas of Five- or More Level Sections

When a five- or more level section is encountered it is possible to compute its area by summing up the areas of the triangles and/or trapezoids which make up the figure, by using a coordinate rule, or by using a planimeter as described in the next few paragraphs.

(e) Measuring Areas of Irregular Cross Sections with a Polar Planimeter

An irregular cross section is defined as one where the ground surface is so irregular that neither a three-level cross section nor a five-level cross section will provide sufficient information to sufficiently describe the area. Intermediate elevations at irregular intervals are necessary between center line and slope stakes to adequately specify the cross section.

When irregular cross sections are involved it may be possible to break them down into convenient figures such as triangles and trapezoids and compute their areas. Sometimes a form of the coordinate method may be used. These methods, however, are rather tedious to apply and it is more common to plot the cross sections on cross-

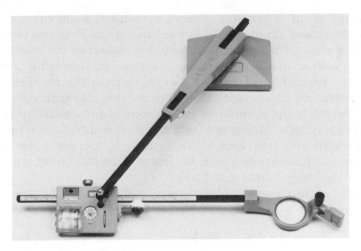

Figure 19-5 Planimeter. (*Courtesy of the Leitz Company.*)

section paper (particularly if the sections are very irregular and also if some of the sides are curved). Once this is done the areas are determined by traversing their perimeters with polar planimeters. In almost every case the area of a cross section can be determined by planimeter with a precision as good as is justified by the precision used in making the field measurements with which the cross section was drawn.

A polar planimeter is a device which can be used to measure the area of a figure on paper by tracing the boundary of the figure with a tracing point. As the point is moved over the figure the area within is mechanically integrated and recorded on a drum and disk. An excellent mathematical proof of the workings of the planimeter is presented in the book by Davis, Foote, and Kelly.[1]

The planimeter is particularly useful for measuring the areas of irregular pieces of land such as those discussed in Section 14-10, as well as the areas of irregular cross sections. If the operator is careful he can obtain results within 1% or better depending on the accuracy of the plotted figures, the types of paper used and the carefulness with which the figures are traced.

The major parts of a planimeter are the tracing point, the anchor arm with its weight and post, the scale bar, and the graduated drum and disk with its vernier. When an area is to be traced the drawing is stretched out flat so there are no wrinkles and the anchor point is pushed down into the paper at a convenient location so that the operator can trace all or a large part of the area desired at that one location.

[1] R. E. Davis, F. S. Foote, and J. W. Kelly, *Surveying Theory and Practice*, 5th ed. (New York: McGraw-Hill Book Company, 1966), pp. 67–69.

The tracing point is set at a distinct or marked point on the drawing and the graduated drum is read (or set to zero). Then the perimeter is carefully traced until the tracing point is returned to the starting point, at which time a final drum reading is taken. In traversing around the figure the operator must be very careful to note the number of times the drum reading passes zero. Counters are available with some planimeters with which the number of revolutions is automatically recorded. The initial reading is subtracted from the final reading and represents to a certain scale the area within the figure. If the perimeter is traced clockwise the final reading will be larger than the initial reading, but smaller if traversed counterclockwise.

If the anchor point is placed outside the area to be traversed the area of the figure will equal

$$A = Cn$$

where C is a constant and n is the difference from the initial to the final readings on the drum. The constant C is usually given on the top of the tracing arm or on the instrument box. It is equal to 10.00 sq. in. for many planimeters. If the user is not sure of C, he can easily construct a figure of known area (as for instance a 5 in. by 5 in. square), run the tracing point around the area and determine what C would have to be to give the correct area when multiplied by the net reading on the drum. For some instruments the value of C is given in SI units.

If the anchor point is placed inside the area to be traversed (as it will often be for large areas) it will be necessary to make a correction to the area computed. It is possible to hold the tracing arm in such a position that the tracing point can be moved completely around 360° and the drum reading will not change. The area of this circle called the *zero circle* or the *circle of correction* must be added to Cn if a figure is traversed in a clockwise direction with the anchor point on the inside. It will be noted that if the area of the figure is less than that of the zero circle the change in the drum reading will be minus for a clockwise traverse.

Example 19-1 which follows illustrates the use of the planimeter for determining areas of figures in a drawing.

Example 19-1

Upon calibration it was found that a given planimeter traversed 10 sq. in. for each revolution of its drum.

(a) If a given map area is traversed with the anchor point outside the area and a net reading of 16.242 revolutions is obtained, what is the map area traversed?

Solution

$$\text{Area} = (10)(16.242) = \mathbf{162.42 \ sq. \ in.} \text{ on the map.}$$

(b) If the same area is traversed with the anchor point located inside the area and the net reading is 8.346 revolutions, what is the area of the zero circle?

Solution

Area of zero circle = $(10)(16.242 - 8.346) =$ **78.96 sq. in.**

(c) Another map area is traversed with the anchor point placed inside the area and a net reading of 23.628 revolutions is obtained. If this area was plotted on the map with a scale of 1 in. = 20 ft, what was the actual area on the ground?

Solution

Area on map = $(10)(23.628) + 78.96 = 315.24$ sq. in.

Area on ground = $(20 \times 20)(315.24) =$ **126,096 sq. ft**

Electronic polar planimeters are available today which provide digital readout in large, bright numbers. These devices are simple to read and may be easily set to zero. They may also be used to handle cumulative adding and subtracting of areas.

19-5 COMPUTATION OF EARTHWORK VOLUMES

Earthwork volumes may be computed from the areas of cross sections by two methods as will be described in this section. The distance between cross sections is dependent on the precision required for the volume calculations. Obviously as the price per cubic yard goes up it becomes more desirable to have the cross sections closer together. For instance, for rock excavation or for underwater excavation costs are so high that cross sections at very close intervals are required perhaps no more than 10 ft. For ordinary road or railroad earthwork the sections are probably taken at 50- or 100-ft intervals. In addition to the cross sections taken at regular stations it is also necessary to take them at the beginning and ending points of curves, at locations where unusual changes in elevations occur and for points where ground elevations coincide with natural grades. These later points are called *grade points*.

The earth between two cross sections forms an approximate *prismoid*. A prismoid is a solid which has parallel end faces (or bases) and sides which are plane surfaces. In this section two methods are presented for estimating the volume of these prismoids these being the average-end-area method and a method using the prismoidal formula.

(a) Average-End-Area Method

A very common technique used for computing earthwork volumes is the average-end-area method. In this approach the volume of earth between two cross sections is assumed to equal the average area of the two end cross sections times the distance between them. The volume

$$V = \left(\frac{A_1 + A_2}{2}\right)\left(\frac{L}{27}\right)$$

where A_1 and A_2 are the end areas in square feet while L is the distance between the cross sections in feet. The expression has been divided by 27 to give an answer in cubic yards.

Should cross sections be taken at 100-ft stations, the expression may be reduced to the following form, the volume again being in cubic yards:

$$V = 1.85(A_1 + A_2)$$

The average-end-area method is very commonly used for computing earthwork quantities because of its simplicity. It is not, however, a theoretically exact method unless the two end areas are equal, but the errors are not usually significant. Should one of the areas approach zero, as on a hillside where the cross section is running from cut to fill, the error will be rather large. For this case it may be well to calculate the volume as a pyramid, with $V = \frac{1}{3}$ the area of the base times the height.

Though the average-end-area method is approximate, the precision obtained is fairly consistent with the precision obtained using the field measurements made for the cross sections. The costs per cubic yard for earthwork are usually low and thus it is not normally justifiable economically to make refinements in earthwork calculations. Also, in most cases the method gives volumes on the high side, which is in favor of the contractor. To improve the accuracy of the average-end-area method it is necessary to decrease the length between stations. This is particularly desirable if the ground surface is very irregular.

Sometimes when a road has a very sharp curve and large cuts or fills, adjustments may be made for curvature in making volume calculations but usually such adjustments are not considered significant.

Example 19-2 illustrates the computation of earthwork between two stations using the average-end-area method.

Example 19-2

Determine the earthwork volume between Stations 100 and 101 using the average-end-area method if the road width is 30 ft.

Station	Left	Center	Right
100 + 00	$\dfrac{C10.0}{34.0}$	$\dfrac{C6.0}{0.0}$	$\dfrac{C9.0}{33.0}$
101 + 00	$\dfrac{C6.0}{28.0}$	$\dfrac{C4.0}{0.0}$	$\dfrac{C7.0}{31.0}$

Solution

$$\text{Area at Station } 100 + 00 = \frac{w}{4}(h_L + h_R) + \frac{C}{2}(d_L + d_R)$$

$$= \left(\frac{30}{4}\right)(10.0 + 9.0) + \left(\frac{6.0}{2}\right)(34.0 + 33.0)$$

$$= 343.5 \text{ sq. ft}$$

$$\text{Area at Station } 101 + 00 = \left(\frac{30}{4}\right)(6.0 + 7.0) + \left(\frac{4.0}{2}\right)(28.0 + 31.0) = 215.5 \text{ sq. ft}$$

$$V = 1.85(A_1 + A_2) = (1.85)(343.5 + 215.5) = \textbf{1034.1 cu. yds}$$

(b) Volume by Prismoidal Formula

Should the ground surface be such that two adjacent end areas are quite different from each other or should a high degree of precision be desired in the calculations, as where rock quantities or volumes of concrete are being determined, the average-end-area method may not be sufficiently accurate.   For such cases the prismoidal formula which follows will often be used.

$$V = \frac{L}{27}\left(\frac{A_1 + 4A_m + A_2}{6}\right)$$

In this expression A_1 and A_2 are the areas of the cross sections at the ends or bases of the prismoid while A_m is the area of a section halfway between the two ends and L is the distance between the two end cross sections.   *The area of the middle section* A_m *is determined by taking cross sections there or by averaging the dimensions of the end cross sections and using these values to calculate the area.   It is not determined by averaging the end areas.*

The prismoidal formula usually yields smaller volumes than does the average-end-area method.   Its use is probably only justified when cross sections are taken at very short intervals and where the area of successive cross sections are quite different.   When earthwork is being contracted for, the method to be used for computing volumes should be clearly indicated in the contract.   Should no mention be made of the method to be used the contractor will in all probability be able to require that the owner use the average-end-area method—whatever the owner's intentions may have been.

Example 19-3 illustrates the application of the prismoidal formula.

Example 19-3

Using the prismoidal formula, compute the earthwork volume between stations 100 and 101 using the data of Example 19-2.

Solution   We determine cross-section dimensions at Station 101 + 50 by averaging dimensions at Stations 100 and 101:

Station	Left	Center	Right	Area by 3-Level Section Formula
100 + 00	$\dfrac{C10.0}{34.0}$	$\dfrac{C6.0}{0.0}$	$\dfrac{C9.0}{33.0}$	343.5 sq. ft
100 + 50	$\dfrac{C8.0}{31.0}$	$\dfrac{C5.0}{0.0}$	$\dfrac{C8.0}{32.0}$	277.5 sq. ft
101 + 00	$\dfrac{C6.0}{28.0}$	$\dfrac{C4.0}{0.0}$	$\dfrac{C7.0}{31.0}$	215.5 sq. ft

$$V = \frac{L}{27}\left(\frac{A_1 + 4A_m + A_2}{6}\right) = \left(\frac{100}{27}\right)\left(\frac{343.5 + 4 \times 277.5 + 215.5}{6}\right)$$

$$= 1030.2 \text{ cu. yds}$$

Though the prismoidal formula does provide the true volumes of prismoids, the average-end-area method is more commonly used because the difference between the two methods is usually quite small except where abrupt changes in cross sections occur. Furthermore it is somewhat tedious to apply the prismoidal formula because of the extra work involved in computing the average dimensions and the area of the center cross sections. Unless rock excavation or concrete quantities are being computed, the use of the prismoidal formula is not justified anyway because of the low precision of the work usually obtained in cross sectioning. Finally, if it is desired to determine the prismoidal volume it is easier to use the average-end-area method and correct the results obtained with the *prismoidal correction* formula. This correction expression is accurate for three-level sections and is reasonably accurate for most other cross sections.

In the expression to follow for C_V, the prismoidal correction, C_1 and C_2 are the center cuts or fills at the end cross sections A_1 and A_2, while w_1 and w_2 are the distances between slope stakes (see Fig. 20-10) at these sections. The correction is in cubic yards.

$$C_V = \frac{L}{12 \times 27}\,(C_1 - C_2)(w_1 - w_2)$$

The correction calculated is subtracted from the volume obtained by the average-end-area method unless the formula yields a minus answer, in which case the correction is added. Example 19-4 shows the application of the prismoidal correction expression to the calculations of Example 19-2, which employed the average-end-area method.

Example 19-4

Correct the average-end-area solution of Example 19-2 to a prismoidal volume using
the prismoidal correction formula.

Solution

$$C_V = \frac{L}{12 \times 27} (C_1 - C_2)(w_1 - w_2)$$

$$= \frac{100}{(12)(27)} (6.0 - 4.0)(67.0 - 59.0)$$

$$= 4.94 \text{ cu. yds}$$

Prismoidal volume = 1034.1 - 4.9 = **1029.2 cu. yds**

19-6 MASS DIAGRAM

For highway and railway construction it is desirable to make a cumulative plot of the earthwork quantities (designating cuts plus and fills
minus) from one point to another (perhaps the beginning to the ending
points). Such a plot, illustrated in Fig. 19-6, is called a *mass diagram*.

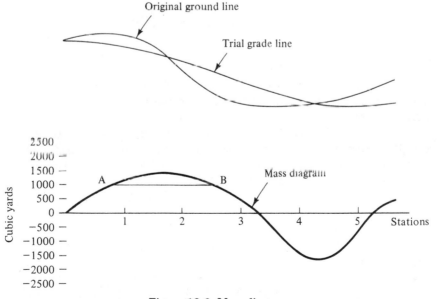

Figure 19-6 Mass diagram.

It is usually plotted directly below the profile of the route. The ordinate at any point on the diagram is the cumulative volume of cut and fill to that point while the abscissa at any point is the distance in stations along the survey line from the starting point.

From a mass diagram it can readily be seen whether the cuts and fills balance throughout the job and how far the earth will have to be hauled. If there is an excess of fill needed beyond the cut quantities it will be necessary to obtain earth from other sources such as borrow pits. If there is more cut than fill, it may be necessary to dump the extra fill as waste, or perhaps widen the planned fills or make the valleys shallower.

When material is excavated and placed in trucks or pans or other earthmoving equipment, it will normally occupy a larger volume than it had in its original position. When solid rock is broken up it may take up as much as two times its original volume.

When the excavated material is rolled down in thin layers in an embankment at an optimum moisture content such that the greatest compaction is achieved, the resulting fill volume will be appreciably less than the cut quantities. Thus for most fills (other than those consisting of rock) it takes more cut volume than fill volume. The excess may vary from 5 to 20% depending on the character of the material involved. If a fill is made in a marshy area the original material underneath will settle appreciably, thus requiring even more fill. As a result, it is necessary to make use of the so called *balance factor* in figuring cut and fill quantities. A value of about 1.2 is frequently used, that is, about 1.2 yards of cut make 1 yard of fill. For rock excavation the balance factor will be less than 1.0 to allow for swell from cut to fill. This factor needs to be accounted for in constructing the mass diagram.

Referring to Fig. 19-6, the existing or original ground profile is plotted and a trial grade line (shown in the figure) is drawn. In drawing the trial grade line an effort is made to visually balance the cut and fill quantities. Earthwork volumes are calculated as described in the previous sections of this chapter and the mass diagram is drawn taking into account the estimated balance factor. If the cuts and fills don't balance well or if the earth has to be transported or hauled long distances the grade line will have to be adjusted and the process repeated.

Earth grading contracts refer to a certain distance as being a *free haul* distance. It may be specified as being 500 ft, 1000 ft, 2000 ft, or some other value. If the earth is not moved more than this distance the contractor will be paid the standard price per cubic yard. If it is moved more than this distance he will be paid extra as specified in the contract. The extra hauling is referred to as *overhaul*. The unit of measurement usually used for overhaul is the *station yard* where one station yard indicates the hauling of 1 cubic yard of material for one station.

From the figure it will be noted that the peaks on the diagram show where there is a change from cut to fill while the valleys show a change from fill to cut. If a horizontal line *AB* is drawn as shown in the figure the cuts and fills between the two points will exactly balance.

19-7 VOLUMES FROM CONTOUR MAPS

It is quite practical to estimate earthwork quantities from a contour map. For this discussion, reference is made to the five-foot contour interval map of Fig. 19-7. It is assumed that the hill shown is to be graded off to elevation 515. The areas within the 525, 520, and 515 contour lines can be easily determined with a planimeter. From these values the volume of earth to be removed can be calculated from the following:

$$V = (5)\left(\frac{A_{525} + A_{520}}{2}\right) + (5)\left(\frac{A_{520} + A_{515}}{2}\right) + \text{earth}$$

volume above the 525 contour line

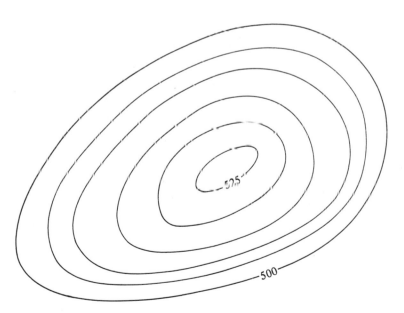

Figure 19-7

PROBLEMS

In Problems 19-1 to 19-4, find the volume in cubic yards of the excavation for the borrow pits shown. The numbers at the corners represent the cuts in feet.

19-1.

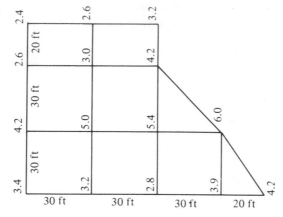

(*Ans.:* 959 cu. yd)

19-2.

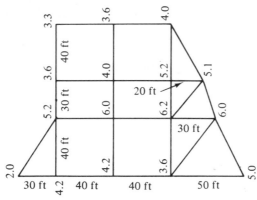

19-3.

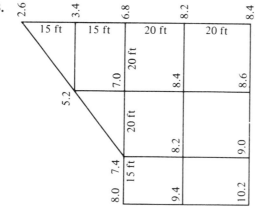

(*Ans.:* 792 cu. yd)

19-4.

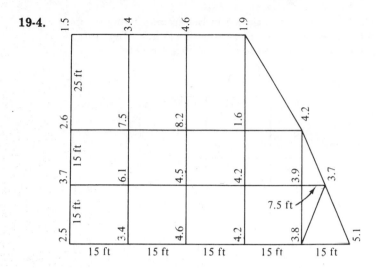

19-5. An area has been laid out in 40-ft squares as shown in the accompanying illustration and the ground elevations are as indicated on the sketch. How many cubic yards of cut are required to grade the area level to elevation 40.0?

(*Ans.:* 2806 cu. yd)

19-6. An area has been laid out as shown in the accompanying illustration. The elevations of the corners are indicated on the sketch. Determine how many cubic yards of cut are required to grade the area level to elevation 60.0.

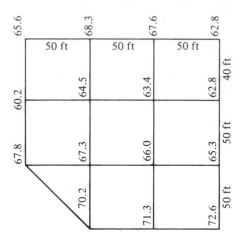

19-7. An area has been laid out as shown in the accompanying illustration and the elevations of the corners are as indicated on the sketch. How many cubic yards of cut are required to grade the area level to elevation 60.0 neglecting the balance factor?

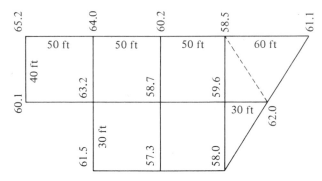

(*Ans.:* 232 cu. yd)

19-8. The following notes are for cross sections at stations 67 and 68. If the width of the roadbed is 30 ft, calculate the area of the two cross sections.

Station	Cross Sections		
67	$\dfrac{C6.2}{28.4}$	$\dfrac{C3.4}{0.0}$	$\dfrac{C1.3}{18.0}$
68	$\dfrac{C10.8}{26.2}$	$\dfrac{C4.8}{0.0}$	$\dfrac{C2.6}{19.4}$

19-9. A given planimeter has a constant of 2, that is, one revolution of the drum equals 2 sq. in.

(a) If a map area is traversed with the anchor point outside the area and a net reading of 16.422 revolutions is obtained, what is the map area?

(*Ans.:* 32.844 sq. in.)

(b) If the same area is traversed with the anchor point inside the area and the net reading is 8.324 revolutions, what is the area of the zero circle?

(*Ans.:* 16.196 sq. in.)

19-10. Upon calibration, it was found that a given planimeter traversed 10 sq. in. for every revolution of its drum and the area of the zero circle was 72.00 sq. in. With the anchor point inside the area the reading before traversing an unknown area was 4.832 revolutions and after traversing the area was 8.986 revolutions. What is the area in sq. in.? If the map scale is 1 in. = 50 ft what is the area on the ground?

19-11. A given planimeter has a constant of 6. A map area is traversed with the anchor point outside the area and a net reading of 16.420 revolutions is obtained. The same area is traversed with the anchor point inside the area and a net reading of 8.320 revolutions is obtained. The scale of the map on which the area is being measured is 1 in. = 50 ft. Determine:

(a) the map area traversed   (*Ans.:* 98.52 sq. in.)

(b) the area of the zero circle   (*Ans.:* 48.60 sq. in.)

(c) the actual ground area   (*Ans.:* 5.65 acres)

19-12. Compute the volume of excavation between stations 67 and 68 of Problem 19-8 using

(a) the average-end-area method and

(b) the prismoidal formula.

19-13. The cross section areas along a proposed dike as obtained with a planimeter are as follows:

Station	End Area (sq. ft)
46	622
47	1466
48	962

Calculate the total volume of fill in cubic yards between stations 46 and 48 using the average-end-area method.   (*Ans.:* 8354.6 cu. yds)

19-14. For a proposed highway the following planimetered areas in sq. ft were obtained. Using the average-end-area method, determine the total volumes of cut and fill between stations 84 and 87.

Station	Cut	Fill
84	86	
85	65	
85 + 50	0	0
86		38
87		64

Construction Surveying

20-1 INTRODUCTION

The construction industry is the largest in the United States and survey-ing is an essential part of that industry. A boundary survey and the preparation of the necessary topographic maps is the first step in the construction process. From these maps the positions of the structures are established and when the final plans for the project are available, it is the duty of the surveyor to set the required horizontal and vertical positions for the structures. In other words, construction surveying involves the transfer of the dimensions on the drawing to the ground so that the work is done in its correct position. This type of surveying is sometimes called "setting lines and grades." The work of the surveyor for construction projects is often referred to as *layout work* and the term "layout engineer" may be used in place of the term "surveyor."

Of necessity, construction surveying begins before actual construc-tion and continues until the project is completed. Surveying is an essential part of the construction process and must be carried out in coordination with other operations in order to have an economical job and to prevent serious mistakes. The reader can readily understand the expense and inconvenience caused if one reinforced concrete foot-ing for a building is placed in the wrong position. As another example, just imagine the problems involved if a few thousand feet of a sewer line are set on the wrong grade.

Construction drawings show the sizes and positions of structures that are to be erected, such as buildings, bridges, roads, parking lots, storage tanks, pipelines, and so on. The job of the construction surveyor is to locate these planned features in their desired positions on the ground. He does this by placing reference marks (such as construction stakes) sufficiently close to the planned work so as to permit the masons, carpenters, and other tradesmen to properly position the work with their own equipment (folding rules, mason's levels, string lines, and so on).

The construction surveyor will find that his work is quite varied. One day he may be doing topographic work for a proposed building, while on another he may be setting stakes for pipeline excavation. It is often necessary for him to make measurements before and after some types of work, e.g., computing quantities of earthwork moved by a contractor. At other times, he will set stakes to guide the construction of foundations; he may be aligning the columns of a steel frame building, or checking a completed structure to see that it is correctly positioned, and so on.

A construction project requires four kinds of surveys for its completion:

1. A property or boundary survey by a registered land surveyor to establish the location and dimensions of the property.
2. A survey to determine the existing conditions such as contours, man-made and natural features, streams, sewers, power lines, roads, nearby structures, and so on. This work may also be done by the land surveyor along with the boundary survey.
3. The construction surveys which determine the position and elevation of the features of the construction work. These surveys include the placing of grade stakes, alignment stakes and other layout control points.
4. Finally there are the surveys which determine the positions of the finished structures. These are the "as-built" surveys and they are used to check the contractor's work and show locations of structures and their components (water lines, sewers, etc.) which will be needed for future maintenance, changes, and new construction.

Only a few of the many possible construction surveying or layout problems can be presented in this chapter. The reader will get a general idea of the problems involved by confronting a range of topics in construction surveying. Included are introductory discussions on building, highway, bridge and pipeline surveys. In spite of the variety, however, the reader will notice that no new surveying principles are involved. All of those needed for construction surveying were presented in the earlier chapters.

20-2 TRADE UNIONS

In most states trade unions do not claim jurisdiction over construction surveying. Nevertheless, surveyors involved in layout work can become involved in disputes where the unions claim that surveying work should be done by union members. For instance, the unions feel that their carpenters are supposed to nail batter boards and layout building partitions. They further claim that their ironworker foremen are supposed to check the positioning and alignment of structural steel and that their concrete finishers are to set screeds for concrete slabs. In certain areas of the country, the union operating engineers have jurisdiction over field survey work done by the contractor. If, however, the surveyor is employed as an office engineer and part of his job is to do field surveying, only the field surveying work would be under the jurisdiction of the operating engineers.[1]

20-3 PROPERTY SURVEY FROM THE CONTRACTOR'S VIEWPOINT

The contractor for a construction job should be furnished with a property survey made by a registered surveyor. In fact the construction contract signed by the contractor should require that the property survey be furnished by the owner. Furthermore the contractor should be able to prove that the survey he used was indeed furnished by the owner.

Even if the contractor or one of his employees is a registered surveyor, he should not make the property survey nor should the contractor even employ a surveyor to make it. If he or one of his employees make the survey the contractor will be taking onto himself the responsibility of locating the structure and if a mistake is made he may find his payment for the entire building project in jeopardy. Numerous lawsuits occur each year when buildings are placed in the wrong positions—frequently being located wholly or partly on other people's land. Even if the structure is only a small distance out of position the result

---

[1] K. Royer, *Applied Field Surveying* (New York: John Wiley & Sons, Inc., 1970), p. 124.

can be a disastrous lawsuit. This kind of mistake could possibly bank-
rupt a contractor.[2]

20-4 PRELIMINARY SURVEYS

To prepare the plans for a building the architect needs a map of the site
so that he can carefully locate the building. These maps are typically
drawn to a large scale such as 1 in. = 10 ft or 1 in. = 20 ft. The information to be included are property lines; elevations for the preparation
of contour lines; locations, sizes and materials of existing buildings on
the site or adjacent to it; location of any immovable objects; the locations of existing streets, curbs, and sidewalks; the locations of fire
hydrants; sizes and locations of gas and water lines and storm and
sanitary sewers including manhole locations and invert elevations
(these are low points on the inside circumference of the pipes); the
locations of power lines, telephone lines, light poles, trees, and other
items.

Before the design of the structure can begin, the above-listed information needs to be furnished to the engineer-architect group. This
data will ordinarily be provided on the site map; for buildings this
drawing is frequently called the *building plot plan.*

The layout of the building will be based upon the information
presented on the plot plan and the proposed building design will be
superimposed on that plan. The final drawing will show the location
of the building with respect to the property lines and with respect to
the streets, utilities, etc. It will probably also show the final contour
lines which are to exist after the construction is complete.

The preliminary survey may very well include a survey of existing
structures on the site and perhaps structures on adjacent property
which might be affected by the new construction. Such surveys should
be performed prior to the start of construction. They should include
the measurement of the vertical and horizontal positions of the foundations of these buildings so that later it can be determined if any lateral
or vertical movement has occurred during construction. In this regard
it should be noted that settlements continue for quite a few years after
a building is constructed and if the existing building or buildings on or
adjacent to the site are relatively new they may still be settling. As a
result settlement which occurs in an existing building when a new one is
erected may or may not be due entirely to the new construction.

[2]K. Royer, *Applied Field Surveying* (New York: John Wiley & Sons, Inc.,
1970), pp. 110–111.

The surveys of existing buildings should also include examinations of both the exterior and interior of those buildings to record their condition. For instance such items as the location and sizes of wall cracks should be noted.

20-5 GRADE STAKES

If a project is to be constructed to certain elevations, it is necessary to set stakes in order to guide the contractor. Once the rough grading has been completed, it is necessary to place these stakes in order to control the final earthwork. A grade stake is a stake that is driven into the ground until its top is at the elevation desired for the finished job or until the elevation of the top has a definite relation to the desired elevation. Grade stakes are necessary for sewers, street pavements, railroads, buildings, and so on.

When appreciable cuts are to be made, it may not be feasible to drive stakes into the ground until their tops are at the desired final elevations because they may have to be driven below the ground or buried. If appreciable fills are to be made, the tops of the grade stakes may have to be located above the ground level. Here it is the custom to drive stakes to convenient elevations above the ground and mark them with the cuts or fills necessary in order to obtain the desired elevations. It is very helpful to the contractor when the stakes are placed at heights such that cuts or fills are given whole numbers of feet.

For a particular point the surveyor determines the required elevation and subtracts it from the HI of the level. This tells him what the rod reading should be when the top of the stake is at the desired elevation. Then, more or less by trial and error, the stake is driven until the required rod reading is obtained when the rod is held on top of the stake.

When earth grading operations are near their desired values it is possible to drive stakes until their tops are at the specified final elevations. For these situations it is customary to drive the stakes to grade and color their tops with blue lumber crayons called *keel*. These stakes, known as *blue tops*, are commonly used for grading operations along the edges of highways, railroads, and so on. Some people require that grade stakes be set at road center lines instead of along shoulders.

20-6 REFERENCING POINTS FOR CONSTRUCTION

All survey stakes, even the very stoutest ones, are vulnerable to disturbance during construction. As a result it is necessary to reference them to other points so they may be reestablished in case they are displaced.

In construction or layout surveying, the terms stakes and hubs are frequently used. A *stake* is usually thought of as an approximately $1 \times 2 \times 18$ in. or longer piece of wood sharpened at one end to facilitate driving in the ground. A *hub* is an approximately 2×2 in. piece of wood of variable length which is driven flush or almost flush with the ground and which has a tack driven into it to mark the precise position of the point. Usually one or more stakes on which identification of the hubs is written are driven partly into the ground by the hubs. The stakes may also be flagged.

For highway construction it is usually common practice to reference every tenth station (1000-ft points) as well as the starting and ending points of curves and the points of intersections of the tangents to curves. For building construction it is conventional to reference building corners and even property corners if they are in danger of being obliterated.

The references used should be fairly permanent. Markers may consist of crosses or marks chiseled into concrete pavements or curbs or sidewalks, or nails and bottle caps driven into bituminous pavements. Other types are tacked wooden stakes, nails in trees, or preexisting features such as the corners of existing buildings. One disadvantage of the latter two types of points is that the tripod may not be set over them. Sometimes references may be marked on a curb or pavement or wall with keel. In addition, directions are often written on pavements describing how to find the marks whether the marks themselves are on the pavement or wall or in some other location.

It is desirable to set a sufficient number of reference points so that if some of them are lost the referenced point can still be reset. It is advisable that important points also be recorded and described in the notes in case the markings are obliterated or the points themselves disturbed.

If the distances between reference points can be kept to less than 6 ft the points may be quickly and easily reestablished with a carpenter's rule and the plumb bob. Should reference points be set at greater distances it will be necessary to use a tape and perhaps a transit to check and reset construction points. It is probably true that almost all construction points will be disturbed at least once and perhaps many times during the construction process. As a result, the surveyor needs to carefully position the reference points so that if they are disturbed they can easily be reestablished. It is always desirable to put guard stakes and flags around important construction points. In spite of these precautions, it is still necessary to make continual position and alignment checks. Otherwise some of the construction work may be im-

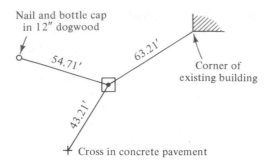

Nail and bottle cap
in 12" dogwood

54.71'

63.21'

Corner of
existing building

43.21'

✝ Cross in concrete pavement

Figure 20-1 Three reference points.

properly positioned, resulting in possible removal and replacement at potentially great expense.

One very common method of referencing survey points is illustrated in Fig. 20-1. In this figure, three reference points are used; if the survey point is disturbed it can be reestablished by swinging an arc with a steel tape (held horizontally) from each of the reference points. The desirability of keeping the reference points within one tape length of the survey points is obvious (6 ft is even better as previously discussed). To protect the hubs and make it easier to locate them, it is desirable to drive slanted stakes over their tops.

A convenient and frequently used method for referencing building corners is shown in Fig. 20-2. It is desirable where possible to set reference points the same distance away from the survey point being referenced so that no notes have to be used to reset it.

It is quicker to set reference points at random as shown in Fig. 20-3 in order to make use of relatively permanent objects such as trees or buildings or pavement or poles in the vicinity. To reestablish survey

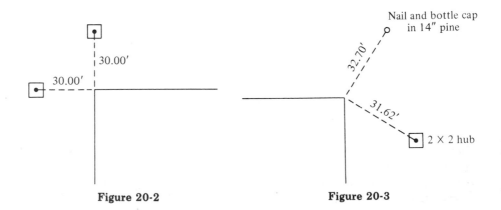

Nail and bottle cap
in 14" pine

30.00'

30.00'

32.70'

31.62'

2 × 2 hub

Figure 20-2   **Figure 20-3**

points from these references it is convenient to use three men so that simultaneous measurement can be made from both reference points.

20-7 BUILDING LAYOUT

The first step in laying out a building is to properly locate the building on the lot. Many cities have building ordinances that provide certain minimum permissible setback distances from the front edge of lots to the fronts of buildings. The purposes of these regulations are to improve appearances and to enhance fire protection. The survey crew sets alignment stakes with distances and directions measured from existing buildings, or streets or curbs, so that the structure will be erected in its desired position.

After the building is properly located it is necessary for it to be laid out with the correct dimensions. This is always important but is particularly so when some or all of the building components are prefabricated and must fit together.

It is obviously critical for a construction job that the various parts of the structure be placed at the desired elevations. To accomplish this goal the construction surveyor will establish one or more bench marks in the general vicinity of the project. These bench marks are placed away from the immediate vicinity of the building so they will not be destroyed by the construction operations, and are used to provide vertical control for the project. Once these are set, the surveyor will then establish a good many less permanent but more accessible bench marks quite close to the project (say, within 100 to 200 ft). The location of these less permanent points should be carefully selected so that turning points will ideally not be needed when elevations have to be set at the project. Such careful selection of the points may result in critical time saving which is so important on construction projects.

For large construction projects it is desirable to establish elevations based on sea-level values (National Geodetic Vertical Datum of 1929). This datum is particularly important for underground utilities. Should an assumed elevation datum be used, its elevations probably should be sufficiently large so that no negative elevations will occur on the project.

Another idea that may save considerable time involves establishing a permanent position for setting up the level so that the instrument can be set up in the same position every day. One way of doing this is to actually have notches for the tripod shoes cut into a concrete sidewalk or pavement or into a specially placed concrete pad. In this way the HI of the level will always be the same if the same instrument and tripod are used and the rod reading for any particular desired location (say, a floor level) will be constant.

Sometimes a special concrete pier can be justified for a large job. Another case where a platform may be advisable is where it is necessary to constantly check nearby existing buildings for possible settlement during the progress of the new construction. This is particularly important during excavation, especially if there is much vibration or blasting. A set of targets can be established on nearby buildings, the level set up at the fixed point, and the telescope used to quickly check for settlements by sighting on the targets. In this regard it is necessary to use the same instrument and tripod (fixed leg not adjustable leg) each day. It is further necessary to frequently check the elevation of the permanent bench mark against surrounding bench marks which are away from the influence of the construction activity.

Later during construction, it may be convenient to establish a bench mark at a certain point in the walls or other part of the building and use that point to measure or set other elevations in the structure. In addition, other reference points may be set within the building from which machinery or other items may be properly located once the building is enclosed. This is often done by setting brass reference points or disks in the floors or other points in the building. Once the walls are erected, the surveyor will be unable to take long sights from his reference points outside the building.

20-8 BASE LINES

Before the actual layout measurements can begin, it is necessary for reference lines or base lines to be carefully established. For large construction projects, the usual procedure is to set a main base line down the center line of the structure and to set stout stakes or hubs (preferably 2 X 2 in. or larger) with tacks in them at intervals not exceeding 100 ft. It is anticipated that these hubs will stay in position for some time during construction. If station numbers are assigned to the points they should preferably be large numbers so that no negative stations will be needed on the extreme ends of the system.

In addition, monuments are set along the center line at each end beyond the area of the construction work. The ends of the line are best marked with heavy cast-in-place concrete monuments with metal tablets embedded in their tops. The monuments along the ends of the line may be occupied by the surveyor and will enable him to check and reset, if necessary, points within the construction area.

A central base line will be so often disturbed during construction that it is common practice to set another or secondary base line parallel to the central one but offset some distance from it. In fact sometimes two base lines, one in front of the building and one in the back, may be established.

Figure 20-4 shows the case where a base line is set to one side of the building. Along such a base line hubs are set where needed to enable the surveyor to set or align corners or other important features of the building. After the corners for a building are set it is essential that their locations be carefully checked. One check that can be used for rectangular buildings is to tape their diagonals to see if they are equal (as shown by the dotted lines shown in Fig. 20-4).

When it is necessary to constantly provide alignments for points in a building where reference markers have been disturbed (a typical situation for the construction of most buildings) the procedure described in the preceding paragraphs and illustrated in Fig. 20-4 has the disadvantage that it is necessary to set up the instrument and lay off 90° angles every time a point is needed. It is preferable to use a system where the important points in the building can be checked or reestablished by merely sighting between two points or by stretching a string or wire between them. Such a system is illustrated in Fig. 20-5. For this building both a base line and an auxiliary base line were established together with the auxiliary points shown. When feasible, it is desirable to attach or paint targets on the walls of existing buildings, instead of setting the auxiliary points in the ground. Such targets are much less likely to be disturbed than stakes in the ground.[3]

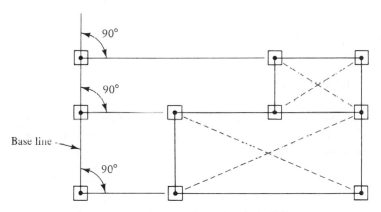

Figure 20-4 Layout for a simple building.

[3]B. A. Barry, *Construction Measurements* (New York: John Wiley & Sons, Inc., 1973), pp. 149–155.

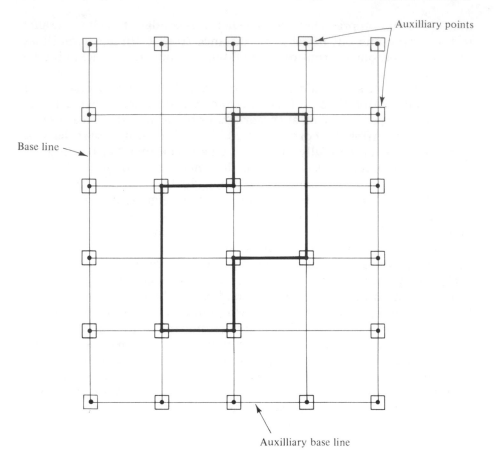

Figure 20-5 Layout of a building with base line, auxiliary base line, and auxiliary points.

20-9 BATTER BOARDS

After hubs are set for the corners of a building they may be secured by means of references set just outside the work area in the form of batter boards. If hubs were placed only at the proposed corners of a building, they would be in the way of excavation and construction operations and they would not last long. For this reason, batter boards are ordinarily used. They consist of wooden frameworks which have nails driven into their tops (or have notches cut into them) from which strings or wires are placed in order to outline the position of the building lines and perhaps the outside of the foundation walls.

Batter boards are used not only for building corners but also for the construction of culverts and sewer lines, for bricklaying and many

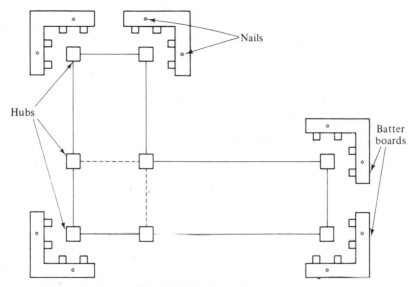

Figure 20-6 Batter boards.

other construction jobs. Examples of batter boards are shown in Figs. 20-6 and 20-7. Batter boards should be placed firmly in the ground and should be well braced. Further, they should be placed a sufficient distance from the excavation so they will be located in undisturbed ground and yet close enough so that strings may conveniently be stretched between them.

To build the batter boards, several stout timbers are driven in the ground probably 4 to 6 ft from the desired excavation. As wire lines are often used between batter boards, they have to be quite stout to withstand the pulling. The vertical posts are 2 × 4s or 4 × 4s and 1 or 2 in. boards are nailed across them. The cross boards are set between the timbers with a level rod and adjusted to the desired elevation with a level. To set the nails which are driven vertically on the cross boards

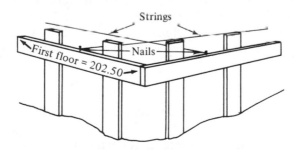

Figure 20-7 Batter boards.

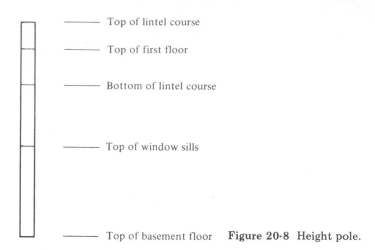

——— Top of lintel course

——— Top of first floor

——— Bottom of lintel course

——— Top of window sills

——— Top of basement floor   **Figure 20-8** Height pole.

to hold the strings for alignment, the transit is used or perhaps strings are stretched across the corner point using plumb bobs as needed.

The main purpose of batter boards is to enable the workmen to measure values from readily accessible reference points without the necessity of having a surveyor always on the spot. Since these boards are used to provide both alignment and elevation they must be set very carefully. It is common to set the batter boards at some prearranged height as a certain number of feet above the foundation or the finished floor grade. A controlling elevation such as that of the first floor level should be clearly marked on the batter boards as illustrated in Fig. 20-7.

Once the batter boards are set and checked the strings or wires can be stretched between the nails (or notches) as shown in Fig. 20-7. The strings can be taken down or replaced as often as desired during construction. For instance they can be put in place and the corner of the structure reset at any time by hanging a plumb bob at the intersection of the strings. The amount of excavation needed at a particular point for the foundation can be determined by measuring down from the string. The strings can be taken down when it is necessary for workmen or machines to excavate for the foundation. They can be put back up for use by the brick masons or other workmen.

In masonry structures a height pole is often used to give the elevations of various points or courses in the walls. Such a pole is shown in Fig. 20-8.

20-10 LAYOUT FOR STRUCTURAL STEEL BUILDINGS

The footings or piers for structural steel buildings can be aligned from the base lines as previously described in Section 20-8. Batter boards are often used to enable the workmen to excavate for and construct the footings.

Steel columns are set on steel base plates and connected to the footings or piers with anchor bolts which are cast into the footings. The anchor bolts will pass through drilled holes in the base plates. The anchor bolts are set into the footing and carefully aligned vertically. Plywood templates are used to hold the bolts in their desired positions. These templates are set on top of the footing forms and must be aligned vertically and horizontally by the surveyor.

One of the most important aspects in successfully erecting a structural steel building is the proper positioning of the anchor bolts and base plates. Except for some very small columns (which may have their base plates welded to their bases in the shop) the base plates are shipped loose to the job and very carefully set to their correct elevations.

The concrete for the footings is placed from $\frac{1}{2}$ to 2 in. lower than its final elevation. The plates are then placed at their correct elevations using shim packs. These packs consist of 3 to 4-in. metal squares with thicknesses varying from $\frac{1}{16}$ to $\frac{3}{4}$ in. each. They are placed under the corners of the base plates and their thicknesses adjusted until the tops of the base plates are at the desired elevation. (An alternative to shim packs involves the use of leveling nuts placed on the anchor bolts under the base plates. They can be adjusted up or down until the plate is at the desired elevation.) At this point a final check of the base plate position must be made using the transit or the batter board strings.

When the plates are in their final position a small wood form is placed a couple of inches away from the edge of the plates and grout is poured or pumped in to fill in the space under the plates. The grout or mortar is preferably of a nonshrinking type made with a metallic aggregate.

As the columns are set in the building they must be properly aligned. The distances between the columns are carefully checked with a tape and their positions may be adjusted by means of jacking or hammering. The columns must be kept vertical by using the transit or plumb bobs. Diagonal bracing cables are very useful here and they are used to pull the columns into position with turnbuckles.

20-11 LAYOUT FOR REINFORCED CONCRETE BUILDINGS

The layout work needed for reinforced concrete buildings follows very closely the alignment work which is used for structural steel buildings. The necessary alignment may be accomplished with transits or with strings. It is necessary to offset these distances so as to be at the correct positions for setting the concrete forms. The reinforcement which is placed in the footings and runs up into the columns or walls must be carefully aligned as are the anchor bolts used for structural steel columns.

The snapping of chalk lines onto concrete floors or footings or on the forms is a very useful method of establishing reference lines from which measurements can be taken. Chalk lines can often take the place of mason's strings or wires, and they don't present tripping hazards or the problems of resetting them when the lines are broken.[4]

It is necessary to do a great deal of spirit leveling in a large reinforced concrete building to set reference marks on concrete floors or columns. As the building is constructed, tapes are often used to control the vertical distances. Because of obstructions, many of these measurements are made in elevator shafts or stairwells or on the outsides of the buildings.

To maintain the vertical alignment of walls and columns it is usually necessary to use transits or plumb bobs. Carpenter's levels held against the formwork do not usually provide sufficient accuracy for maintaining vertical alignment.

20-12 LAYOUT NOTES FOR CONSTRUCTION

No particular standard forms are used for keeping field notes for construction layouts, but a sketch should be made showing the actual angles and lengths used. Often the construction drawings themselves can be used and the dimensions thereon circled, checked or marked in some way. Should the actual construction dimensions be different from the values shown in the drawings, the dimensions used should be carefully noted. For instance, if a distance is shown on the drawing as being 320.00 ft and the actual field measurement is 320.04 ft, the actual value should be recorded. If every measured field dimension agreed exactly with the planned dimension there could very well be a suspicion that the surveyor didn't bother to check the values. A complete set of "as-built plans" is often required or should be furnished as part of a construction job.

20-13 SLOPES AND SLOPE STAKES

In working with cuts and fills for highway construction, the side slopes are generally based on the material involved. The slope is given as a ratio of so many units horizontally to so many units vertically. For

---

[4]B. A. Barry, *Construction Measurements* (New York: John Wiley & Sons, Inc., 1973), p. 159.

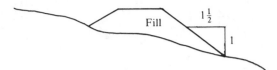

Figure 20-9

instance, a 2-to-1 slope means that the bank in question goes 2 ft horizontally for each 1 ft vertically (see Fig. 20-9). The $1\frac{1}{2}$ to 1 slope is
probably the most common one used, but if the material consists of
solid rock, the slopes may be much steeper. For very loose material,
such as sand, they will be much flatter.

After the grade of the road has been established and the material
slopes decided for cuts and fills, it is necessary for the surveyor to stake
out the work. The cut or fill to be made at a particular station along
the center line of the road equals the difference between the ground
elevation there and the final elevation from the grade line. The surveyor will need to stake the road center line and set *slope stakes* at the
intersection of the natural ground line and the side slopes of the cut or
fill as shown in Fig. 20-10.

A few descriptive comments are made here regarding the trial-and-
error problem of setting slope stakes because it may give the student a
little trouble. If the ground surface is horizontal, the distance from the
road center line to the intersection of the ground surface and the cut or
fill can easily be computed, as shown in Fig. 20-11.

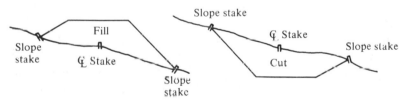

Figure 20-10 Slope stakes.

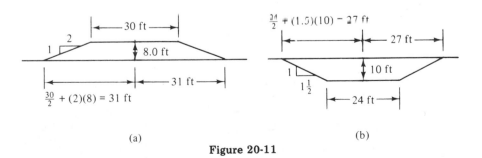

(a)   (b)

Figure 20-11

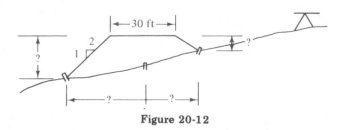

Figure 20-12

If the ground surface is sloping, as is the case in Fig. 20-12, the problem becomes a trial-and-error one because the vertical amounts of cut or fill as measured from the slope stake positions are unknown, as are the horizontal distances from the center line to the slope stakes.

The usual practice is to set slope stakes with the 50-ft woven tape and the engineer's level. The zero end of the tape is placed at the center line. The instrumentman roughly estimates the distance to the slope stake and sends the rodman that distance from the center line. The instrumentman then takes a reading on the level rod and calculates to see if the rodman is at the desired point. For this discussion, see Fig. 20-13 in which the instrumentman estimated the cut at the lower slope stake as 8.0 ft and then computed the horizontal distance from the center line as $(30/2) + (2)(8) = 31$ ft.

The rodman was sent out 31 ft from the center line and the level rod reading was taken. It was found that the difference in elevation from the top of the fill to the point in question was 10.0 ft. The horizontal distance from the center line for that difference in elevation should have been $(30/2) + (2)(10) = 35$ ft, so the rodman was not at the correct point.

Next, the rodman was sent out to the 40-ft point and the elevation difference was found to be 12.0 ft. For this point; the horizontal distance should have been $(30/2) + (2)(12) = 39$ ft. The rodman moved to the 39-ft point where the elevation difference was 11.9 ft. The calculated horizontal distance equaled $(30/2) + (2)(11.9) = 38.8$ ft, which was just about right.

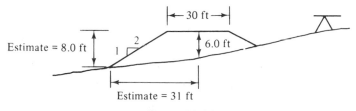

Figure 20-13

Once the trial horizontal distance is within 1 or 2 tenths of a foot of the computed horizontal distance, the slope stake is driven. An experienced man usually sets the stake within 2 or 3 trials at most, but the beginner may require a few more attempts.

Slope stakes are normally set sloping outward from the road for fills and inward for cuts. The station numbers are usually written on the outside of the stakes and the cuts or fills are given on the inside, often with the distance to the center line of the road given. Sometimes after the correct position of the slope stakes is determined, they are offset by 2 to 5 ft in order to preserve them during grading operations.

20-14 BRIDGE SURVEYS

For short bridges with no offshore piers, the first step is to lay out the center line of the roadway. Then the station of the abutment is staked on the center line and the angle of the abutment face across the center line is established. This should be done by establishing points at the end of the cross line and beyond and referencing them carefully. Should the abutment be out in the water, the cross line will be offset on the shore. In a similar manner the governing lines for the wing walls are established at and beyond the limits of the excavation, preferably at both ends of each wing wall.

When the faces of the walls are battered, it is customary to stake one line for the bottom of the batter and another for the top.

Once the foundation concrete has been set, the surveyor will need to provide the lines needed for setting forms. As the walls are constructed elevation marks are established on the forms or on the hardened concrete. In addition, the alignment is marked on the completed parts of the structure.

For long bridges the measurement of distances is extraordinarily important. They should be measured with a precision even better than that of first-order work because of the necessity of having the parts of the superstructure fit together within very small limits or tolerances. The pier locations are established with a somewhat lower precision than that used for the fitting of the superstructure. When the entire structure is completed, permanent survey points are usually established at various points so that any lateral and vertical movements of the bridge may be detected.[5]

[5] R. E. Davis, F. S. Foote, and J. W. Kelly, *Surveying Theory and Practice*, 5th ed. (New York: McGraw-Hill Book Company, 1966), pp. 734–736.

Pipelines can be of two types, pressure or gravity. In the pressure pipe-line the liquid or gas that fills the pipes moves along uphill or down under pressure. In gravity pipelines (such as sanitary sewers or storm drains) the liquid flows along by gravity and the grades of the lines have to be controlled very carefully.

The *pipe inverts* are used to control the elevations and grades of the lines. Because of manufacturing imperfections there is some varia-tion in pipe dimensions and, therefore, the pipe invert is the logical point for elevation control because any dimension irregularities will be thrown to the top of the pipe where they will not affect the flow.

The old method of controlling pipeline excavation and construc-tion by using batter boards and stringlines is described in this section, while more modern procedures are presented in Section 20-16.

In order to stake out a sewer line, the stakes are set at the desired stations (commonly 50-ft intervals) and offset from the planned pipe center line and their elevations are determined. The station number, offset distance, and cut required are marked on each of the stakes. Alignment is handled with the transit, but the very important elevations are normally obtained with a level. After excavation is completed, the stakes will be used to give alignment and grade for the pipe-laying operation.

The offset stakes are normally driven to a depth such that their tops will be at certain desired elevations above the pipe inverts. During excavation the depth of the trench is often checked from these stakes to be sure that the proper grade is being obtained. The bottom of the trench will obviously be a few tenths of a foot below the pipe invert because of pipe thickness, gravel or sand beds for the pipes, and so on. These values must be considered in the elevations. As shown in Fig. 20-14, it is easy to check trench depths from the offset stakes by using a board, a carpenter's level, and a level rod or some type of height pole.

After the trench excavation is completed, a stringline is placed directly above the desired center line of the pipe and is made parallel to the desired grade at the inverts. In order to accomplish this, a cross batter board is placed at each elevation stake, as shown in Fig. 20-15.

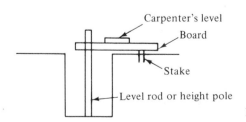

Figure 20-14

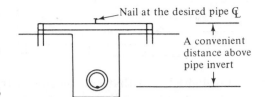

Figure 20-15

The top of each of these batter boards will be the same vertical distance above the desired position of the inverts. A nail is placed (as shown in the figure) at the top of the batter board at the desired center line of the pipe. A string run between these nails and tightly stretched will have a grade equal to that desired for the pipe inverts.

Starting at the lower end of the trench, the pipe is laid joint by joint. The invert elevation of both ends of the first section and its alignment under the string are checked. As the pipe is laid along the line, each additional section of the pipe is set by using a height pole of the proper length from the string to check the elevation of the forward end.

20-16 TRANSIT AND LASER METHODS OF LAYING PIPE

A transit can be set up on a platform across the trench and the telescope inclined at the desired grade. With the line of sight set a convenient distance above the proposed bottom of the trench, the grade can be controlled for excavation and later for laying the pipe, as shown in Fig. 20-16. It will be noted that the instrumentman will have to make frequent readings with the transit and will often have to check to see if his instrument has settled or been otherwise disturbed.

Another useful device for controlling pipeline excavations and laying pipe is the laser. So many applications are being found for the laser that some construction people are saying that eventually it will be the only tool needed for the layout and control of construction projects. It can be quickly, accurately, and economically used for many purposes, such as distance measurement, alignments for tunnel borings, setting of pipes with desired grades, and setting line and grade for many types of construction.

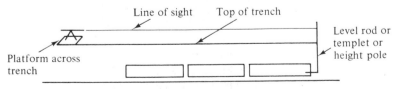

Figure 20-16

Figure 20-17

The laser is an intense light beam that can be concentrated into a narrow ray containing only one color (red) or one wavelength of light. The resulting beam can be projected for short or long distances and is clearly visible as an illuminated spot on a target. It is not disturbed by wind or rain, but it will not penetrate fog. A laser can be set up on a bracket or even attached to a transit telescope. The beam is lined in the proper direction at the desired grade and can be left relatively unattended.

Today, instead of using batter boards and strings, the laser can be used to control the alignment for excavating a trench and setting a pipe. The laser can be set so that it shines on the boom of a backhoe so that the operator can clearly see the illuminated spot. By its position he can closely control the depth to which he is digging. For laying the pipe the laser is set in the proper direction at the desired distance above the pipe invert. With the aid of an L-shaped pole or templet, as shown in Fig. 20-17, the workmen can control the invert elevation. It may also be possible to direct the laser beam from the inside of manholes through the pipes being laid and control the grade without any interference from the backfill operations. This can be done even if the pipes are too small for human access.

The use of lasers for pipe-laying operations provides several advantages: less labor required; quicker work; better alignments and grades; and no batter boards to be erected and later removed.

PROBLEM

20-1. For the accompanying sketch and readings with the level placed at the center line of the road, would no. 1, no. 2, or no. 3 be at the correct location for the slope stake?   (*Ans.:* no. 2)

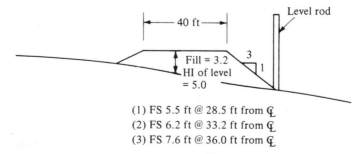

(1) FS 5.5 ft @ 28.5 ft from ₵
(2) FS 6.2 ft @ 33.2 ft from ₵
(3) FS 7.6 ft @ 36.0 ft from ₵

21-1 INTRODUCTION TO HORIZONTAL CURVES

The center lines of highways and railroads consist of a series of straight lines connected by curves. The curves for fast traffic are normally circular, although spiral curves may be used to provide gradual transitions to or from the circular curves. Three circular curves are shown in Fig. 21-1. The *simple curve* consists of a single arc. The *compound curve* consists of two or more arcs with different radii. The *reverse curve* consists of two arcs that curve in different directions. Only the simple curves are discussed here.

Several definitions relating to curves are presented in the next few paragraphs and are illustrated in Fig. 21-2. A curve is initially laid out with two straight lines or tangents. These lines are extended until they intersect and that *point of intersection* is called the *P.I.* The first tangent encountered is called the *back tangent* and the second one is called the *forward tangent.*

The curve is laid out so that it joins these tangents. The *points on tangents (P.O.T.s)* are the points where the curves run into the tangents. The first of these points is on the back tangent at the beginning of the curve and is called the *point of curvature (P.C.).* The second one is at the end of the curve on the forward tangent and is called the *point of tangency (P.T.).* In another notation the point of curvature may be written as *T.C.*, indicating that the route changes from a tangent to a curve; the point of tangency may be written as *C.T.*, indicating that the route goes from curve back to tangent.

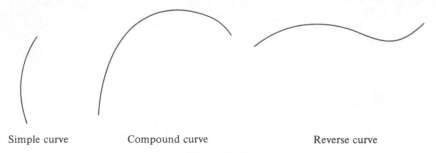

Simple curve   Compound curve   Reverse curve

Figure 21-1

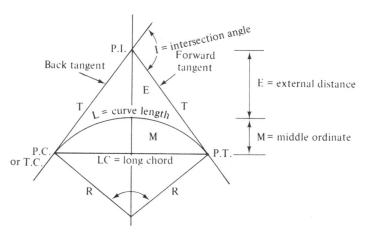

Figure 21-2 Curve notation.

The angle between the tangents is called the *intersection angle* and is labeled *I*. The *radius of the curve* is *R*, while *T* is the *tangent distance* and equal to the length of the back or forward tangents. The distance from the *P.I.* to the middle point of the curve is called the *external distance* and is denoted by *E*. Finally, the chord of the arc from the *P.C.* to the *P.T.* is called the *long chord* (*L.C.*) and the distance from the middle of the curve to the middle of the long chord is labeled *M*, the *middle ordinate*, and *L* is the actual curve length.

21-2 DEGREE OF CURVATURE AND RADIUS OF CURVATURE

The sharpness of a curve may be described in any of the following ways.

Radius of curvature. This method is often used in highway work where the radius for the curve is probably given as some multiple of 100 ft. The smaller the radius, the sharper the curve.

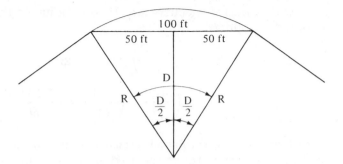

Figure 21-3 Degree of curvature, chord basis.

Degree of curvature, chord basis.   In this method, the degree of curvature is defined as the central angle subtended by a chord of 100 ft as illustrated in Fig. 21-3. The radius of such a curve may be computed with the following expression:

$$R = \frac{50}{\sin \frac{1}{2}D}$$

Degree of curvature, arc basis.   As shown in Fig. 21-4, the degree of curvature on the arc basis is the central angle of a circle which will subtend an arc of 100 ft. It will be noticed that a sharp curve has a large degree of curvature and a flat curve a small degree of curvature. For a

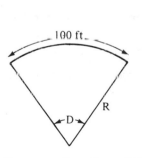

Sharp curve (large D, small R)

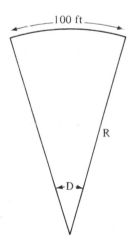

Flat curve (small D, large R)

Figure 21-4 Degree of curvature, arc basis.

particular curve with a degree of curvature D, the radius of the curve, R, can be computed as follows:

$$\text{Circumference of circle} = \left(\frac{360°}{D}\right)(100) = 2\pi R$$

$$R = \left(\frac{360°}{D}\right)\left(\frac{100}{2\pi}\right) = \frac{5729.58}{D}$$

The arc basis is used for the calculations presented in this chapter. Actually both the chord and arc methods are used extensively in the United States. The method selected often depends on the experience of the surveyor involved. For long gradual curves, which are common in railroad practice, the chord basis (where the lengths along the arcs are considered to be the same as along the chords) is normally used. For highway curves and curved property boundaries, the arc basis is more common.

It will be noted that the difference between the chord and arc bases is normally not large. For instance, for a $1°$ curve, R on the arc basis is 5729.58 ft, and 5729.65 ft on the chord basis. Corresponding values for a $4°$ curve are 1432.39 ft and 1432.69 ft, respectively.

21-3 CURVE EQUATIONS

The formulas needed for circular curve computation are presented in this section, with reference to Fig. 21-2. The radius of curvature was previously given as

$$R = \frac{5729.58}{D}$$

The tangent distance T is the distance from the *P.I.* to the *P.C.* or *P.T.* and can be computed from

$$T = R \tan \tfrac{1}{2} I$$

The length of the long chord is

$$L.C. = 2R \sin \tfrac{1}{2} I$$

The external distance E is

$$E = R \left(\sec \frac{I}{2} - 1\right) = R \text{ ex. sec } \frac{I}{2}$$

where

$$\text{ex. sec} = 1 - \sec$$

The middle ordinate M equals

$$M = R - R \cos \frac{1}{2} I = R \text{ vers } \frac{I}{2}$$

where

$$\text{vers} = 1 - \cos$$

The length of the curve is

$$L = \frac{100I}{D}$$

For surveyors who frequently work with curves, the use of versines and external secants simplifies the calculations. Various texts provide tables of their values.[1]

Curves are staked out using straight-line chord distances. If the degree of curvature for a particular curve is $3°$ or less, the curve can be staked out using 100-ft chords as shown in Fig. 21-5 and yet keep the arc length and chord length values sufficiently close to each other so as to be within the precision of tape measurements. As the usual curve will be of an odd length (that is, not a whole number of hundreds of ft), there will certainly be subchord lengths at the ends of the curve as indicated in the figure. For curves from $3°$ to $7°$, it is necessary to go to 50-ft chords, and for those from $7°$ to $14°$, to 25-ft chords to maintain satisfactory precision.

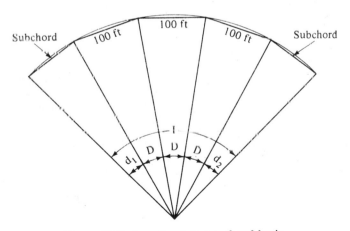

Figure 21-5 Length of curve, chord basis.

[1] For example, T. F. Hickerson, *Route Location and Design*, 5th ed. (New York: McGraw-Hill Book Company, 1967), pp. 559–588.

For flat curves, say $2°$ or $3°$, the chord distances and arc distances are almost the same to two places beyond the decimal, whereas for sharper curves the difference is more pronounced. For a 100-ft long $2°$ curve, the chord is 99.995 ft, and for a 100-ft $3°$ curve, it's 99.989 ft. For a $6°$ curve, the chord is 99.954 ft, and for a $10°$ curve, it's 99.873 ft.

21-4 DEFLECTION ANGLES

The angle between the back tangent and a line drawn from the *P.C.* to a particular point on a curve is called the *deflection angle* to that point. Circular curves are laid out almost entirely by using these angles. From the geometry of a circle, the angle between a tangent to a circular curve and a chord drawn from that point of tangency to some other point on the curve equals one-half of the angle subtended by that chord. Thus for a 100-ft chord, the deflection angle is $D/2$, and for a 50-ft chord is $50/100 \times D/2$. These values are illustrated in Fig. 21-6 where D is $3°$. It will be noted that the deflection angle from the *P.C.* to each succeeding 100-ft station can be calculated by adding $D/2$ to the last deflection angle, or to each succeeding 50-ft station by adding $D/4$ to the last deflection angle.

21-5 SELECTION AND STAKING OUT OF CURVES

Before a horizontal curve can be selected, it is necessary to extend the tangents until they intersect, measure the intersection angle I, and

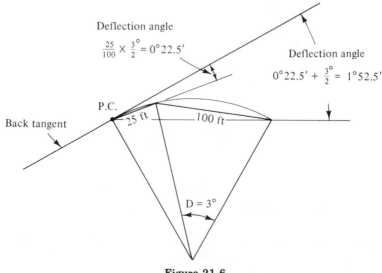

Deflection angle

$$\frac{25}{100} \times \frac{3°}{2} = 0°22.5'$$

Deflection angle

$$0°22.5' + \frac{3°}{2} = 1°52.5'$$

P.C.

Back tangent

25 ft   100 ft

$D = 3°$

Figure 21-6

select D, the degree of curvature. For any pair of intersecting tangents, an infinite number of curves can be selected, but the field conditions will narrow down the choices considerably. For high-speed highways, the degree of curvature is generally kept below certain maximums, while for twisting mountain roads, the lengths of tangents may be severely limited by the topography. Should a road be running along the bank of a river, the external distance E may be restricted.

Actually, the curve to be used can be selected by assuming a value for D, R, T, E, or L. This permits the calculation of the other four since they are dependent upon each other. Usually the radius of the curve or the degree of curvature is the value assumed. Then the necessary computations can be made as illustrated in Example 21-1. In addition to the determination of R, T, $L.C.$, E, M, and L, these computations include the calculations of the deflection angles and the setting up of the fieldbook notes for staking out the curve.

The positions of the $P.C.$ and $P.T.$ are determined by measuring the calculated tangent distance T from the $P.I.$ down both tangents. The reader should note that the station of the $P.C.$ equals the station of the $P.I.$ minus the distance T, and that the station of the $P.T.$ equals the $P.C.$ plus the curve length L (not the $P.I.$ plus T). The transit is set up at the $P.C.$ or $P.T.$ and the curve staked out. As the work proceeds, the transit can be set up at intermediate points as well. This latter procedure will be described after Example 21-1.

Example 21-1 illustrates the calculations and field book notes needed for the staking out of a horizontal circular curve. It will be noted that these notes are arranged and the stations numbered from the bottom of the page upward. This practice, which is common for route surveys such as this one, enables the surveyor to look forward along the route and follow his notes while going in the same direction. In the same manner, as he looks at his sketch page and forward along the route, items to the right of the project center line on the ground are to the right on the sketch page and vice versa.

Example 21-1

For a horizontal circular curve, the $P.I.$ is at station $64 + 32.2$, I is $24°20'$, and a D of $4°00'$ has been selected. Compute the necessary data and set up the field notes for 50-ft stations.

Solution   The values of R, T, $L.C.$, E, M, and L are computed by the formulas previously presented although tables are given in many books for simplifying the calculations.[2]

$$R = \frac{5729.58}{4} = 1432.39 \text{ ft}$$

[2]T. F. Hickerson, *Route Location and Design*, 5th ed. (New York: McGraw-Hill Book Company, 1967), pp. 396–458.

$$T = (1432.39)(0.21560) = 308.82 \text{ ft}$$

$$L.C. = (2)(1432.39)(0.21076) = 603.78 \text{ ft}$$

$$E = (1432.39)(0.02298) = 32.92 \text{ ft}$$

$$M = (1432.39)(0.02246) = 32.17 \text{ ft}$$

$$L = \frac{(100)(24.33)}{4} = 608.25 \text{ ft}$$

From this data, the stations of the $P.C.$ and $P.T.$ can be calculated as follows:

$$
\begin{array}{rl}
P.I. = & 64 + 32.2 \\
-T = & -3 + 08.8 \\
\hline
P.C. = & 61 + 23.4 \\
+L = & 6 + 08.2 \\
\hline
P.T. = & 67 + 31.6
\end{array}
$$

As the degree of curvature is between $3°$ and $7°$, the curve will be staked out with 50-ft chords. The distance from the $P.C.$ (61 + 23.4) to the first 50-ft station (61 + 50) is 26.6 ft, and the deflection angle to be used for that point is $(26.6/100) \times (D/2) = (26.6/100) \times (4°/2) = 0°32'$. For each of the subsequent 50-ft stations from 61 + 50 to 67 + 00, the deflection angles will increase by $D/4$ or $1°00'$. Finally, for the $P.T.$, the deflection angle will be $11°32' + (31.6/100)(4°/2) = 12°10'$. This value equals $\frac{1}{2} I$, as it should.

The field notes are set up as shown in Fig. 21-7.

21-6 FIELD PROCEDURE FOR STAKING THE CURVE

The tangent distances are taped from the $P.I.$ down both tangents to locate the $P.C.$ and $P.T.$ For this discussion, the transit is assumed to be set up at the $P.C.$ and the curve and numbers of Example 21-1 used. It should be noticed however that if the entire curve can be seen from the $P.T.$, it is possible to avoid one instrument setup by setting up there, because after the curve is completed, the survey can proceed down the forward tangent from the same setup.

The transit is set up at the $P.C.$ with the A vernier reading zero while sighting on the $P.I.$ The first deflection angle $(0°32')$ is turned and the chord distance of 26.6 ft is taped to locate station 61 + 50. The next deflection angle $(1°32')$ is turned and 50 ft is taped beyond station 61 + 50 to stake station 62 + 00. This procedure is continued until the $P.I.$ is reached. These points should check very carefully. For long curves it is considered better to run in the first half of the curve from the $P.C.$ and the second half back from the $P.T.$ so that small errors that occur can be adjusted at the middle of the curve where

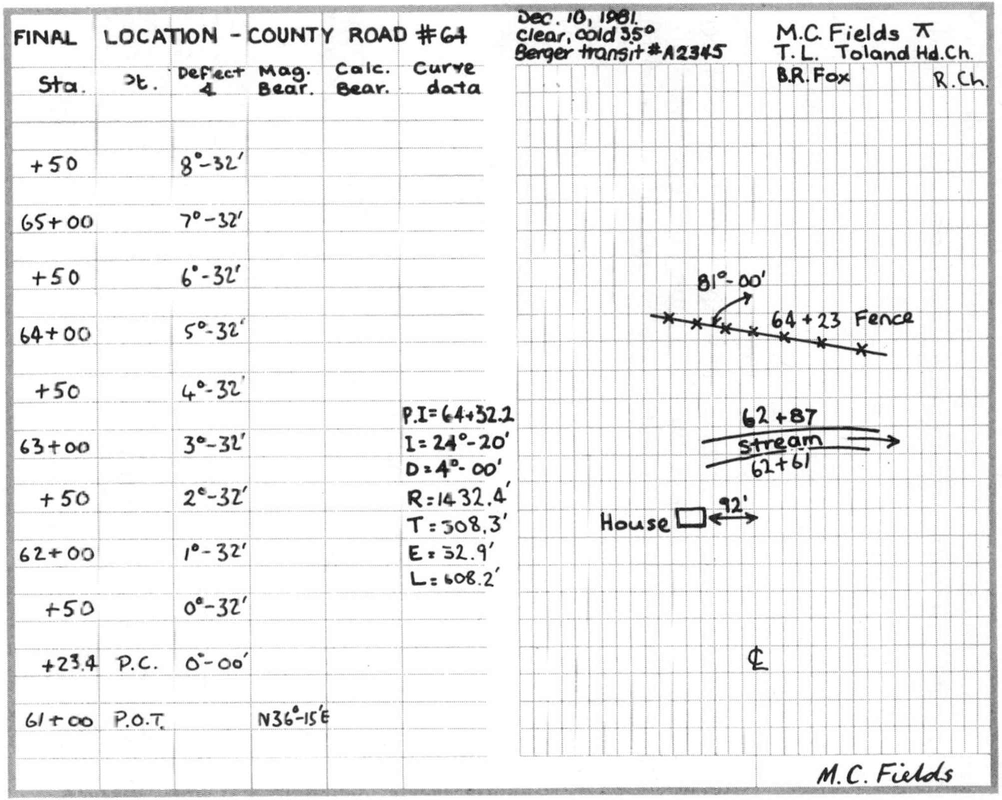

FINAL LOCATION - COUNTY ROAD #64

Sta.	Pt.	Deflect. ∢	Mag. Bear.	Calc. Bear.	Curve data
+50		8°-32'			
65+00		7°-32'			
+50		6°-32'			
64+00		5°-32'			
+50		4°-32'			P.I=64+32.1
63+00		3°-32'			I=24°-20'
+50		2°-32'			D:4°-00'
62+00		1°-32'			R:1432.4'
+50		0°-32'			E:52.9'
+23.4	P.C.	0°-00'			L:608.2
61+00	P.O.T.		N36°-15'E		

Dec. 10, 1981
Clear, cold 55°
Berger transit #A2345

M.C. Fields ⊼
T.L. Toland Hd.Ch.
B.R.Fox   R.Ch.

81°-00'
64+23 Fence
62+87
Stream
62+61
House 92'
℄

M.C. Fields

Figure 21-7 Field notes for a horizontal curve.

385

a little variation is not as important as it would be near the points of tangency to the curve.

Very often it may not be possible to set all of the points on the curve from one position of the instrument. The transit may have to be moved up to one of the intermediate stations on the curve and the work continued. To do this, the transit is set up on the intermediate station, and is backsighted on the *P.C.* with the telescope inverted and the *A* vernier reading $0°00'$. The telescope is reinverted, the deflection angle that would have been used for that next station turned, and the process continued.

As an illustration for Example 21-1, it is assumed that the instrument is moved up to station 64 + 00 and backsighted on the *P.C.* with the telescope inverted. The telescope is reinverted and turned to an angle of $6°32'$, the distance of 50 ft taped, and station 64 + 50 set.

Another horizontal curve situation is considered in Example 21-2. A surveyor is assumed to be working along the boundary of a piece of private property which adjoins a curved section of a state highway. He would like to determine the information necessary to describe correctly the curved boundary of the land.

Example 21-2

As shown in Fig. 21-8, a surveyor has determined that the intersection angle for a particular highway curve is $28°00'$. In addition, he has measured the external distance *E* from the *P.I.* to the highway center line for this curve and found it to be 73.6 ft.

(a) Determine the values of *D*, *R*, and *T* for the highway center line.

(b) If the highway right of way is 50 ft from the highway center line as shown in the figure, determine for the property on the inside of the curve the values of *R* and *D*.

Solution

(a) $R = \dfrac{73.6}{0.03061} = 2404.4$ ft

$D = \dfrac{5729.58}{2404.4} = 2.383° = 2°23'$

$T = (2404.4)(0.24933) = 599.5$ ft

(b) *R* of property line or of right of way line = 2404.4 − 50 = 2354.4 ft

D of property line curve = $\dfrac{5729.58}{2354.4} = 2.434° = 2°26'$

The surveyor who is staking out new lots or resurveying old ones along a highway will often find that parts of the lot boundaries adjacent to the right of way (ROW) of the highway will be curved. His work should reflect these curved property lines.

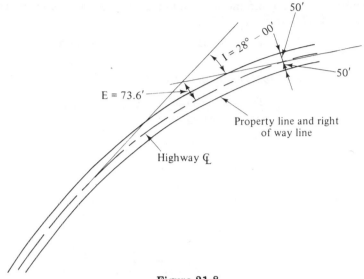

Figure 21-8

From the equations given earlier in this chapter we have seen that only two of the elements $(R, T, L.C., E, M,$ and $L)$ of a curve are necessary to define the curve. It is often desirable, however, for the surveyor to provide three or even more of these elements. If more than two elements are provided they must be compatible with each other. Among the curve elements most commonly provided are the radius, intersection angle, curve length, and sometimes the length and bearing of the chord.

Example 21-3 illustrates the information necessary to stake out 100-ft lots along a right-of-way curve for a highway.

Example 21-3

It is assumed that the curve considered in Example 21-1 represents the center line of a highway with a 50-ft right of way on each side. It is desired to lay off 100-ft lots on the inside of the curved right of way as shown in Fig. 21-9. Compute the

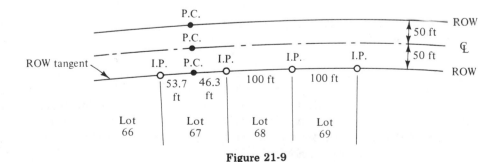

Figure 21-9

necessary data for laying off the lots assuming that the last corner on the ROW back tangent is 53.7 ft from the *P.C.*

Solution   The radius of the ROW curve is 50 ft less than the radius of the center-line curve. Therefore

$$R = 1432.39 - 50.00 = 1382.39 \text{ ft}$$

$$D = \frac{5729.58}{R} = \frac{5729.58}{1382.39} = 4.144° = 4°08.6'$$

$$L = \frac{100I}{D} = \frac{(100)(24.333)}{4.144} = 587.19 \text{ ft}$$

Setting up the instrument at the *P.C.* on the ROW curve the deflection angle to the first lot corner on the curve (a distance equal to 100.0 - 53.7 = 46.3 ft) is equal to

$$\left(\frac{46.3}{100}\right)\left(\frac{4.144°}{2}\right) = 0.959° = 0°57.5'$$

From the same position of the instrument the deflection angle to the next lot corner is

$$\left(\frac{146.3}{100}\right)\left(\frac{4.144°}{2}\right) = 3.031° = 3°01.9'$$

Or it can be calculated as

$$0°57.5' + \frac{D}{2} = 0.959° + \frac{4.144°}{2} = 3.031° = 3°01.9'$$

The deflection angle for the *P.T.* can be determined by taking the initial deflection angle (0°57.5') and adding $D/2$ to it for each 100 ft plus the angle value for the partial lot width before the *P.T.*; or it equals

$$\left(\frac{587.19}{100}\right)\left(\frac{4.144°}{2}\right) = 12.167° = 12°10'$$

which is half of *I* of 24°20'.

Note: Since the degree of curvature is between 3° and 7°, the curve should be laid out with 50-ft chords with the iron pin corner monuments placed at the 100-ft points.

21-7 HORIZONTAL CURVES PASSING THROUGH CERTAIN POINTS

Sometimes it is necessary to lay out a horizontal curve which passes through a certain point. For instance it may be desired to establish a curve which passes no closer than a certain distance to some feature such as a building or stream. The distance from the *P.I.* to the point in question and the angle from one of the tangents can be measured. Such a situation is considered in Example 21-4 and illustrated in Fig. 21-10.

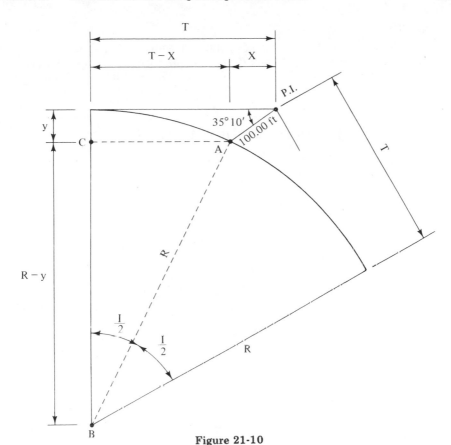

Figure 21-10

To solve the problem the distances x and y shown in Fig. 21-10 can be computed by trigonometry. With reference to the figure it can be seen that the radius of the curve R can be determined by considering the right triangle ABC. For this triangle the following quadratic equation applies:

$$R^2 = (R - y)^2 + (T - x)^2$$

The value of T can be expressed in terms of R as $R \tan \frac{1}{2} I$ and substituted into the preceding equation leaving only R as an unknown. Then R can be determined with the quadratic equation, by trial and error with a calculator or by completing the squares. If the quadratic equation is used, the solution will yield two answers but one of them will be seen to be unreasonable. Hickerson[3] provides a solution to this type of problem which does not involve the use of a quadratic equation.

[3] T. F. Hickerson, *Route Location and Design*, 5th ed. (New York: McGraw-Hill Book Company, 1967), pp. 90–91.

Example 21-4

A horizontal curve is to be run through point A as shown in Fig. 21-10. From the *P.I.* the distance to point A is 100.00 ft and the angle from the back tangent to a line from the *P.I.* to point A is $35°10'$. If I is $65°00'$, determine the values of R, D, T, and L.

Solution

$$y = (100.00)(\sin 35°10') = 57.596 \text{ ft}$$

$$x = (100.00)(\cos 35°10') = 81.748 \text{ ft}$$

$$T = (R) \ \frac{\tan 65°00'}{2} \ = 0.63707R$$

$$R^2 = (R - y)^2 + (T - x)^2$$

$$R^2 = (R - 57.596)^2 + (0.63707R - 81.748)^2$$

Squaring the terms and simplifying,

$$R^2 - 540.46R + 24{,}639.137 = 0$$

Using the quadratic equation with $a = 1.00$, $b = -540.46$ and $c = 24{,}639.137$:

$$R = \frac{+540.46 \pm \sqrt{(-540.46)^2 - (4)(1.00)(24{,}639.137)}}{(2)(1.00)}$$

$$R = 490.20 \text{ ft} \quad \text{or} \quad 50.26 \text{ ft (not feasible)}$$

$$D = \frac{5729.58}{490.20} = 11.688°$$

$$T = (490.20) \left(\tan \frac{65°00'}{2} \right) = 312.29 \text{ ft}$$

$$L = \frac{(100)(65.00)}{11.688} = 556.13 \text{ ft}$$

21-8 VERTICAL CURVES

The curves used in a vertical plane to provide a smooth transition between the grade lines of highways and railroads are called *vertical curves*. These curves are parabolic rather than circular.

Figure 21-11 shows the nomenclature used for vertical curves. When moving along the road, the first of the grade lines that will be encountered is called the *back tangent*. The other one is called the *forward tangent*. To distinguish the points of tangency and their intersection from the similar terms used for horizontal curves, the letter V (for vertical) is added to their abbreviations. For instance, the point of intersection for the tangents is called the *P.V.I.* (point of vertical inter-

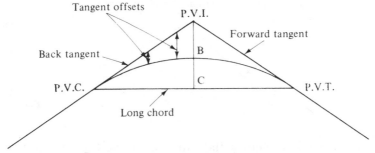

Figure 21-11 Vertical curve nomenclature.

section) and the points of tangency are referred to as the *P.V.C.* (point of vertical curvature at start of curve) and *P.V.T.* (point of vertical tangency at end of curve). The long chord is a chord from the *P.V.C.* to the *P.V.T.* and the *tangent offsets* are distances measured from the tangents in a vertical direction to the curve.

Several different methods are available for making the necessary calculations, but only one method is presented here. It is a very simple and satisfactory one that uses tangent offsets from the grade lines. The parabola has three mathematical properties that make its application very convenient. These are as follows:

1. The curve elevation at its midpoint will be halfway from the elevation at the *P.V.I.* to the elevation at the midpoint of the long chord.
2. The tangent offsets vary as the square of the distance from the point of tangency.
3. For points spaced at equal horizontal distances, the second differences are equal. This is a useful property for checking vertical curve calculations as is illustrated in Example 21-5. The differences between the elevations at equally spaced stations are called the *first differences*. The differences between the first differences are called the *second differences*.

21-9 VERTICAL CURVE CALCULATIONS

The station of the vertex or *P.V.I.* can be determined from the grade lines. Then the length of the curve is selected—usually as some whole number of hundreds of feet or stations. The length of a vertical curve is defined as the horizontal distance from the *P.V.C.* to the *P.V.T.* This length is normally controlled by the specification being used. For instance, the length required for a highway vertical curve must be suffi-

Station 65 + 00
Elevation 264.20

Figure 21-12

cient to provide a sufficiently flat curve over the top of a hill so that certain minimum sight distances of oncoming vehicles or objects in the road are provided. The AASHTO (American Association of State Highway and Transportation Officials) provides such information depending on the design speed of the highway.[4]

Example 21-5 illustrates the calculations required for determining the elevations along a vertical curve. For this example the given grade lines intersect at station 65 + 00 at an elevation of 264.20 ft, and the curve is assumed to be 800 ft long. From this information the stations and elevations of the *P.V.C.* and *P.V.T.* are determined, as well as the elevations of the intermediate stations along the grade lines or tangents. The elevation at the midpoint of the long chord is equal to the average elevation of the *P.V.C.* and the *P.V.T.*

Next, the elevation of the curve midpoint (point *B* in Figs. 21-11 and 21-13) is determined by averaging the elevations of point *C* and the

Stations	61	62	63	64	65	66	67	68	69
Elevations on grade lines	248.20	252.20	256.20	260.20	264.20	261.20	258.20	255.20	252.20

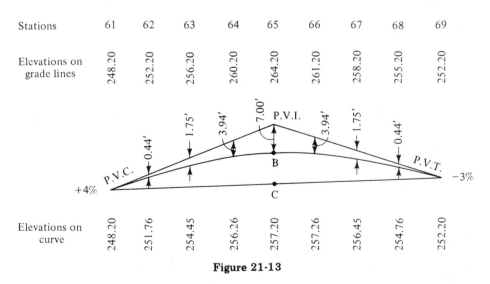

Elevations on curve	248.20	251.76	254.45	256.26	257.20	257.26	256.45	254.76	252.20

Figure 21-13

[4] AASHTO, *A Policy on Geometric Design of Rural Highways* (Washington, D.C., 1965), pp. 203–211.

P.V.I. Finally, the tangent offset distances from the grade lines are calculated, thus giving the elevations of the intermediate stations on the curve.

Example 21-5

Figure 21-12 shows the known data for a vertical curve. Assume a curve length of 800 ft, and compute the elevation of each full station by using the tangent offset method.

Solution   From the length of the curve the stations of the *P.V.C.* and *P.V.T.* are determined (61 + 00 and 69 + 00, respectively) and from the tangent grades the station elevations along the tangents are obtained. Then the elevation of point *B*, the mid-point of the curve, is calculated as follows:

$$\text{Elevation of } P.V.I. = 264.20 \text{ ft}$$

$$\text{Elevation of } C \text{ (midpoint of long chord)} = \frac{248.20 + 252.20}{2} = 250.20 \text{ ft}$$

$$\text{Elevation of } B = \frac{264.20 + 250.20}{2} = 257.20 \text{ ft}$$

The tangent offset values are calculated as follows:

$$\text{Difference in elevation from } P.V.I. \text{ to } B = 264.20 - 257.20 = 7.00 \text{ ft}$$

$$\text{Tangent offset at station } 62 + 00 = \frac{(1)^2}{(4)^2}(7.00) = 0.44 \text{ ft}$$

$$\text{Tangent offset at station } 63 + 00 = \frac{(2)^2}{(4)^2}(7.00) = 1.75 \text{ ft}$$

TABLE 21-1

Station	Point	Elevation on Grade Lines	Tangent Offset	Curve Elevation	1st Difference	2nd Difference
61	P.V.C.	248.20	0.00	248.20		
					+3.56	
62		252.20	0.44	251.76		0.87
					+2.69	
63		256.20	1.75	254.45		0.88
					+1.81	
64		260.20	3.94	256.26		0.87
					+0.94	
65	P.V.I.	264.20	7.00	257.20		0.88
					+0.06	
66		261.20	3.94	257.26		0.87
					-0.81	
67		258.20	1.75	256.45		0.88
					-1.69	
68		255.20	0.44	254.76		0.87
					-2.56	
69	P.V.T.	252.20	0.00	252.20		

The elevations of the stations are computed by subtracting the tangent offsets from the grade elevations. The resulting values, along with the second difference math check, are recorded in Table 21-1.

21-10 MISCELLANEOUS ITEMS RELATING
     TO VERTICAL CURVES

Elevation of intermediate points on curves. It is often necessary to compute the elevation of points on vertical curves at closer intervals than full stations, for example, at 50 ft or 25 ft, as well as the values for some occasional intermediate point. For such cases as these, the tangent offset method will work as well as it did at full stations, although the numbers are not quite so convenient. As an illustration, the elevation of station 63 + 75.2 for the curve of Example 21-5 equals

$$248.20 + (.04)(275.2) - \frac{(2.752)^2}{(4)^2} (7.00) = 255.90 \text{ ft}$$

Highest or lowest point on curve. When the two grade lines for a vertical curve have opposite algebraic signs, there will be either a low point or a high point between the $P.V.C.$ and the $P.V.T.$ This point will probably not fall on a full station, and yet its position will often be of major significance as, for example, where drainage facilities are being considered. The high or low point (where the slope of the curve is zero) may be obtained by substituting into the following equation:

$$X = -\frac{g_1}{r}$$

where X is the distance in stations from the $P.V.C.$, g_1 is the percentage of slope of the back tangent and g_2 is that of the forward tangent, r is the rate of change of grade per station = $(g_2 - g_1)/L$, and L is the length of the curve in stations.

For the curve of Example 21-5, the high point is as follows:

$$r = \frac{g_2 - g_1}{L} = \frac{-3 - 4}{8} = -\frac{7}{8}$$

$$X = -\frac{g_1}{r} = -\frac{+4}{-7/8} = +4.57 \text{ stations} = 65+57$$

The elevation of the high point equals

$$264.20 - (0.03)(57) - \frac{(3.43)^2}{(4)^2} (7.00) = 257.34 \text{ ft}$$

Example 21-6 provides another vertical curve example. The calculations are made exactly as they were for the curve of Example 21-5. For all vertical curve problems it is necessary for the person making the calculations to be very careful to use the correct signs for determining the elevations along the tangents. He or she must be equally careful with the signs of the offset distances from these tangents, which are used to determine the curve elevations. In this particular case the offsets are measured down from the tangent or grade lines. Space will not be taken here to show the second difference check, which is nevertheless always desirable.

Example 21-6

A −3% grade line intersects a −5% grade line at station 62+00 where the elevation is 862.30 ft as shown in Fig. 21-14. Determine the elevations at full 100-ft stations along the curve if a 600-ft long curve is to be used.

Solution

1. The elevations along the grade lines are computed and shown in the figure.

2. Elevation of midpoint of $L.C.$ = $\dfrac{871.30 + 847.30}{2}$ = 859.30 ft

3. Elevation at midpoint of curve = $\dfrac{862.30 + 859.30}{2}$ = 860.80 ft

4. The tangent offsets are computed and subtracted from the grade-line elevations, giving the final curve elevations as shown in the figure.

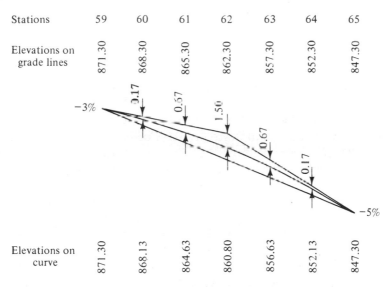

Stations	59	60	61	62	63	64	65
Elevations on grade lines	871.30	868.30	865.30	862.30	857.30	852.30	847.30

−3%   0.17   0.57   1.50   0.67   0.17   −5%

Elevations on curve	871.30	868.13	864.63	860.80	856.63	852.13	847.30

Figure 21-14

21-11 SIGHT DISTANCE

As previously mentioned, it is necessary for vertical curves to be so constructed that drivers will have certain minimum sight distances for seeing cars or other objects on the road. The driver should be able to see an object of a given height at no less than the estimated distance he would travel while reacting to put his foot on the brake pedal plus the distance required for the car to stop. The following formula for the minimum sight distance S

$$S^2 = \frac{8Lh}{g_1 - g_2}$$

applies when S is less than the curve length L in stations and h is the height of the driver's eye above the pavement as specified by the AASHTO.

For an 800-ft curve with $g_1 = 4\%$ and $g_2 = -3\%$ (Example 21-5) and with $h = 3.75$ ft,

$$S = \sqrt{\frac{(8)(8.00)(3.75)}{4.00 - (-3.00)}} = 5.8554 \text{ stations} = 585.54 \text{ ft}$$

Should the preceding expression yield a value greater than L, the object or vehicle sighted will be off the vertical curve and on the tangent and it will be necessary to use the following equation, which involves both a parabola and a straight line.

$$S = \frac{L}{2} + \frac{4h}{g_1 - g_2}$$

21-12 VERTICAL CURVE PASSING THROUGH CERTAIN POINT

A problem commonly faced in working with vertical curves is that of passing a curve through a definite point. For instance, as shown in part (a) of Fig. 21-15 it may be desired to have a vertical curve pass a certain distance over the top of a culvert. The problem is to determine the correct length of a curve which will pass through the point in question. In a similar fashion a vertical curve on a bridge may need to pass a certain distance for clearance over an existing roadway, railroad, or navigable stream as shown in part (b) of the figure. To solve the problem, the tangent offset expression is written from both sides of the curve using the length L as one unknown and the tangent offset at the $P.V.I.$ as the other one. The two equations may be solved simultaneously for the unknown values as illustrated in Example 21-7.

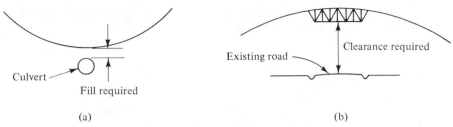

Figure 21-15

Example 21-7

A −6% grade line intersects a +5% grade line at station 11 + 00 (elevation 43.00 ft). The top of a culvert oriented at 90° to the road center line at station 10 + 00 is to be located at elevation 53.00 ft. Determine the vertical curve length in whole stations such that there will be from 1.0 to 3.0 ft of cover material over the top of the culvert.

Solution   Assume the material covering the top of the culvert is 2.0 ft thick and determine the theoretical curve length required. Thus the desired elevation on the curve at that station (10 + 00) is 53.00 + 2.00 = 55.00 ft.

With reference to Fig. 21-16, expressions for the offsets at station 10 + 00 are written with respect to both tangents and labeled y_{10}.

1. The distance from the back tangent to the desired curve elevation = 55.00 − 49.00 = 6.00 ft = y_{10}.

2. The distance from the forward tangent (extended to station 10 + 00) to the desired curve elevation = 55.00 − 38.00 = 17.00 ft = y_{10}.

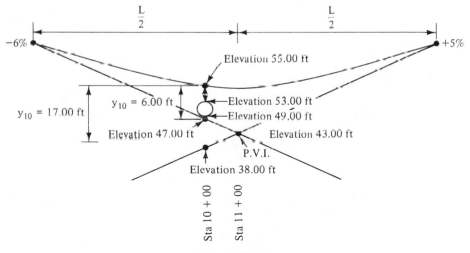

Figure 21-16

In the following two offset expressions, P is the tangent offset at the $P.V.I.$ and L the desired curve length in stations.

For the back tangent:

$$\frac{6.00}{P} = \frac{\left(\dfrac{L}{2} - 1.00\right)^2}{\left(\dfrac{L}{2}\right)^2} \qquad\qquad P = \frac{6.00\left(\dfrac{L}{2}\right)^2}{\left(\dfrac{L}{2} - 1.00\right)^2} \qquad (1)$$

For the forward tangent:

$$\frac{17.00}{P} = \frac{\left(\dfrac{L}{2} + 1.00\right)^2}{\left(\dfrac{L}{2}\right)^2} \qquad\qquad P = \frac{17.00\left(\dfrac{L}{2}\right)^2}{\left(\dfrac{L}{2} + 1.00\right)^2} \qquad (2)$$

Equating the values for P [Eqs. (1) and (2) above] and solving for L:

$$L = 7.88 \text{ stations}$$

Use $L = 8$ stations and calculate elevations along the curve making sure the culvert cover is between 1 and 3 ft.

PROBLEMS

21-1. Two highway tangents intersect with a right deflection angle of $31°10'$. If a $3°30'$ horizontal circular curve is to be used to connect the tangents compute R, T, L, and E for the curve.

   (*Ans.*: $R = 1637.02$, $T = 456.55$, $L = 890.48$, and $E = 62.47$ ft)

21-2. A horizontal circular curve having a radius of 600.00 ft is to connect two highway tangents. If the chord length $L.C.$ is 700.00 ft, compute E, L, M, and the intersection angle I.

In Problems 21-3 to 21-5, prepare the field book notes for laying out the curves with stakes at full stations for the data given.

21-3. $P.I.$ at $64 + 32.7$, $I = 20°00'$, and $D = 3°00'$

   (*Ans.*: $R = 1909.9$, $T = 336.8$, $L = 666.7$, Defl. $\sphericalangle$ @ $64 + 00$ is $4°34'$)

21-4. $P.I.$ at $84 + 21.9$, $I = 28°22'$, and $D = 4°00'$

21-5. $P.I.$ at $102 + 10.3$, $I = 18°32'$, and $D = 2°30'$

   (*Ans.*: $R = 2291.8$, $T = 373.9$, $L = 741.3$, Defl. $\sphericalangle$ @ $102 + 00$ is $4°33'$)

21-6. It is necessary to have a horizontal curve pass through a specified point. The point was located by distance and angle from the $P.I.$ of the curve (150.00 ft and $32°00'$ to the left of the back tangent with the instrument located at the $P.I.$). If the intersection angle between the back and forward tangents is $50°00'$ to the right, determine the curve radius required.

21-7. A 1000-ft vertical curve is to be used for joining a +5.0% grade line and a -3.0% grade line. The vertex or intersection of the grade lines is at station

62 + 00 and the elevation there is 742.55 ft. Compute the elevations at 100-ft stations throughout the curve.

(*Ans.:* Curve elevation at stations 58, 60 and 64 = 722.15, 728.95, 732.95)

21-8. Repeat Problem 21-7 if the curve is 800 ft long and the first grade is +3.6% instead of +5.0%.

21-9. A +4.0% grade meets a −2.1% grade at station 46 + 50 where the elevation is 682.24 ft. Compute the elevation of all full stations if the curve is to be 800 ft long.

(*Ans.:* Curve elevations at stations 44, 46 + 50, 49 = 671.38, 675.57, and
676.13)

21-10. Repeat Problem 21-9 if the second grade is +1.4% instead of −2.1%.

21-11. A −2.6% grade line meets a +3.8% grade line at station 103 + 00. Elevation = 1626.88 ft. Determine the elevation of all full and half stations if a 650-ft curve is to be used.

(*Ans.:* Curve elevations at stations 102 and 104 + 50 = 1631.97 and
1634.09)

21-12. Compute the location and elevation of the high point of the curve of Problem 21-7.

21-13. At what stations is the elevation of the curve of Problem 21-7 equal to 729.55 ft?   (*Ans.:* Stations 60 + 24 and 66 + 26)

21-14. A grade of +4.0% passes station 36 + 00 at an elevation of 622.80 ft. A grade of −5.0% passes station 52 + 00 at elevation 623.40 ft. Compute the station and elevation of the vertex or *P.V.I.* of these grades.

In Problems 21-15 to 21-20, compute the elevation at 100-ft stations throughout the given curve.

	Initial Grade	Final Grade	Station of *P.V.I.*	Elevation of *P.V.I.* (ft)	Length of Curve (ft)
21-15.	−4%	+2%	62+00	642.70	800
21-16.	−3%	+5%	103+00	310.20	1000
21-17.	−5%	−3%	31+00	1422.80	1200
21-18.	3%	−5%	43+00	831.90	600
21-19.	+3%	+6%	91+00	1641.60	600
21-20.	+5%	+2%	52+50	291.70	800

(*Ans. 21-15:* Curve elevations at stations 59, 63, and 65 = 655.08,
648.08 and 649.08)

(*Ans. 21-17:* Curve elevations at stations 26, 30, and 35 = 1447.88,
1429.88 and 1411.13)

(*Ans. 21-19:* Curve elevations at stations 89, 92, and 93 = 1635.85,
1648.60 and 1653.85)

21-21. A −5% slope meets a +2% grade at elevation 162.20 ft at station 61 + 00. Assuming an 800-ft vertical curve, determine the elevation on the curve at each 100-ft station and find the station and elevation at the low point on the curve.

(*Ans.:* Curve elevation at station 59 = 173.95,
low point on curve is 167.92 at station 62 + 71.4)

21-22. A −3.2% grade intersects a +2.1% grade at station 103 + 00 at an elevation of 153.20 ft. If a 1200-ft long vertical curve is to be used to connect the tangent grades determine the elevation at stations 99 + 00 and 104 + 00.

21-23. A −4% downgrade intersects a +3% upgrade at elevation 120.00 ft at station 67 + 00. It is desired to pass a vertical curve through a point of elevation 128.00 ft at station 68 + 00. Determine the curve length required.

(*Ans.:* 928.3 ft)

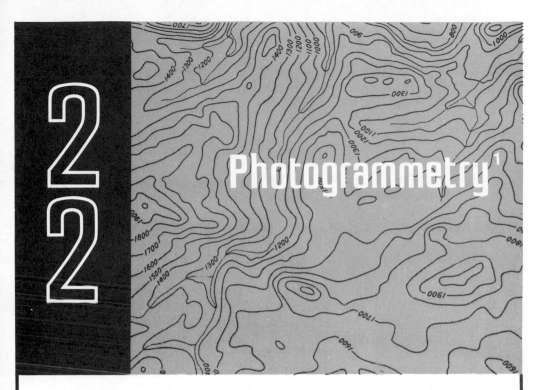

22 1 INTRODUCTION

Photogrammetry is a field of study that is very closely related to surveying. Photogrammetry is a tool that can be used effectively as an economical substitute for certain field surveying operations. Also, a certain amount of field surveying is necessary to establish the system of horizontal and vertical control points that is required for many photogrammetric operations. Photogrammetry has been shown to be a faster and less costly method of producing topographic base mapping for many large engineering projects. Photogrammetry has almost completely replaced field methods of topographic mapping for large areas. However, topographic maps for small areas (less than 100 acres) can be produced at lower cost by the field surveying techniques previously described in Chapter 16. Photogrammetry is a tool that is not only useful to surveyors and engineers but also to many other disciplines such as forestry, geology, archaeology, urban planning, geography, and agriculture. The purpose of this chapter is to provide surveying students with a basic introduction to the field of photogrammetry and its relationship to surveying.

---

[1] Written by Dr. Donald B. Stafford, Associate Professor, Department of Civil Engineering, Clemson University, Clemson, S.C.

There is some degree of confusion regarding the definition of the term photogrammetry. This confusion occurs because photogrammetry has been used to define a very broad area of study that involves the extraction of quantitative measurement data and qualitative interpretation data from aerial photographs, as well as a more restricted area of study that applies only to the quantitative use of aerial photographs. This confusion can be overcome by using terms to identify specific types of photogrammetry. However, this practice has not been generally accepted or used widely by the engineering profession and other disciplines that use photogrammetry.

In the broad context, photogrammetry is defined by the American Society of Photogrammetry as the art, science, and technology of obtaining reliable information about physical objects and the environment through processes of recording, measuring, and interpreting photographic images and patterns of radiant energy and other phenomena. This broad definition can be subdivided into two components: *metric photogrammetry* and *interpretative photogrammetry*. Metric photogrammetry is generally understood to refer to the use of measurements made on aerial photographs to obtain quantitative data about the earth's surface. In many instances, the end product that results from metric photogrammetry operations is a map of a project area.

The term interpretative photogrammetry is used to describe the situation in which aerial photographs or images produced by electronic sensors are carefully studied to produce an interpretation of the conditions existing in the area covered by the photographs or images. The product of interpretative photogrammetry operations is usually a map showing the distribution of soil types, rock types, land use, vegetation species, or other environmental factors. For many years, this field of study was known as aerial photographic interpretation and this term is still appropriate when only aerial photographs are being interpreted. Since the mid-1960s, the term *aerial remote sensing* or simply *remote sensing* has been used to identify a broader area of application in which the product being interpreted may be an image produced by an electronic sensing device or an aerial photograph. Thus, aerial remote sensing includes aerial photographic interpretation and also the interpretation of a variety of other sensor products such as infrared imagery, radar imagery, multispectral imagery, satellite photography, satellite imagery, and other products.

A much simpler definition of photogrammetry is widely used by those involved in the practical applications of photogrammetry and by many users of photogrammetric products. This definition states that photogrammetry is the art and science of measuring elevations, distances, and directions from aerial photographs and a minimum amount of field work or known ground control. In this context, the ultimate objective of most photogrammetric operations is the production of a base map. In the remainder of this chapter, the term photogrammetry is used in the context of this simple definition or as a shorter phrase for the term metric photogrammetry.

The use of aerial photographs is a fundamental element in photogrammetric operations. An aerial photograph is defined as a photograph made from an aerial platform which shows a portion of the earth's surface to some scale. The aerial platform is usually an airplane, although it may be a balloon or a spacecraft. The scale of an aerial photograph is a very important characteristic because it influences the accuracy that can be obtained from many photogrammetric operations. An understanding of the characteristics of aerial photographs is a prerequisite to the study of photogrammetry.

22-3 TYPES OF AERIAL PHOTOGRAPHS

Aerial photographs can be divided into two groups: vertical aerial photographs and oblique aerial photographs. Vertical aerial photographs are taken with the optical axis of the aerial camera lens pointed along a line perpendicular to the earth's surface. The geometry of vertical aerial photographs is relatively simple and therefore vertical aerial photographs are usually preferred for photogrammetric applications. Although few aerial photographs strictly fulfill the definition of vertical aerial photographs, aerial photographs that have less than three degrees of tilt are considered to be vertical for many applications. Figure 22-1 shows a vertical aerial photograph of the Washington, D.C., area.

Oblique aerial photographs are photographs taken with the optical axis of the camera lens purposely tilted at an angle other than 90° to the earth's surface. The angle of tilt usually ranges between 30° and 60° for oblique aerial photographs. It is common practice to designate oblique aerial photographs as either low or high obliques on the basis of the nature of the image of the earth's surface depicted in the photograph. Low oblique aerial photographs do not have an image of the horizon in the photograph because small values of the tilt angle are used. High oblique aerial photographs show the earth's horizon because the angle of tilt is relatively large.

Although oblique aerial photographs can be used for certain photogrammetric applications, these photographs are not used extensively because of the complex geometry of the photograph. The primary problem with oblique aerial photographs is that the scale of the photograph

Figure 22-1 Vertical aerial photograph of the Washington, D.C., area. *(Photograph courtesy of the U.S. National Ocean Survey.)*

varies throughout the photograph. This scale variation means that measurements on the oblique photograph cannot be easily converted to distances on the ground. Oblique aerial photographs can be used effectively to portray visual information about terrain features because the oblique view is more easily understood by the public than is the plan view contained in vertical aerial photographs.

Other classifications of aerial photographs can be made based on a variety of factors such as film type, type of aerial camera, altitude of the aerial platform, and other parameters. However, these classifications are not sufficiently important to warrant detailed coverage in a basic introduction to photogrammetry.

22-4 SCALE OF AERIAL PHOTOGRAPHS

As mentioned earlier, the scale of an aerial photograph is a basic parameter. The scale of the photograph is the property that relates distance measurements made on the aerial photograph to ground distances. One approach to expressing the scale of an aerial photograph is to use a dimensionless ratio of distance on the photograph to a corresponding distance on the ground. The scale expressed in this manner is usually referred to as a *representative fraction*. This form of the scale has the advantage that any system of units can be used as long as the same units are used for both the distance on the photograph and the distance on the ground. Examples of the scale expressed in this manner are typical aerial photograph scales of 1/6000 (or 1 : 6000) and 1/20,000 (or 1 : 20,000).

Another approach to expressing the scale of an aerial photograph is the method commonly used by engineers in which the number of feet on the ground that corresponds to a one-inch distance on the photograph is given. This form of the scale is commonly referred to as an *engineer's scale*. Although the engineer's scale is usually computed in the form of x feet per inch, the scale can be expressed as 1 inch equals x feet.

The simplest way to look at the scale of an aerial photograph is to assume that the ground surface depicted in the photograph is a plane, as shown in Fig. 22-2. This condition would be representative of the situation in which the ground surface being photographed consists of level terrain. By using similar triangles, the following equations for aerial photograph scales can be developed:

$$S_{RF} = \frac{\text{Photo distance}}{\text{Ground distance}} = \frac{f}{H'}$$

$$S_{ENGR} = \frac{\text{Ground distance}}{\text{Photo distance}} = \frac{H'}{f}$$

where S_{RF} is the scale expressed as a representative fraction, S_{ENGR} is the scale expressed as an engineer's scale, f is the camera focal length, and H' is the height of the camera above the ground surface.

In computing the scale as a representative fraction, the same units must be used for both numerator and denominator so that a dimension-

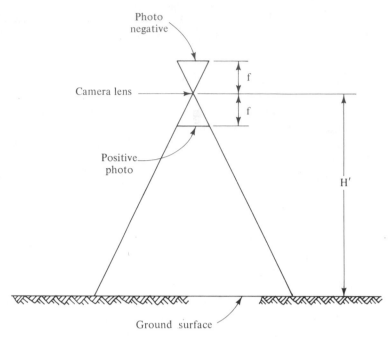

Figure 22-2 Simplified cross section of the geometry of a vertical aerial photograph in flat terrain.

less fraction will be obtained. In computing the scale as an engineer's scale, the numerator must be in feet and the denominator must be in inches. Conversion from one form of the scale to the other form can readily be made by converting the units and inverting the values. Special care must be exercised in computing aerial photograph scales to ensure that errors in units are not made.

There are instances where the user of an aerial photograph does not have information on camera focal length or aircraft altitude above the ground. In this situation, the aerial photograph scale can be computed by using known or measured ground distances between points that can be identified on the aerial photograph. For example, the distance between highway intersections or public land survey lines can be measured on the aerial photograph and related to the corresponding ground distance. The ground distance can be obtained by ground measurement, maps, or known public land survey distances. Example 22-1 illustrates this procedure.

Example 22-1

The distance between two adjacent street intersections on an aerial photograph of an urban area has been measured as 2.25 in. The block lengths are known to be 900 ft long in the area shown on the photograph. Compute the scale of the aerial photograph as an engineer's scale and as a representative fraction.

Solution

$$S_{ENGR} = \frac{\text{Ground distance}}{\text{Photo distance}} = \frac{900 \text{ ft}}{2.25 \text{ in.}} = 400 \text{ ft/in.}$$

$$S_{RF} = \frac{\text{Photo distance}}{\text{Ground distance}} = \frac{2.25 \text{ in.}}{900 \text{ ft } (12 \text{ in./ft})} = \frac{1}{4800}$$

   Aerial photograph scales are influenced by topography or changes in elevation. Special care must be taken in computing the scale of aerial photographs in areas of rolling terrain because of this effect. Figure 22-3 is a schematic diagram that can be used to show the effect of elevation on scale. The equations for photograph scale now contain an elevation term as follows:

$$S_{RF} = \frac{f}{H - h}$$

$$S_{ENGR} = \frac{H - h}{f}$$

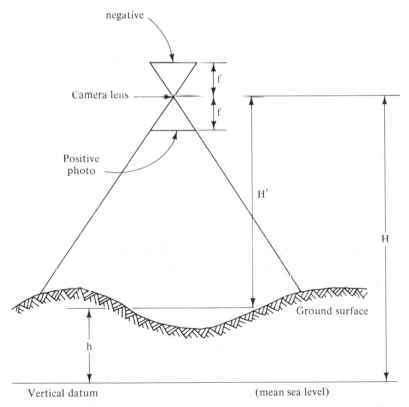

Figure 22-3 Cross section of a vertical aerial photograph in rolling terrain.

where h is the elevation of the point for which the scale is being computed, and H is the height of the aircraft above a vertical datum such as
mean sea level. S_{RF}, S_{ENGR}, and f are as defined previously.

These equations show that for all points, the aerial photograph
scale is dependent on the elevation. In other words, a particular scale
is only applicable for points at the same elevation. However, for many
applications, the average scale can be computed for an aerial photograph by using the average elevation of the terrain. The user of aerial
photographs should recognize that an error is introduced when the
average scale is used for points having an elevation significantly different from the average elevation of the terrain.

22-5 RELIEF DISPLACEMENT

The geometry of aerial photographs is different from the geometry of a
topographic or planimetric map. This is related to the fact that aerial
photographs are central perspectives while maps are orthographic
perspectives. One important effect of this difference is that aerial
photographs exhibit *relief displacement.* Relief displacement is the
condition in which the image of a point is displaced from its true map
position. The magnitude of the relief displacement of a point is directly related to the relief or elevation difference of the point and some
elevation reference. Relief displacement is best illustrated by observing
the photograph of a tall thin object such as a flagpole, as shown in
Fig. 22-4.

On an aerial photograph, the image of the top of the flagpole
appears at a different location than the image of the bottom of the
flagpole. The difference in the location of the two images represents
the amount of relief displacement. It can be shown that the relief displacement is always along a radial line from the center of vertical
photographs. Points above the average terrain elevation are displaced
outward from the center of the photograph and points below the
average elevation are displaced inward toward the center. The amount
of the relief displacement can be computed by using the following
equation:

$$d_r = \frac{rh}{H'}$$

where d_r is the relief displacement, r is the radial distance from the
center of the photograph to the displaced image point, h is the height
of the object, and H' is the height of the aircraft above the ground. In
using this equation, the units of the relief displacement will be the same
as the units of the radial distance measurement, while h and H' can be
in any units as long as the same units are used.

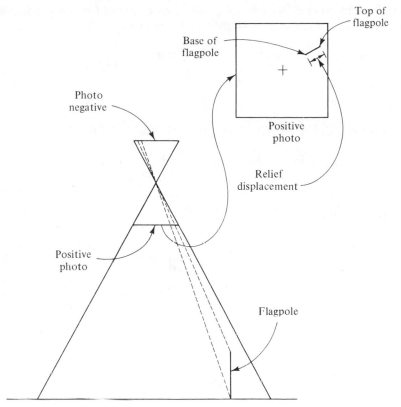

Figure 22-4 A schematic aerial photograph of a flagpole illustrating the concept of relief displacement.

The relief displacement equation can be used to determine the height of objects for certain situations in which the relief displacement can be measured on the aerial photograph. This condition exists when the image of both the top and bottom of an object can be observed on the photograph and the top of the object occurs vertically above the bottom. Features such as flagpoles, utility poles, some transmission towers, some smokestacks, and corners of buildings fulfill these criteria. The use of this approach to determine the height of objects is illustrated in Example 22-2.

Example 22-2

The relief displacement of the corner of a building was measured on a vertical aerial photograph and found to be 0.18 in. The top of the building was measured to be 2.95 in. from the center of the photograph. The photograph was taken from a height of 3000 feet above the ground. Compute the height of the building in feet.

Solution   The relief displacement equation can be solved for h to give the expression: $h = d_r H'/r$. Therefore, the height of the building is:

$$h = \frac{d_r H'}{r} = \frac{(0.18 \text{ in.})(3000 \text{ ft})}{2.95 \text{ in.}} = 183 \text{ ft}$$

22-6 STEREOSCOPIC VIEWING OF AERIAL PHOTOGRAPHS

One of the most important concepts of photogrammetry is the concept of stereoscopic viewing of aerial photographs. When two aerial photographs taken under the proper conditions are viewed in a certain specified manner, the viewer perceives a three-dimensional image of the terrain in the photograph. The three-dimensional view permits the study of the shape of the terrain as it relates to engineering projects and also provides the basis for topographic mapping by photogrammetric techniques.

There are three requirements for stereoscopic viewing:

1. The two aerial photographs must provide two views of the terrain taken from different positions.
2. The two aerial photographs must be oriented properly for viewing.
3. The viewer must have normal binocular vision.

The first requirement is usually fulfilled by the procedure that is used to take aerial photographs of project areas. As the aircraft flies along a particular line (the flight line) over the area being photographed, aerial photographs are taken at periodic intervals such that there is at least 50% overlap between adjacent photographs along the flight line, as shown in Fig. 22-5. In order to ensure that at least 50% overlap is achieved, the time interval between photographs is usually selected so that about 60–65% overlap is obtained under normal conditions.

The second requirement is fulfilled by orienting the two aerial photographs being viewed stereoscopically in the same relative position that the camera existed in the field at the time the aerial photographs were taken. This means that the two photographs must be overlapped along the flight line with the overlapping images superimposed. Then the two overlapping images are separated along the flight line by the proper amount to satisfy the requirements for the type of stereoscopic viewing device being used.

The third requirement is fulfilled by anyone who has vision in both eyes, even if the vision must be corrected by glasses. Vision in

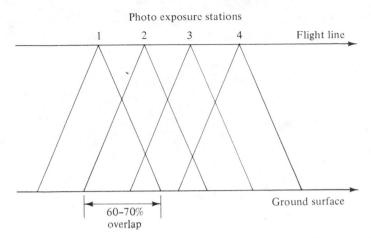

Figure 22-5 Overlapping of aerial photographs along the flight line to permit stereoscopic viewing.

both eyes is necessary so that the two different images of the terrain contained in the separate overlapping aerial photographs can be transmitted to the brain by the two eyes of the viewer. The stereo model of the terrain is then perceived by the brain based on the slight differences in the two views of the terrain recorded on the aerial photographs.

A special viewing device is commonly used as an aid in the stereoscopic viewing of aerial photographs. The primary purpose of stereoscopic viewers is to insure that the left eye sees only the aerial photograph on the left and the right eye sees only the aerial photograph on the right. Two common types of stereoscopic viewers are the small lens or pocket stereoscope that use refraction to produce the desired effect and the larger mirror stereoscope that use reflection to produce the desired result. Aerial photograph users can also teach themselves to see the stereoscopic view without the aid of a stereoscope by learning to fix the image perceived by each eye on the two separate overlapping aerial photographs, a process called *stereoscopic fusion*. However, considerable practice is necessary to achieve this capability.

An example stereogram that has been prepared from a pair of overlapping aerial photographs is shown in Fig. 22-6. The stereogram can be viewed with a pocket stereoscope or by stereoscopic fusion to observe the three-dimensional shape of the terrain covered by the photographs. The stereoscopic view that is obtained from this stereogram illustrates a common characteristic of such views, the vertical exaggeration that occurs when aerial photographs are used to produce a three-dimensional image.

Figure 22-6 Stereogram prepared from a pair of overlapping large scale aerial photographs. (*Photographs courtesy of the Department of Civil Engineering, Clemson University.*)

22-7 PLANNING AERIAL PHOTOGRAPHY MISSIONS

Careful planning must be exercised in developing the specifications and flight plans for aerial photography missions. The parameters that are chosen for the aerial photographs are very important because these values influence the accuracy of the various photogrammetry operations which use the photographs. There are a number of interrelated variables that must be selected to define the characteristics of the aerial photographs that will be produced by a particular aerial photography mission. Also, a detailed flight plan must be prepared to define the area of coverage and the spacing between flight lines and between photographs.

The scale of the aerial photographs must be selected based on the use that is to be made of the photographs. For example, if topographic maps at a particular scale are to be prepared by photogrammetric techniques using a specific type of stereoscopic plotter (see Section 22-10), the scale of the aerial photographs to be used will be generally defined. Once the scale is selected, the combination of aerial camera focal length and flying altitude above the ground can be determined based on a number of considerations.

Two important parameters that must be selected are end-lap between adjacent aerial photographs along the flight line and side-lap between adjacent flight lines. These values are usually expressed in percents and can vary over a relatively narrow range depending on several considerations. End-lap values of between 60 and 70% are common and side-lap values of 20 to 40% are typical. The end-lap is an important variable in determining the photograph spacing along the flight line and the side-lap is important in determining the spacing between flight lines.

Figure 22-7 shows a typical flight plan for an aerial photography mission. Project areas are usually flown in a series of parallel north–south flight lines. One exception to this general rule is the case of rectangular project areas that are long in the east–west direction and short in the north–south direction in which case east–west flight lines are used. For very irregular project areas, other flight line orientations may be adopted. The cost of flying the mission is lowest when the number of flight lines is minimized so that time and aircraft operational expense involved in the turning operation between flight lines is reduced.

The base map that is used for the flight map must be chosen carefully. The primary purpose of the flight map is to insure that the flight crew can maintain the flight line alignment and start and stop the camera at the proper location. Therefore, the flight line base map must show a sufficient amount of natural and man-made detail that can be

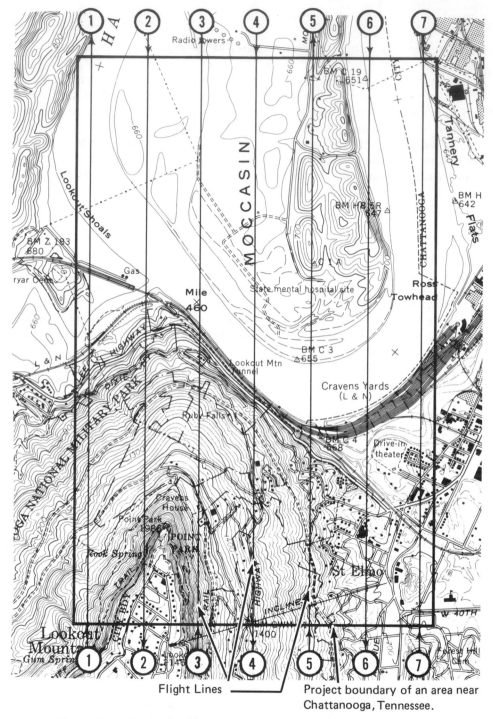

Flight Lines ———

Project boundary of an area near Chattanooga, Tennessee.

Figure 22-7 Example of a flight line map of an area near Chattanooga, Tennessee. (*Base map courtesy of the Tennessee Valley Authority.*)

identified from the air to orient the crew adequately. In developed
areas, the flight crew can usually orient the flight lines with reference
to highways, railroads, buildings and other man-made features. However, in undeveloped areas the flight crew may have trouble relating
the base map features to the view observed from the air. In these instances, a flight line base map that consists of an aerial mosaic or other
available small-scale aerial photographs may be more suitable.

Another important aspect of flight planning is the preparation of a
set of specifications for the aerial photographs. These specifications
define the parameters that the aerial photographs should have and provide information on the limits in these values that will be permitted
before the aerial photographs would be rejected. The specifications
cover a number of aerial photograph characteristics including scale,
camera focal length, percent of end-lap, percent of side-lap, tilt, and
other parameters. The tilt values are particularly important since the
assumption of vertical aerial photographs may not be applicable if the
tilt values are too high.

There are a number of other considerations in flight planning that
are beyond the scope of this chapter. Several of the references at the
end of this chapter provide more detailed information on flight planning for aerial photography missions.

22-8 PHOTOGRAMMETRIC CONTROL SURVEYS

Probably the most direct relationship between surveying and photogrammetry is the situation in which field surveying techniques are used
to provide the necessary ground control data for photogrammetric
operations. Many photogrammetric operations require that the coordinates and elevations of a certain number of points be known. Ground
control data requirements are usually given in terms of the number of
horizontal control points for which the map coordinates are needed
and the number of vertical control points for which the elevations are
needed. Although a primary advantage of photogrammetric techniques
is the reduction in total project costs by reducing the amount of field
surveying required, there will always be a need for some minimum
amount of surveying to generate the needed ground control data. The
cost of photogrammetric control surveys generally ranges from 20%
to 50% of the total cost of photogrammetric mapping projects. Therefore, special attention must be given to the planning and supervision
of photogrammetric control surveys to minimize the cost of these
operations.

The surveying techniques used to develop horizontal ground control points require the measurement of distances and angles so that the

coordinates of the points can be computed. The most common coordinate system that is used for horizontal ground control points is the state plane coordinate system (see Chapter 24), although latitude and longitude or other coordinate systems can be used. The types of surveying operations required to produce the horizontal ground control data can consist of triangulation, traverses, trilateration (see Chapter 23), or other similar work. The surveying technique selected for the horizontal control surveys depends on a number of factors including the topography, the extent of the highway and railroad network in the project area, and density of vegetation. It is important to select the surveying technique that will provide the horizontal control data with the required accuracy at the lowest possible cost. Electronic distance measurement equipment has become widely accepted and used for horizontal control surveys.

Differential or spirit leveling is the usual surveying technique used for vertical control surveys. The leveling circuit should begin and end at bench marks with known elevations so that error adjustments can be made. In rugged terrain, trigonometric leveling is sometimes employed. For small-scale mapping projects in which the accuracy of the vertical control points is not critical, barometric leveling and other techniques have been used satisfactorily. New electronic devices that can provide three-dimensional coordinates of points based on signals from orbiting satellites have been developed which could revolutionize photogrammetric control surveys.

It is critical that surveyors who are performing photogrammetric control surveys understand how the ground control data is to be used. With a knowledge of how the ground control data is applied in various photogrammetric operations, the surveyor can locate the ground control points in the optimum location, insure that the required accuracy is provided, and select points in such a manner that errors can be minimized. It is particularly important that the surveyor understand the critical influence that the ground control data has on the accuracy of photogrammetric operations. The equipment and surveying techniques employed should be selected to provide the necessary accuracy at the lowest possible cost. Providing an unrealistic level of accuracy in the ground control data is uneconomical.

The photogrammetrist should also have an understanding of surveying equipment and techniques used for control surveys. By having a knowledge of surveying procedures, the photogrammetrist can avoid requesting unnecessary data and making unrealistic accuracy demands that would significantly increase the cost of ground control surveys. By working closely and cooperatively, the surveyor and photogrammetrist can ensure that the best and most economical set of ground control data is obtained.

One very important consideration in ground control surveys is to insure that the ground control points can be located and defined very accurately on the aerial photographs being used. This usually means that clearly defined objects with sharp boundaries should be used for control points. In many areas there are sufficient numbers of man-made objects such as sidewalk and highway intersections, utility poles, fence corners, isolated trees, and other similar objects so that the required number of ground control points can be located in the proper location on the photographs. In undeveloped areas artificial targets are often used to define the control points. This procedure, known as pre-marking or paneling, involves placing special shapes on the ground at the proper location before the aerial photographs are taken. Although this procedure is somewhat expensive, it can be very effective in improving the system of ground control points and minimizing errors in locating ground control points that would reduce the accuracy of the photogrammetric operations.

Surveyors who plan to become involved in photogrammetric control surveys should consult a good photogrammetry textbook to learn about the control data requirements for various photogrammetry operations. A photogrammetry textbook will describe the number of horizontal and vertical control points required and the preferred location of the points within the aerial photographs for different photogrammetry operations.

22-9 MOSAICS AND ORTHOPHOTOS

One photogrammetric product that is useful for a variety of purposes is a *mosaic*. A mosaic is defined as an assembly of overlapping aerial photographs into a continuous picture of an area. Mosaics are often prepared of certain geographic areas of interest such as cities or project areas so that the area of concern is shown conveniently in a single image rather than having to handle several photographs. The characteristics of aerial mosaics vary over a wide range depending on the time and effort that is expended in preparing the aerial photographs and assembling the moasic.

Aerial mosaics are commonly divided into three classes: uncontrolled, semi-controlled, and controlled. The accuracy of measurements made on mosaics are lowest for uncontrolled mosaics, intermediate for semi-controlled mosaics, and highest for controlled mosaics.

Uncontrolled mosaics are prepared by assemblying contact print aerial photographs while matching the photographic detail of adjacent photographs as carefully as possible. Because of errors due to scale variation between photographs, tilt, aerial camera lens distortion, and other sources, the photographic detail does not match exactly and errors are introduced into the mosaic. No attempt is made to match specific features on the aerial photographs to known ground control points in an effort to improve the accuracy of uncontrolled mosaics. Uncontrolled mosaics are satisfactory for many applications in which high accuracy is not of critical importance.

Semi-controlled mosaics represent the next highest level in accuracy and cost of preparation. Semi-controlled mosaics can be prepared by using aerial photographs that have been rectified to correct for tilt and projection printed to a uniform scale to correct for scale variation between photographs. Alternatively, semi-controlled mosaics can be prepared by matching features on nonrectified and nonratioed aerial photographs to plotted ground control points. Semi-controlled mosaics represent a significant improvement in accuracy over uncontrolled mosaics but the cost of preparation is considerably higher.

Controlled mosaics represent the highest quality of preparation and the highest level of accuracy is produced. First, the individual aerial photographs are rectified to remove errors due to tilt and the photographs are ratioed or printed to a uniform scale to remove scale variation errors. Also, the coordinates of known ground control points are plotted on the base map and the features corresponding to these points are matched to the plotted points. The result is a mosaic in which the photographic detail matches very closely between photographs. Therefore accurate measurements are possible on controlled mosaics. However, the cost of preparing controlled mosaics is also high. It should be noted that the accuracy of mosaics is limited by the fact that each individual photograph contains relief distortions.

Only the central portion of each aerial photograph is used in mosaic preparation in order to improve the matching of photographic detail between adjacent photographs. By trimming and discarding the outer portions of each photograph, the portions of the photographs where tilt errors, lens distortion errors, and other errors are most pronounced are eliminated. The individual photographs are glued to a rigid material such as plywood or masonite board to provide support. The resulting continuous photograph of an area can be used for a variety of purposes.

A special type of photogrammetric product that has become more common in recent years is the *orthophoto*. Orthophotos show the photographic images of objects in the terrain in the form of an orthographic perspective as contrasted with the central perspective view provided in aerial photographs. The most significant characteristic of orthophotos is that the relief distortions have been removed. Therefore, accurate measurements of distances, directions, and areas can be made directly on the orthophotos because image displacements caused by relief distortions do not occur. The preparation of orthophotos involves the differential rectification of aerial photographs using techniques that are beyond the scope of this discussion.

Orthophotos can be used individually or more commonly assembled into mosaics of several orthophotos. Such mosaics are known as *orthophotomosaics*. Since orthophotos and orthophotomosaics show the photographic detail in its true map position, contours (also produced by photogrammetric techniques) showing the shape of the terrain can be superimposed on the mosaic to offer another dimension of terrain information. Orthophotos and orthophotomosaics have the important advantages of showing the complete photographic detail in correct position and including the superimposed contours for terrain elevation data. These products are beginning to be widely accepted as a map base for a variety of applications. Many engineering projects such as route selection, drainage design, and transmission line layout are now accomplished using orthophoto products for base maps. This product will continue to replace the more common topographic mapping techniques for a wide variety of mapping projects because of the inherent advantages of orthophotos.

22-10 STEREOSCOPIC PLOTTERS AND TOPOGRAPHIC MAPPING

The most important application of photogrammetry is the production of mapping products from overlapping aerial photographs. Although photogrammetric techniques are sometimes used to produce planimetric maps which show the true map positions of natural and man-made features, the most common photogrammetric product is topographic maps which show the shape of the terrain by the use of contours and also include planimetric detail. The photogrammetric plotters that are used for map production are sophisticated instruments which employ the concept of stereoscopic viewing of aerial photographs to provide a three dimensional view of the terrain to allow contour lines of equal elevation to be plotted. Figure 22-8 schematically illustrates the concept of a stereoscopic plotter.

The widespread substitution of photogrammetric techniques using stereoscopic plotters for field topographic mapping methods has occurred because of the lower cost of the photogrammetric approach on large mapping projects. Many aerial survey firms, engineering consulting firms, and government agencies have developed topographic mapping programs by purchasing stereoscopic plotters and training employees in photogrammetric mapping techniques. Several United States government agencies that have topographic mapping functions

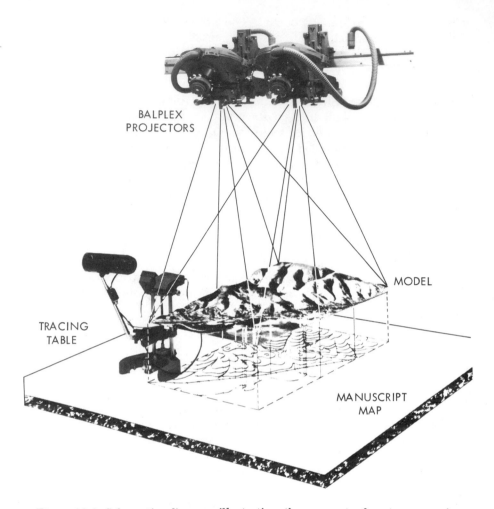

BALPLEX
PROJECTORS

MODEL

TRACING
TABLE

MANUSCRIPT
MAP

Figure 22-8 Schematic diagram illustrating the concept of a stereo scopic plotter. (*Courtesy of Bausch & Lomb.*)

such as the U.S. Geological Survey and the Tennessee Valley Authority have been particularly aggressive in adopting and improving photogrammetric mapping procedures.

There are a number of different types of stereoscopic plotters that have been developed by various equipment manufacturers in the United States and Europe. These plotters vary considerably in cost, accuracy, and operating procedures. However, a relatively standard type of plotter can be used to describe the common characteristics of stereoscopic plotters and photogrammetric mapping techniques. Typical plotters used widely by aerial survey firms and government agencies in the United States are the Kelsh plotter and the ER-55 plotter and Balplex plotter manufactured by Bausch and Lomb.   Figure 22-9 shows a modern Kelsh plotter.

A basic component of all photogrammetric mapping techniques is a pair of overlapping aerial photographs.   The photographs are usually taken with about 60–70% overlap along the flight line so that they can be viewed stereoscopically in the plotter.   For most plotters the aerial photographs are printed in the form of positive transparencies on a glass plate.   This form of aerial photograph is known as a diapositive and is used to provide a stable and accurate photograph.   The glass plates are very flat and are not affected significantly by changes in temperature and moisture conditions.   Some plotters use diapositives that are the same size as the original aerial photograph while other plotters use reduced-size diapositives that are produced by projection printing.

A typical stereoscopic plotter is composed of three primary systems: the projection system, the viewing system, and the measuring and plotting system.   The projection system is composed of the lighting and optical components that are necessary to project images of the aerial photographs contained in the diapositives in such a manner that a stereoscopic view of the terrain is produced.   The projection system

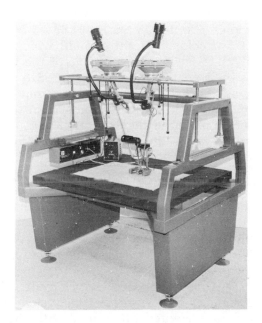

Figure 22-9 Modern Kelsh plotter, Model 5030-B, a common type of stereoscopic plotter for topographic map production. (*Courtesy of the Kelsh Instrument Division of Danko Arlington, Inc.*)

usually consists of a light source placed above the diapositive and a lens system to ensure that the projected image is focused properly.

The viewing system is the mechanism that permits the operator to observe the stereoscopic view produced by the projection system. The primary function of the viewing system is to force the left eye of the operator to view the image projected from the left diapositive and the right eye to view the image projected from the right diapositive. This permits the operator to perceive the three-dimensional shape of the terrain, generally referred to as a stereomodel. Several different viewing systems are available and considerable changes in viewing systems have occurred in recent years.

For many years, the most common type of viewing system was the anaglyphic system. In this viewing system, a red filter is placed in the left projector and a blue-green filter is placed in the right projector so that a red image is projected from the left diapositive and a blue-green image is projected from the right diapositive. When the operator wears a pair of glasses composed of a red filter glass over the left eye and a blue-green filter glass over the right eye, the operator perceives the stereomodel. Because of the colors imparted to the projected images, the anaglyphic viewing system cannot be used with color aerial photographs. Also, the filters reduce the amount of light projected and serve to darken the model. However, this viewing system is relatively simple and inexpensive and has been used in a variety of stereoscopic plotters over the years.

In recent years two new viewing systems have been developed and widely adopted because of the inherent advantages of permitting the use of color diapositives and having much less light loss than the anaglyphic viewing system. One of the modern viewing systems is known as the stereo-image alternator or SIA system. This system uses rotating synchronized shutters which alternatively block and permit the passage of light from the projection system to the operator. One pair of shutters is placed in the projection system and the operator views the projected image through a second pair of shutters. By synchronizing the rotating shutters such that the left projection shutter is open when the left viewing shutter is open, the operator observes first the image from the left diapositive and then the image from the right diapositive. The shutters open and close so rapidly that the operator is not aware of the alternating images and the stereomodel is perceived. Stereoscopic plotters that previously used anaglyphic viewing systems can be retrofitted with SIA viewing systems and many such conversions have been made.

Another modern viewing system is the polarized platen viewing system or PPV system. This system uses polarized light filters rather than the colored filters used with the anaglyphic viewing system. A pair of polarized filters of opposite polarity are used in the projectors and the operator views the projected images on the platen through a pair of eyeglasses with polarized filters over each eye. The PPV viewing system can also be retrofitted on plotters that originally used the anaglyphic viewing system and many such modifications have been made.

The measuring and plotting system provides a means of locating a specific point in the stereomodel and plotting this position on a map manuscript. The system permits planimetric features to be plotted in the correct map position and allows contours to be plotted to define the topography of the terrain. The most common type of measuring and plotting system consists of a tracing table which is composed of a platen, a floating dot reference mark in the center of the platen, a geared device for recording the vertical position of the platen (equivalent to elevations in the stereoscopic model), and a pencil to plot the position of the tracing table on the map manuscript. The platen is a white disk on which the diapositive images are projected so that a small area of the stereomodel can be viewed. The floating mark which consists of a very small dot of light can be made to rise or drop until it is in contact with the three-dimensional surface of the stereomodel by raising or lowering the platen. By setting the elevation dial at a specific even value and keeping the floating dot in contact with the surface of the stereomodel, a particular contour can be located and plotted by moving the tracing table with the plotting pencil extended to contact the manuscript. In a similar manner, the true map position of planimetric features can be plotted by following the outlines of the features while maintaining the floating dot in contact with the three-dimensional surface of the stereomodel and keeping the plotting pencil in contact with the manuscript. There are many other detailed aspects of stereoscopic plotters that cannot be covered herein because of space limitations. The student interested in learning more about photogrammetric techniques for topographic mapping should consult one of the photogrammetry textbooks in the list of references at the end of this chapter.

The use of stereoscopic plotters to produce topographic maps is a very versatile approach. A wide range of map scales can be accommodated by taking aerial photographs at various altitudes with different focal length cameras. For instance, topographic maps with scales as large as 1 in. equals 20 ft and a contour interval of one foot are often prepared in densely developed urban areas. On the other hand, topographic maps with scales as small as 1 in. equals 2000 ft and contour intervals of 10 or 20 ft are common for the $7\frac{1}{2}$-minute quadrangle series of maps. More common scales for engineering projects range from 1 in. equals 100 ft with a two-foot contour interval to 1 in.

equals 500 ft with a ten-foot contour interval. It is important to recognize that the proper contour must be matched with the proper scale when producing topographic maps by photogrammetric techniques.

22-11 SOURCES OF AERIAL PHOTOGRAPHS

Although many photogrammetric projects require that a new set of aerial photographs be taken to fulfill the specific needs of the project, other projects can be accomplished with existing aerial photographs. In particular, aerial photographic interpretation projects can often use existing aerial photographs. If existing aerial photographs with the proper characteristics can be identified and used, a considerable savings can usually be realized. If a new set of aerial photographs must be taken, the total cost of the aerial photography mission must be borne by the purchaser. On the other hand, existing aerial photographs can usually be purchased for the cost of reproduction or only a slightly higher cost.

The user of aerial photographs should realize that a number of aerial survey firms and government agencies take aerial photographs for a variety of purposes. Therefore, most areas in the United States have been photographed several times by different organizations. In many instances one or more of the existing coverages will have a combination of dates, scales, film type, and overlap that are suitable for a particular application.

The organization with the largest inventory of aerial photographs is the Agricultural Stabilization and Conservation Service (ASCS) of the U.S. Department of Agriculture. This agency and its predecessor agencies produced aerial photographs of most counties in the U.S. that had a significant amount of agricultural land use on approximately a five-year repeat cycle over the past 40 years. Other U.S. Department of Agriculture agencies that have also produced significant amounts of aerial photography are the Soil Conservation Service and the U.S. Forest Service. Various U.S. government agencies that have significant aerial photography inventories are listed in Table 22-1.

Many state governments also have agencies with active aerial photography programs. The departments of transportation and highway agencies in many states have a photogrammetry department that has produced aerial photographs of large areas in the state. In some states, the state natural resources agency may have a large inventory of aerial photographs. Many county or municipal governments have a planning, engineering, or public works agency that maintains files of aerial photographs.

Many aerial survey firms have extensive inventories of aerial

TABLE 22-1 United States Government Agencies
with Large Inventories of Aerial Photography

U.S. Department of Agriculture
  Agricultural Stabilization and Conservation Service
  Soil Conservation Service
  Forest Service
U.S. Geological Survey
National Ocean Survey (formerly U.S. Coast and Geodetic Survey)
National Aeronautics and Space Administration (NASA)
Tennessee Valley Authority
Bureau of Land Management
Bureau of Reclamation
National Park Service
U.S. Army Corps of Engineers

photographs in their files. These inventories vary widely in scale, film type, and other pertinent characteristics. In particular, aerial survey firms tend to accumulate large quantities of aerial photographs in the general vicinity of their home office and branch offices.

One problem that is frequently encountered in trying to determine the amount and characteristics of available aerial photographs is that many different agencies must be contacted. An organization that can be very helpful in this effort is the National Cartographic Information Center, an agency of the U.S. Geological Survey. This agency maintains current records of aerial photography coverage (as well as map coverage) that can be used as a source of information on which government agencies and aerial survey firms have aerial photographs of a specific geographic area. This program is known as the Aerial Photography Summary Record System and the use of this central data source can save considerable time and effort in gathering information on existing aerial photography. Inquiries to the National Cartographic Information Center can be made by writing to the agency at 507 National Center, 12001 Sunrise Valley Drive, Reston, Virginia 22092.

22-12 PHOTOGRAMMETRY ORGANIZATIONS

The primary professional organization in the photogrammetry field in the United States is the American Society of Photogrammetry. The national headquarters for this organization is located at 105 North Virginia Avenue, Falls Church, Virginia 22046. One of the important functions of the American Society of Photogrammetry is the publishing of the monthly technical journal entitled *Photogrammetric Engineering and Remote Sensing*. This magazine contains articles on all aspects of photogrammetry and aerial remote sensing, advertisements for photogrammetry equipment manufacturers and aerial survey firms, news articles of interest to the photogrammetry profession, and announcements of job opportunities in the photogrammetry field. Over the past several years, the American Society of Photogrammetry has prepared and published a number of important professional manuals including the *Manual of Photogrammetry*, *Manual of Remote Sensing*, *Manual of Photographic Interpretation*, and *Manual of Color Aerial Photography*. A student membership grade is available which allows students to receive the monthly journal and obtain discounts on other publications produced by the organization. Other professional organizations that provide information on photogrammetric techniques and the applications of photogrammetry are the Surveying and Mapping Division of the American Society of Civil Engineers, the American Congress on Surveying and Mapping, the Photogrammetric Society which has its headquarters in London, and the International Society of Photogrammetry, an association of 60 national societies of photogrammetry.

SELECTED REFERENCES

American Society of Photogrammetry. *Manual of Photogrammetry*, Fourth Edition. American Society of Photogrammetry, Falls Church, Virginia, 1980.

Avery, Thomas Eugene. *Interpretation of Aerial Photographs*, 3rd ed. Burgess Publishing Co., Minneapolis, Minnesota, 1977.

Church, Earl F. *Elements of Photogrammetry*. Syracuse University Press, Syracuse, NY, 1944.

Church, Earl F., and Alfred O. Quinn. *Elements of Photogrammetry*. Syracuse University Press, Syracuse, NY, 1948.

Colwell, Robert N., editor-in-chief. *Manual of Photographic Interpretation*. American Society of Photogrammetry, Falls Church, Virginia, 1980.

Crone, D. R. *Elementary Photogrammetry*. Frederick Ungar Publishing Co., New York, 1968.

Hallert, B. *Photogrammetry*. McGraw-Hill Book Company, New York, 1960.

Moffitt, Francis H. *Photogrammetry*. 2nd ed. International Textbook Company, Scranton, Pennsylvania, 1967.

——, and Edward M. Mikhail. *Photogrammetry*, 3rd ed. Harper & Row Publishers, New York, 1980.

Reeves, Robert G., editor-in-chief. *Manual of Remote Sensing*. American Society of Photogrammetry, Falls Church, Virginia, 1975.

Schwidefsky, K. *An Outline of Photogrammetry*. New York, Pitman Publishing Corp., 1959.

Smith, John T., Jr., editor-in-chief. *Manual of Color Aerial Photography*. American Society of Photogrammetry, Falls Church, Falls Church, Virginia, 1968.

Spurr, Stephen H. *Photogrammetry and Photo-Interpretation.* 2nd ed. The Ronald Press Company, New York, 1960.

Thompson, Morris M., editor-in-chief. *Manual of Photogrammetry*, 3rd ed. (two volumes). American Society of Photogrammetry, Falls Church, Virginia, 1966.

U.S. Dept. of Transportation, Federal Highway Administration. *Reference Guide Outline, Specifications for Aerial Surveys and Mapping by Photogrammetric Methods for Highways.* Washington, D.C., 1968.

Wolf, Paul R. *Elements of Photogrammetry.* McGraw-Hill Book Company, New York, 1974.

Zeller, D. M. *Textbook of Photogrammetry.* H. K. Lewis and Co. Ltd., London, 1952.

PROBLEMS

22-1. Compute the scale (as a representative fraction) of a set of aerial photographs known to have an engineer's scale of 1 in. equals 2500 ft.

(*Ans.:* 1/30,000)

22-2. Carefully define the terms metric photogrammetry, interpretative photogrammetry, and aerial remote sensing.

22-3. Compute the scale (as a representative fraction) of a set of aerial photographs taken from an altitude of 3750 ft with a camera having a nominal focal length of 6 in. over a project area that has an average elevation of 750 ft.   (*Ans.:* 1/6,000)

22-4. Describe the primary characteristics of vertical and oblique aerial photographs.

22-5. Compute the engineering scale of a vertical aerial photograph that shows two public land survey lines known to be one mi apart on the ground to be 5.28 in. apart on the photograph.   (*Ans.:* 1000 ft/in.)

22-6. Compute the scale (as a representative fraction) of a vertical aerial photograph that contains an image of a drag strip that is known to be one-quarter mi in length if the image of the drag strip is 6.6 in. long.

22-7. Compute the engineer's scale of a set of aerial photographs taken with an aerial camera having a nominal focal length of $3\frac{1}{2}$ in. from an altitude of 1800 ft over a flat area that has an elevation of 400 ft.

(*Ans.:* 400 ft/in.)

22-8. A set of aerial photographs has been taken from an altitude of 15,000 ft with an aerial camera having a nominal 8.25 in. focal length lens. What is the maximum terrain elevation that can exist in the area if the aerial photographs scale at any point must meet the scale specification of 1 : 20,000 ± 5 percent?

22-9. An aerial photograph of the central business district in a city shows the image of a large parking lot in which the painted parking stalls are clearly visible. One parking bay that contains 50 parking spaces which are known to be 9 ft wide on the ground measures 2.25 in. on the photograph. What is the engineer's scale of the aerial photograph?   (*Ans.:* 200 ft/in.)

22-10. List the three requirements that must be fulfilled to accomplish stereoscopic viewing.

22-11. The vertical component of an electric power transmission tower has a relief displacement that measures 0.21 in. on a vertical aerial photograph. The photograph was taken from a height of 4400 ft above the ground and the top of the tower is located at a distance of 4.86 in. from the center of the photograph. Compute the height of the transmission tower.

(*Ans.:* 190 ft)

22-12. Describe the three primary types of aerial mosaics.

22-13. Compute the relief displacement that would be expected to exist for a corner of the World Trade Center on an aerial photograph taken from an altitude of 10,000 ft above the ground if the top corner of the building is located at a distance of 3.5 in. from the center of the photograph. The World Trade Center is 1350 ft high.   (*Ans.:* 0.47 in.)

22-14. List and briefly describe the three primary components of a stereoscopic plotter.

22-15. An aerial photograph has the image of the top of a vertical flagpole located at a distance of 4.2 in. from the center of the photograph and the relief displacement is measured to be 0.38 in. The photograph was taken from an altitude of 1500 ft above the ground. What is the height of the flagpole?

(*Ans.:* 136 ft)

22-16. Describe the primary characteristics of orthophotos.

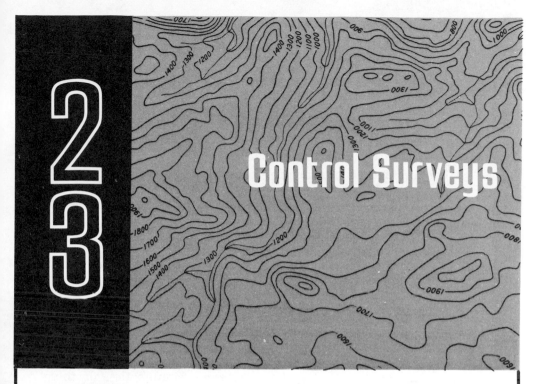

23

Control Surveys

23-1 INTRODUCTION

The determination of the precise position of a number of stations, usually spread over a large area, is referred to as *control surveying*. Control surveys can be divided into two general types: *horizontal* and *vertical*.

The objective of horizontal control surveys is the establishment of a network of triangulation stations. The horizontal position of a station is given in latitude and longitude referenced to the North American 1927 Datum and state plane coordinates (see Chapter 24). The objective of vertical control surveys is the establishment of a network of reference bench marks. The elevations of these bench marks are determined with respect to mean sea level as defined by the National Geodetic Vertical Datum of 1929.

Horizontal and vertical control systems are layed out in the form of nets covering the areas to be surveyed. The control points for the systems are established at locations where other surveys can be conveniently and accurately tied into them.

The results of horizontal and vertical control surveys are used as a basis from which surveys of smaller extent can be originated. Boundary surveys, construction route surveys, topographic and hydrographic surveys and others may be involved.

The establishment of a horizontal control system for a state or other large geographical area (which is the general topic of this chapter

and the next) results in quite a few important advantages. These include the following:

1. All types of surveys are referred to a common datum.
2. Common numerical values are provided for the corners and lines
   of adjoining tracts of land.
3. A common basis is provided for restoring lost corners.
4. The system reduces error accumulation.
5. Blunders can easily be discovered.
6. A common basis is provided for tying together public works
   projects.
7. It's very helpful for computer programming in that all land tracts
   are placed on a consistent system.

23-2 HORIZONTAL CONTROL

Horizontal control can be carried out by precise traversing, by triangulation, by trilateration and perhaps by some combination of these
methods. The exact method or methods used depends on the terrain,
equipment available, information needed and economic factors.

With *traversing* (which has been described in previous chapters), a
series of horizontal distances and angles are measured. With respect to
the other two methods it is generally cheaper and can be extended in
any direction. It is not limited by the lack of intervisibility between
widely spaced stations as are the other methods. As a result it can be
accomplished under somewhat less favorable weather conditions than
triangulation and trilateration. Traversing, however, has the disadvantages that there are fewer checks available for locating mistakes in
the work and the whole system can rather easily sway or bend. To
check a traverse it is necessary to form a loop returning to the starting
point or to tie it into previously established control points.

With *triangulation* the angles in a series of connected triangles are
measured as is the length of at least one side of one triangle, after which
the missing lengths are computed. Triangulation has several particular
advantages. There are more redundancies or checks in the measurements. In other words, more than one route can be used in moving
through the system to calculate desired lengths. There is little tendency
for the system to bend or sway—that is, azimuths can be easily and
accurately carried or established throughout the system. Another
special advantage of triangulation is that outstanding landmarks such as
steeples, water tanks, etc., can be located by establishing directions to
them from different stations. A particular disadvantage of triangulation

is that long range intervisibility is needed between stations and that usually means that special towers are necessary. Intervisibility also means reasonably good weather is required. Triangulation is probably the most expensive of the control methods.

With *trilateration* the lengths of the sides of a series of triangles are measured (usually with EDM equipment) and the angles are computed from the lengths. Trilateration is the most accurate of the three methods because it is possible today to measure distances more accurately than angles. Moreover, it is generally less expensive than is triangulation. Like triangulation, it has the advantage that checks can be made in the calculations while moving through the system by more than one route. On the other hand specially constructed towers are generally needed for trilateration and relatively good weather is required to permit the needed intervisibility of stations. Another disadvantage of this method is that it's not easy to position transmission towers, steeples, water tanks, etc., by electronic distance measurements because to do so requires the placing of reflectors on those landmarks. They of course can be located if angle measurements are made.

The remainder of this chapter is devoted to a discussion of triangulation and trilateration and the presentation of simple numerical examples for those control methods.

23-3 TRIANGULATION

In the past, triangulation was the most important method used for establishing horizontal control for large areas. Distances and directions are determined by using triangles and making a maximum number of angle measurements and a minimum number of distance measurements. The triangulation method was originally adopted because it eliminated the need for a great deal of difficult taping of long distances perhaps over difficult terrain. The sides whose lengths are measured are referred to as *bases* or *baselines*.

Though triangulation is not ordinarily used for surveys of small areas, it may be needed for construction projects where a high degree of precision is required, say 1/25,000 or better such as for bridges, tunnels and similar projects.

A simple form of triangulation was previously mentioned in Chapter 6 where the determination of distances across rivers, ravines, or other relatively inaccessible areas was discussed. Such a situation is shown in Fig. 23-1 where it is desired to determine the distance AB across a river. A base line AC is established and carefully measured. Its distance should be at least half as long as that of AB. The angles at A and C are

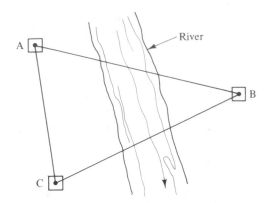

Figure 23-1 Triangulation across a river.

desirably measured by repetition with a repeating instrument or from several positions with a direction instrument. It is also well to measure the angle at B as a check. After these measurements are taken, the length AB can be computed by applying the sine law as follows with reference being made to Fig. 23-1.

$$\frac{l_{AC}}{\sin \sphericalangle B} = \frac{l_{AB}}{\sin \sphericalangle C} = \frac{l_{BC}}{\sin \sphericalangle A}$$

23-4 ACCURACY STANDARDS AND SPECIFICATIONS FOR CONTROL SURVEYS

In this section we consider survey or accuracy standards and specifications. *Survey* or *accuracy standards* are generally defined as being the minimum accuracies deemed necessary to obtain certain specific objectives. *Survey specifications* can be defined as the field operations or "recipes" needed to achieve the particular standards desired. For instance, to achieve a position closure with an accuracy of 1/15,000 it may be specified that distances should be accurate to ±0.002 ft per 100 ft, that angles be accurate to ±5″, that angular closure be equal to ±8″ $\sqrt{n}$ or better where n is the number of angles measured, etc.

If the standard desired is to be achieved, there can be few deviations from the specifications. One of the great weaknesses in past and present land surveys in the U.S. is the lack of surveying specifications and the failure to meet specifications that are provided. If a survey accuracy standard for a particular closed traverse is required to be 1/10,000, the average surveyor will attempt to obtain a precision closure of 1/10,000. Yet the closing precision may not have a great deal to do with accuracy. For instance a surveyor may take a 101-ft steel tape which he thinks is 100 ft long and obtain a 1/10,000 closure precision. The probable result is an accuracy of less than 1/100.

The accuracy standard selected for a particular survey depends on its anticipated use. For most control surveys the cost of using a standard a little higher than what is thought to be necessary is very small. Over the years control systems are generally used for all sorts of purposes often far in excess of what was generally anticipated. As a result it is often thought desirable to select a standard a little higher than what is thought to be needed.

The Federal Geodetic Control Committee has established a set of accuracy standards for horizontal and vertical control surveys. There are in descending order three major classifications. These are first-order, second-order and third-order. For horizontal control the latter two classifications are subdivided into two categories labeled Classes I and II, while for vertical control the first two orders are subdivided into Classes I and II.

In Table 23-1 the minimum accuracies required for horizontal control are briefly summarized. This information is taken from the publication "Classification, Standards of Accuracy, and General Specifications of Geodetic Control Surveys," published in February 1974 by the United States Department of Commerce, Rockville, Md. The minimum relative accuracy between two directly connected points is specified. For vertical control the accuracy standards are specified in terms of K, the distance between bench marks in kilometers. The minimum accuracies required for vertical control were given in Table 9-1. Also included in the tables is information concerning the recommended uses for each classification.

There is an extensive network of triangulation stations across the U.S. that were established with different orders of accuracy. Most of the higher-order work has been done by the National Geodetic Survey while other federal agencies such as the U.S. Geological Survey, the Army Corps of Engineers and others have extended the system, particularly with third-order work.

The primary control network consists of first-order east–west triangulation arcs spaced at about 60 or 70 miles and north–south arcs with approximately the same spacing. There is next a secondary control system in areas which are surrounded by the primary control network. The secondary system which is conducted to second-order Class I standards increases the network density and is used particularly in areas of high land values.

There is also supplemental control executed to second-order Class II standards which enhances control in lightly developed areas. It is particularly used along coastlines and for extensive mapping on construction projects. Finally there is local control used for local construction projects and small-scale topographic maps which are done to third-order Classes I or II. Figure 23-2 shows the status of the control system

TABLE 23-1 Standards of Classification—Horizontal Control (Federal Geodetic Control Committee)

	First-Order	Second-Order		Third-Order	
		Class I	Class II	Class I	Class II
Recommended uses	Primary national network. Metropolitan area surveys. Scientific studies.	Area control which strengthens the national network. Subsidiary metropolitan control.	Area control which contributes to, but is supplemental to the national network.	General control surveys referred to the national network. Local control. Surveys.	
Base measurement standard error not to exceed	1 part in 1,000,000	1 part in 900,000	1 part in 800,000	1 part in 500,000	1 part in 250,000
Relative accuracy between directly connected adjacent points (at least)	1 part in 100,000	1 part in 50,000	1 part in 20,000	1 part in 10,000	1 part in 5,000
Triangle closure. Average not to exceed	1.0"	1.2"	2.0"	3.0"	5.0"
Maximum— seldom to exceed	3.0"	3.0"	5.0"	5.0"	10.0"

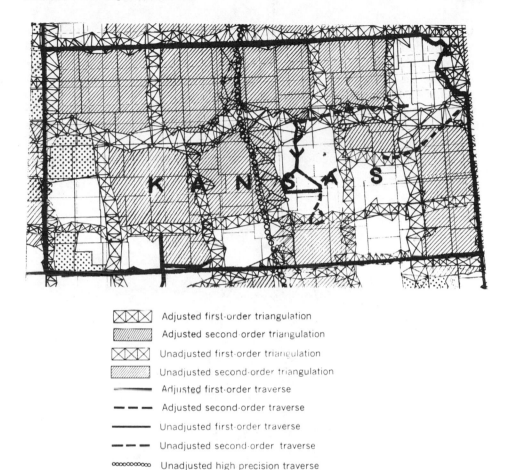

XXXX Adjusted first-order triangulation

▨▨▨ Adjusted second order triangulation

XXX Unadjusted first-order triangulation

▨▨▨ Unadjusted second-order triangulation

——— Adjusted first-order traverse

— — — Adjusted second-order traverse

——— Unadjusted first-order traverse

— — — Unadjusted second-order traverse

∞∞∞∞∞∞ Unadjusted high precision traverse

▦▦▦ Unadjusted USGS second and third-order control

▦▦▦ Adjusted USGS second and third-order control

Figure 23-2 Status of horizontal control in Kansas, 1979.

in the state of Kansas as of October 1, 1979. This information is taken from a U.S. Department of Commerce map of that date entitled "Status of Horizontal Control, United States."

23-5 TRIANGULATION STATIONS

The National Geodetic Survey has a network of stations throughout the U.S. A large percentage of NGS surveys have been conducted by the triangulation method in which an accurately measured base line serves as one side of a group of triangles formed by a number of widely spaced points. The points are selected on the basis of their visibility as for example on the tops of hills or church steeples or radio towers or water tanks. As a result the points are not uniformly spaced. Some of the points are obviously inaccessible and it is necessary to establish eccentric stations from them and to determine the distances and directions from the main points to the eccentric ones.

Sometimes it is necessary to build special towers for making the

Figure 23-3 A Bilby steel tower that consists of two entirely separate structures. The outer tower supports the observers while the inner tower supports the instrument only and is thus free from vibrations caused by movements of the observers. (*Courtesy National Geodetic Survey.*)

observations.   These towers in effect consist of one tower built inside
another because the structural members which support the observers are
not the same members which support the instrument.   Lines of sight
should be kept at least 10 ft above the ground because of possible
refraction effects.

For first-order triangulation the stations are spaced at large distances apart—perhaps 20, 40, or even 100 miles.   Second-order triangulation stations are placed in between the first-order stations.   It is the
ultimate goal of the National Geodetic Survey to have stations of
second-order no further apart than 7 miles.

23-6 STRENGTH OF FIGURES

The strength of figures is an important subject in triangulation.   This
term refers to the effect of the proportions of a triangle on the accuracy with which the lengths of the sides can be computed.   When small
errors in angle measurement affect the computed distances very little
the figure is said to be *strong*.

In triangulation the lengths of triangle sides are computed with the
law of sines.   When triangles are used which contain small angles the
best results may not be obtained because of the fact that the rate of
change of the sines of angles near 0 or $180°$ is quite large as compared
to the rate of change for angles near $90°$.   Thus angles near $90°$ are the
optimum ones to use with those from $30°$ to $150°$ being acceptable.

It is not correct to say that small angles should always be avoided—
rather they should not be used where they will weaken the system.   For
instance, in calculating the length of a given side of a triangle there are
two angles which are used, namely the angle opposite the known side
and the angle opposite the side whose length is to be determined.   In
Fig. 23-4 the length b is assumed to be known and it is desired to determine the length c.   With the sine law only the angles B and C are used

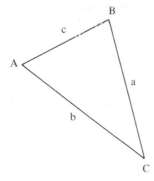

Figure 23-4

and as a result the angle A has no direct effect on the calculations and thus can be quite small without affecting the strength of the figure.

A comprehensive discussion of the strength of figures is presented in "Special Publication No. 247," U.S. Department of Commerce, Coast and Geodetic Survey, by F. R. Gossett, published in 1950, revised in 1959. Included in the material is a discussion of the frequency needed for base lines.

23-7 TRIANGULATION SYSTEMS

There are several different triangulation systems which can be used for a particular survey. In each case, a set of triangles which adjoin or overlap each other are used. In Fig. 23-5, four types of systems which have been used are presented. These systems include a chain of single independent triangles, a chain of quadrilaterals formed with overlapping triangles, a chain of central point figures and a chain of central point figures each with an extra diagonal.

The use of a single chain of triangles such as the one shown in part (a) of Fig. 23-5 does not provide the most accurate results. Such a system is usually employed in rather long and narrow surveys of low precision such as for a valley or a narrow body of water. There is only one route through this type of system while the other systems provide at least two routes. The single triangle system is not as satisfactory as the other triangulation systems and it is necessary to frequently measure base lines as checks for the work. The only other checks on the work are the summations of the angles in each triangle.

Figure 23-5(b) shows the most common triangulation system—a chain of quadrilaterals. Systems of quadrilaterals are best adapted to rather long and narrow surveys where a high degree of precision is required. With a system such as this one the lengths of the sides can be computed with different routes as well as different triangles and angles, offering excellent checks on the computations. Most of the major triangulation arcs in the U.S. consist of quadrilaterals.

When horizontal control is to be extended over a rather wide area involving a rather large number of points, as might be the case in metropolitan areas, a chain of polygons or central point figures such as the one shown in Fig. 23-5(c) may be used. These figures are very strong and are often quite easy to arrange. The central point figure can itself be further strengthened by using a diagonal of the type shown in Fig. 23-5(d). Other systems are available which are combinations of the types mentioned here. In this book, only the chain of single triangles and the chain of quadrilaterals are considered further.

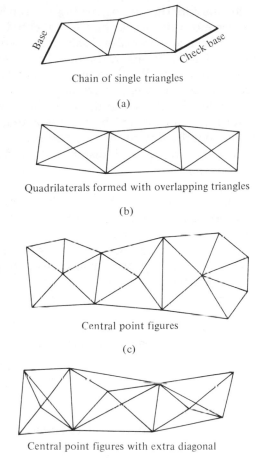

Chain of single triangles

(a)

Quadrilaterals formed with overlapping triangles

(b)

Central point figures

(c)

Central point figures with extra diagonal

Figure 23-5 Triangulation systems.   (d)

23-8 MEASUREMENT OF ANGLES AND BASELINES

The instruments to be used for measuring angles for triangulation depend on the accuracy desired in locating the positions of the triangulation stations.   If first-order work is desired, directional theodolites
should be used with which directions can be read directly to 0.2″. For
second-order work, it is necessary to use instruments capable of being
read to 1″.   For third-order triangulation, engineer's transits which
can be read to 20″ or 30″ may be used if the angles are measured by
repetition.

When the directional theodolite is used, it is set up over a particular station and pointed to each of the desired stations. For first-order

triangulation the set of readings is repeated from 8 to 16 times while for second-order work they are repeated from 4 to 8 times.

For many decades base lines were measured with precise Invar tapes. The locations of the base lines were selected in relatively smooth open areas to facilitate taping. Today, however, while Invar-taped sides are still acceptable, electronic distance-measuring devices are almost always used. As a result, base lines can be located in much rougher country. The longer the lines selected, the more accurate will be the other lengths determined in the system. Slope distances have to be reduced to horizontal distances, and those distances reduced to sea-level distances.

To correct a distance to sea level, it is possible to write a simple proportion. For this purpose R is considered to be the radius of the earth (usually taken as 20,906,000 ft or 6,372,200 m), L the measured distance at some elevation H, and C is the correction to be made. With reference to Fig. 23-6, it can be seen that for a certain angle θ the arcs shown are directly proportional to their radii. Thus the following ratio can be written

$$\frac{L - C}{L} = \frac{R}{R + H}$$

from which the sea-level distance is

$$L - C = \frac{RL}{R + H}$$

Example 23-1

A base line was measured at an elevation of 1462 ft and found to be 3692.320 ft. Convert this value to a sea-level distance

Solution

$$L - C = \frac{RL}{R + H} = \frac{(20{,}906{,}000)(3692.320)}{20{,}906{,}000 + 1462} = \textbf{3692.062 ft}$$

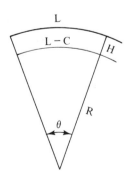

Figure 23-6

23-9 WORK INVOLVED IN TRIANGULATION

The work involved in triangulation usually includes the following steps:

1. The selection of the stations.
2. The construction and placement of monuments, the erection of the signals and in many cases the construction of towers on which the signals and perhaps the instruments are to be placed.
3. The measurement of the required angles.
4. Astronomical observations to establish the true azimuths of the lines.
5. The measurement of the base lines.
6. The necessary office calculations including the adjustment of the angles, the calculation of the lengths of triangle sides, and the coordinates of the station.

23-10 ADJUSTMENT OF CHAIN OF SINGLE TRIANGLES

The system of triangles is placed around the exterior of the area being considered and the triangles are established so as to be as nearly equilateral as possible. All of the angles of the triangles are measured, as are the lengths of at least two sides. As an illustration of the distance measurements, part (a) of Fig. 23-5 shows the base and the check base.

Before the length computation can begin, it is necessary to make the sum of the angles around each point total exactly 360° (called the *station adjustment*), and the sum of the measured angles in each triangle total exactly 180° (called the *figure adjustment*). If precise triangulation is being used, these adjustments are made all in one operation by the method of least squares but for triangulation of ordinary precision it is possible to use a simpler process as described herein.

For the angles about a point the difference between the sum of the angles and 360" is balanced equally between the number of angles. In the same fashion the difference between the sum of the angles in each triangle and 180° is balanced equally between the angles. This procedure can be adjusted somewhat if it is known that some of the angles were measured with a higher degree of precision than some of the others.

Example 23-2 shows the adjustment of the angles of a chain of two triangles. It is assumed that in each case the angles were measured with equal precision.

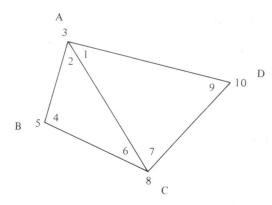

Figure 23-7

Example 23-2

For the two triangles shown in Fig. 23-7, the measured angles are given in the table in the solution part of the problem. Make the station and figure adjustments to the angles.

Solution

Station Adjustment (total of angles around each station adjusted to 360°00'00")

Station	Angle No.	Measured Angle	Adjusted Angle
	1	41°16'10"	41°16'05"
	2	53°36'20"	53°36'15"
A	3	265°07'45"	265°07'40"
	Σ	360°00'15"	360°00'00"
	4	91°16'10"	91°16'20"
B	5	268°43'30"	268°43'40"
	Σ	359°59'40"	360°00'00"
	6	35°07'00"	35°06'50"
	7	78°42'30"	78°42'20"
C	8	246°11'00"	246°10'50"
	Σ	360°00'30"	360°00'00"
	9	60°01'05"	60°00'55"
D	10	299°59'15"	299°59'05"
	Σ	360°00'20"	360°00'00"

Figure Adjustment (total of angles in each triangle adjusted to $180°00'00''$)

Triangle	Angle No.	Angle Value after Station Adjustment	Angle Value from Figure Adjustment
ABC	2	$53°36'15''$	$53°36'27''$
	4	$91°16'20''$	$91°16'32''$
	6	$35°06'50''$	$35°07'01''$
	Σ	$179°59'25''$	$180°00'00''$
ACD	1	$41°16'05''$	$41°16'18''$
	7	$78°42'20''$	$78°42'34''$
	9	$60°00'55''$	$60°01'08''$
	Σ	$179°59'20''$	$180°00'00''$

23-11 ADJUSTMENT OF A QUADRILATERAL

For a quadrilateral to be properly adjusted there are two conditions which must be satisfied. The first of these is the *geometric condition* that the sum of the interior angles must equal $(n - 2)(180°)$, where n equals the number of sides of the figure. Secondly there is the *trigonometric condition* by which the sine of each angle must be proportional to the length of the opposite side of that triangle.

The adjustment of a quadrilateral is illustrated in Fig. 23-8, where it is assumed that the base line distance *AD* has been determined as have the 8 lettered angles. It is desired to determine the length *BC*.

To make the adjustments and to compute the desired length, the following steps are taken:

1. The angles around each point are balanced to a total of $360°$ by distributing the error equally (or approximately so) among the angles.

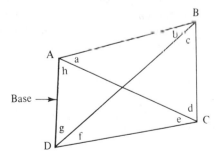

Figure 23-8

2. The sum of all the angles (a through h in Fig. 23-8) is adjusted to equal $(n - 2)(180°) = 360°$. This is accomplished by adding the angles together and balancing the discrepancy from $360°$ between the angles.

3. The opposite angles at the intersection of the diagonals should also be equal. In other words, the following condition should apply:

$$a + b = e + f$$

The values of these angles which were previously adjusted in Step 1 are compared and the difference between them is divided by 4 and distributed to the angles, adding to the smaller pair of angles and subtracting from the larger ones. The following relation should also exist:

$$c + d = g + h$$

A similar procedure is used to adjust these angles as was used for the previous four.

4. For Fig. 23-8, the length of BC is desired. It can be determined by two different routes. Working with triangle ABD the length of side BD can be determined with the sine law. Then from triangle BCD and the sine law the length of side BC can be determined.

In a similar fashion, side CD can be determined by using triangle ACD and then side BC can be determined from triangle BCD. With perfect trigonometric adjustment the length of BC is the same no matter which route is taken. If the results are close to each other they may be averaged for work of ordinary precision. For more precise work there are various methods available by which the trigonometric adjustment may be made. Such adjustments are described in advanced surveying textbooks.

If a chain of quadrilaterals is being used, the adjustment in each quadrilateral is made as described for the single quadrilateral.

The angles of a triangle on the earth's curved surface are spherical. As a result the sum of the spherical angles in such a triangle is slightly larger than $180°$. This value is called the *spherical excess* and equals approximately $1''$ for a triangle covering about 75 square miles of the earth's surface. This excess is neglected in the example problems of this chapter.

Examples 23-3 and 23-4 illustrate the calculations necessary to adjust the angles of a quadrilateral and to compute the lengths of its sides.

Example 23-3

The angles in the quadrilateral of Fig. 23-8 have been measured with the following results:

$$a = 36°12'24''$$
$$b = 43°58'40''$$
$$c = 47°29'54''$$
$$d = 52°18'46''$$
$$e = 39°08'28''$$
$$f = 41°02'40''$$
$$g = 59°12'16''$$
$$h = 40°36'36''$$
$$\Sigma = 359°59'44''$$

Adjust the angles so they satisfy the geometric condition that the sum of the interior angles must equal $(n - 2)(180°)$.

Solution   The total value of the interior angles in the four-sided figure should be $(n - 2)(180°) = 360°$. The sum of angles a through h is $16''$ less than $360°$. As a first step each of the angles is increased by $2''$ to make their total exactly $360°$. The corrected angles are

$$a = 36°12'26''$$
$$b = 43°58'42''$$
$$c = 47°29'56''$$
$$d = 52°18'48''$$
$$e = 39°08'30''$$
$$f = 41°02'42''$$
$$g = 59°12'18''$$
$$h = 40°36'38''$$
$$\Sigma = 360°00'00''$$

As the next step in the adjustment the following angle relations should be true

$$a + b = e + f$$
$$a + b = 80°11'08''$$
$$e + f = 80°11'12''$$

Since the sum of these four angles is $4''$ in error, $1''$ is added to angles a and b and $1''$ is subtracted from angles e and f.

Finally the following angle relation should be true:

$$c + d = g + h$$
$$c + d = 99°48'44''$$
$$g + h = 98°48'56''$$

Since the total of these angles is in error by $12''$, they are adjusted by adding $3''$ to each of angles c and d and subtracting $3''$ from each of angles g and h. The geometrically balanced angles are as follows:

$$
\begin{aligned}
a &= 36°12'27'' \\
b &= 43°58'43'' \\
c &= 47°29'59'' \\
d &= 52°18'51'' \\
e &= 39°08'29'' \\
f &= 41°02'41'' \\
g &= 59°12'15'' \\
h &= 40°36'35'' \\
\hline
\Sigma &= 360°00'00''
\end{aligned}
$$

Example 23-4

It is assumed that the length of side AD of the quadrilateral of Fig. 23-8 has been measured and found to be 864.52 ft. Determine the length of side BC by two different routes through the figure using the adjusted angles from the solution of Example 23-3.

Solution   Using the sine law and triangle ABD, the length of side BD is determined as follows:

$$\frac{AD}{\sin b} = \frac{BD}{\sin (a + h)}$$

$$BD = \frac{(864.52)(\sin a + h)}{\sin b} = \frac{(864.52)(0.9736475)}{0.69438979}$$

$$= 1212.20 \text{ ft}$$

Then from triangle BCD the length of side BC is determined as follows:

$$\frac{BD}{\sin (d + e)} = \frac{BC}{\sin f}$$

$$BC = \frac{BD \sin f}{\sin (d + e)} = \frac{(1212.20)(0.65664792)}{0.99967733}$$

$$= \mathbf{796.25 \text{ ft}}$$

Using a similar procedure and triangles ACD and ABC, the lengths AC and BC are determined.

$$AC = \frac{AD \sin (f + g)}{\sin e} = \frac{(864.52)(0.98404415)}{0.63123624}$$

$$= 1347.71 \text{ ft}$$

$$BC = \frac{AC \sin a}{\sin (b + c)} = \frac{(1347.71)(0.59071129)}{0.99966715}$$

$$= \mathbf{796.37 \text{ ft}}$$

23-12 TRILATERATION

Trilateration is a method of horizontal control in which the lengths of all the lines in geometric figures are measured directly and with the angles of the figures being computed subsequently. The acceptance which this method has gained has been due to the tremendous advances made in recent years with EDM instruments. Trilateration possesses some advantages over triangulation because the measurement of the distances with EDM equipment is so quick, precise, and economical while the measurement of the angles needed for triangulation may be more difficult and expensive.

Originally triangulation was adopted because it enabled the National Geodetic Survey to quickly extend their horizontal control for long distances over rough and forbidding terrain without extensive taping. EDM instruments enable the surveyor to accomplish very much the same objectives. It may very well be that in the future triangulation and trilateration will be used in combination.

To obtain suitable accuracy with trilateration it is necessary to follow certain guidelines such as using certain minimum lengths and geometric configurations. The geometric figures used for trilateration are not as standard as those used for triangulation, but in general the figures are similar. It is better in trilateration to use approximately square figures because slender figures are weaker in the short directions transverse to the figures. Should relatively long narrow figures have to be used because of terrain or other reasons, it is well to strengthen the network by measuring some horizontal angles. The previously mentioned publication "Classification, Standards of Accuracy, and General Specifications of Geodetic Control Surveys" presents standards for trilateration surveys as it does for triangulation and traversing.

Once the distances are obtained for each triangle, the angles may be calculated with the law of cosines. With reference being made to

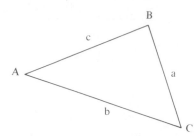

Figure 23-9

Fig. 23-9 the law of cosines can be expressed with the following equation in which a, b, and c are the lengths of the sides and A, B, and C are the opposite interior angles.

$$\cos \measuredangle A = \frac{b^2 + c^2 - a^2}{2bc}$$

The sum of the interior angles should be exactly $180°$. In trilateration as in any other form of surveying, it is wise to make checks on the work as it progresses. For instance the measurement of occasional angles, the calculation of azimuths and the comparison of them with observed azimuths are very helpful. Example 23-5 illustrates the calculation of the angles in a triangle for which the distances are given.

Example 23-5

The following sea-level distances in meters were obtained for a triangle by a trilateration survey. Compute the interior angles of the figure with the law of cosines.

$$a = 1226.423 \text{ m}$$
$$b = 1354.677 \text{ m}$$
$$c = 1416.224 \text{ m}$$

Solution

$$\cos \measuredangle A = \frac{(1354.677)^2 + (1416.224)^2 - (1226.423)^2}{(2)(1354.677)(1416.224)}$$

$$= 0.6089901$$

$$\cos \measuredangle B = \frac{(1416.224)^2 + (1226.423)^2 - (1354.677)^2}{(2)(1416.224)(1226.423)}$$

$$= 0.48208416$$

$$\cos \measuredangle C = \frac{(1226.423)^2 + (1354.677)^2 - (1416.224)^2}{(2)(1226.423)(1354.677)}$$

$$= 0.40133838$$

From which

$$\angle A = 52°29'\ 0.5''$$
$$\angle B = 61°10'42.2''$$
$$\angle C = \underline{66°20'17.3''}$$
$$180°00'00.0''$$

PROBLEMS

23-1. Determine the distance AB across the river shown in the accompanying illustration using the distance AC and the angle values given.

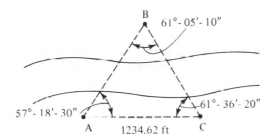

(Ans.: AB = 1240.75 ft)

23-2. For the two triangles shown in the accompanying illustration, the measured values of the angles are as follows:

$$1 = 98°16'32''$$
$$2 = 261°43'38''$$
$$3 = 48°51'26''$$
$$4 = 43°36'08''$$
$$5 = 267°33'11''$$
$$6 = 85°18'19''$$
$$7 = 274°42'01''$$
$$8 = 51°06'18''$$
$$9 = 32°51'42''$$
$$10 = 276°02'04''$$

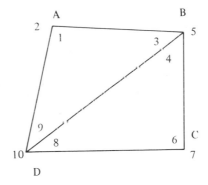

Make the station and figure adjustments for the angles.

23-3. For the quadrilateral shown, the angles have been measured with the following results:

$a = 49°17'05''$

$b = 37°14'24''$

$c = 40°07'31''$

$d = 53°20'32''$

$e = 43°08'12''$

$f = 43°23'45''$

$g = 56°00'44''$

$h = 37°28'11''$

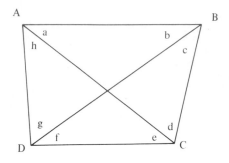

(Ans.: $a = 49°17'09''$, $d = 53°20'42''$)

Adjust the angles to their most probable values.

23-4. The distance AB of the quadrilateral of Problem 23-3 has been measured with EDM equipment and found to be 649.36 ft. Using the adjusted angles from the solution of that problem, determine the length of distance CD with two independent routes.

23-5. The following distances in feet were determined for a triangle for a trilateral survey. Compute the interior angles of the triangle.

$AB = 1642.32$ ft

$BC = 1583.95$ ft

$CA = 1296.84$ ft

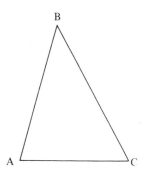

(Ans.: $A = 63°57'29''$, $B = 47°21'33''$, $C = 68°40'58''$)

State–Wide Plane Coordinates

24-1 INTRODUCTION

In earlier chapters we discussed the metes and bounds system and the U.S. public land system for describing property. Each of these methods has its shortcomings and many persons in the surveying profession feel very strongly that a better system is needed for the future. In the U.S., a good state plane coordinate system may be the answer. The use of such a system has been recommended by the American Society of Civil Engineers, the American Congress on Surveying and Mapping, and the American Bar Association.

If a coordinate system were adopted in each state it would make possible the mathematical determination of the location of each property corner. Thus the position of a particular corner would be known in relation to all the other corners in that vicinity whose coordinates were known. If a corner of known coordinates were lost it would be possible to reestablish it by means of the coordinates of the other points in that area. If a number of corners for a large area were destroyed they could again be laid out from the coordinates of points outside the area.

The National Geodetic Survey (NGS) has developed a system for each state in which plane or grid coordinates are provided at various control stations. The surveyor can merely run traverse lines (lengths and bearings) from those points to other points whose coordinates are desired. He will need to adjust his measured distances to sea level and apply a scale correction to them depending on the method used by the

451

NGS in that state. The x and y components of the traverse lines (latitudes and departures) can then be computed and the plane or grid coordinates of the points in question determined. Though the work of the NGS was very complicated, the use of plane coordinates by the practicing surveyor is quite simple.

The reader should clearly understand that a state plane coordinate system cannot replace local monument control. In other words, if a monument exists and no evidence is available to show that it has been disturbed, it represents the corner and any measurements (metes and bounds, coordinates, etc.) are just secondary evidence.

24-2 COMMENTS ON COSTS OF ESTABLISHING AND USING A STATE PLANE COORDINATE SYSTEM

If the land surveyor is required to tie his property corners into a coordinate system with control monuments some distance away, there will be appreciable extra costs to the landowners. As a result, surveyors will object and unless laws are passed requiring the use of the system, it will seldom be used. Thus the density or closeness of the control monuments is a very important factor in obtaining voluntary usage of a coordinate system. For instance if monuments are spaced 10 miles on center and the surveyor has to measure 6 miles in one direction and 4 or more in another to tie into the system, the cost will be high and the surveyor probably will not use the system. If the monuments are spaced every half mile, the likelihood that the system will be used will be greatly increased.

Should the monuments be widely spaced the uncertainty of the surveyor's measurements will also be a major problem. If he has to measure for 4 miles to tie into a control monument there may very well be an uncertainty in his work of ±2 ft or more. As a result, a system such as metes and bounds tied into neighboring property monuments may be just as good or better than a coordinate system based on widely spaced control monuments.

The establishment of a closely spaced set of coordinate positions (say, $\frac{1}{2}$ mile on center) would cost a great deal of money—far more than private surveyors could afford to bear. If, however, such a system were established on the ground and paid for by government agencies it would be quite reasonable to require land surveyors to tie into the system. In some urban areas such as Los Angeles a closely spaced state plane coordinate network is in place. For such cases all surveys should definitely be tied into the system.[1]

[1] C. M. Brown and W. H. Eldridge, *Evidence and Procedures for Boundary Location* (New York: John Wiley & Sons, Inc., 1962), pp. 376-377.

24-3 TYPES OF COORDINATE SYSTEMS

The purpose of coordinate systems is to take advantage of the very precise work of the National Geodetic Survey and use it in a simple fashion to control ordinary surveying work. In other words it is desired to use the mathematics of plane surveying and yet take into account the earth's curvature.

The idea is to project points from the earth's spheroid to some imaginary surface which can be rolled out flat without substantially destroying its shape or size. A plane rectangular grid system is then superimposed onto that flat surface and the location of points specified with x and y components. Several of these so-called map projections have been devised through the years but only three will be discussed here. These are the tangent plane projection, the Lambert projection and the transverse Mercator projection.

24-4 DISADVANTAGES OF STATE PLANE COORDINATES

State plane coordinates have been available for almost 50 years and yet despite their many advantages they are not commonly used throughout the country for the following reasons:

1. The control monuments are too far apart.
2. The surveys are too expensive.
3. Quite a few states (12 to 14) do not have legislation permitting their use.
4. A large percentage of surveyors and attorneys do not understand the system.
5. So many surveys of the past and the present have such a low degree of precision that nothing would be gained by using the coordinates.

24-5 TANGENT PLANE PROJECTION

For the survey of many urban areas it has been customary to refer points to a rectangular coordinate system. A point within the area, such as one whose geographic coordinates have been established by the National Geodetic Survey, is selected as the origin for the system. The coordinates of all points in the area are calculated as though they are on a plane tangent to the earth at the origin. The true meridian through the origin is taken as the y axis while the true east–west line at the origin is taken as the x axis. Sometimes the y axis is referred to as grid north. Obviously the trouble with this method is that the further a particular point is from the origin or point of tangency the greater will be the error of the work. It is usually not used much further than about 20 miles in each direction from the origin.

For average-sized cities, tangent plane coordinates are quite satisfactory but there has been little need for them since the state plane coordinate systems were introduced. These latter systems are the subject of the remainder of this chapter. For areas which have satisfactorily used tangent plane coordinates there is no need to discard them once state plane coordinates are adopted. They can easily be transferred into the new system. The use of tangent plane coordinates is made relatively simple by Publication No. 71 of the U.S. Coast and Geodetic Survey (now the National Geodetic Survey) entitled "Relation Between Plane Rectangular Coordinates and Geographic Position."

24-6 COMMENTS ON EARTH'S CURVATURE

In the early days of the U.S., all surveys were plane surveys; however, the true relative positions of different points on the earth's surface cannot be given unless the earth's curvature and thus spherical coordinates are used. While the average surveyor is acquainted with plane trigonometry, geodesy is usually unfamiliar to him or her.

Today a large amount of surveying is of such an extensive nature that the true shape of the earth must be considered. This shape, as we noted previously, is close to an oblate spheriod of revolution in which the axis at the equator is about 27 miles larger than is the polar axis. In proportion to the overall size of the earth this difference is small, and for all but the most precise geodetic work an average diameter can be used.

In 1933 the North Carolina State Highway Commission asked the U.S. Coast and Geodetic Survey to design a statewide system of coordinates by which plane surveying coordinates could be used and yet take advantage of the very precise geographic coordinates of that organization. A system was designed for the state of North Carolina as well as for the other states of the country. There are actually two basic systems or projections employed: the Lambert conformal conic projection and the transverse Mercator projection. The first of these methods uses an imaginary cone as its developable surface while the second one uses an imaginary cylinder. The mathematics for these systems is described in the publication *Manual of Plane Coordinates Computation* by O. S. Adams and C. N. Claire, Special Publication No. 193 of the National Geodetic Survey.

24-7 LAMBERT CONFORMAL CONIC PROJECTION

As the name implies, this method involves the projection of a section of the earth onto the surface of an imaginary cone. The term *conformal* means that true angular relations (or very nearly so) are retained for all points. There are different forms of the Lambert projection, but the one used for state plane coordinates consists of a cone such as the one shown in Fig. 24-1 whose axis *OE* coincides with the polar axis of the earth. In the figure it will be noted that one element of the cone cuts through the earth's surface at points *B* and *C*. All of the elements of the cone cutting through the earth create the two circles called *standard parallels* which are shown in the figure. A section of the cone between the parallels and outside of them is rolled out flat and the points from the sphere projected on to it. The scale of the projected plane will clearly be exact along these parallels.

Figure 24-2 shows a little more detail of the intersection picture for the element of the cone which intersects the earth at points *B* and *C* in Fig. 24-1. It can be seen from this second figure that points on the earth's surface are projected along radial lines from the earth's center to the surface of the imaginary cone.

When the cone surface is outside of the spheroid, a projected distance on the cone (as *ab* in the figure) is greater than the actual distance (*a'b'*) on the spheroid. When the cone surface is inside the spheroid,

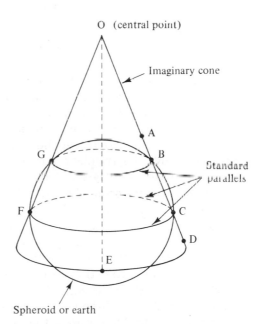

Figure 24-1 Lambert conformal conic projection.

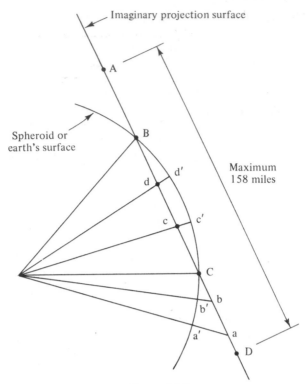

Figure 24-2

a projected distance (as cd) is less than the distance $(c'd')$ on the spheroid.

Figure 24-3 shows the Lambert projection developed into a plane surface for the frustrum of the cone *ADHI*. For this figure the longitude of any point as *D* or *H* can be determined by adding or subtracting the angle θ at point *O* to the central meridian. The longitude at any point on the plane *ADHI* will be exact but the latitude of any point above or below the standard parallels will be slightly in error. The magnitude of the error will not be very large if the height of the plane *AD* is kept within certain limits. The heights have been limited to a maximum of 158 miles, with the standard parallels located at two-thirds of that distance. Unless these values are exceeded there will be no point where the discrepancy between a sea-level distance and a grid distance will be greater than 1/10,000.

The Lambert projection can be extended indefinitely in the east–west direction without affecting the accuracy of the work but it must be limited in the north–south direction as described. The method is therefore most suitable for areas that are wide but not deep.

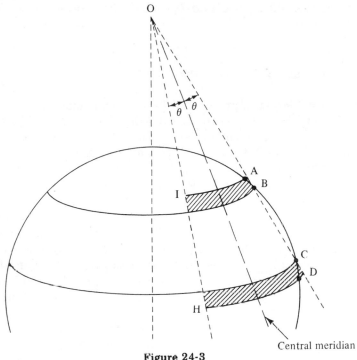

Figure 24-3

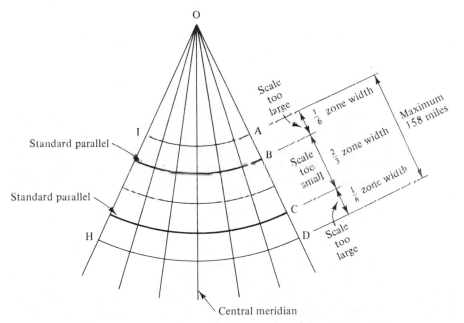

Figure 24-4 Lambert projection showing rectangular grid.

Figure 24-4 shows the grid system superimposed onto the plane surface of the cone.

24-8 TRANSVERSE MERCATOR PROJECTION

The transverse Mercator projection uses an imaginary cylinder which has its axis perpendicular to the polar axis of the earth. The diameter of the cylinder is a little smaller than that of the earth with the result that it intersects the earth with two parallel circles equidistant from the central meridian as shown in Fig. 24-5. In part (a) of the figure the cylinder is shown passing through the earth while the projected area to be used is diagrammed in part (b).

In the projection the scale is exact on the two lines of intersection but between the lines a distance on the projection is smaller than the corresponding distance on the sea-level surface. Outside the two lines the projected distance is larger. Notice in the figure that neither lines of latitude nor longitude (except the central meridian) will appear as straight lines on the projection. The projection can be extended indefinitely in the north–south direction without affecting its accuracy but in the east–west direction its width is limited to 158 miles with the exact intersection lines separated by about two-thirds of the distance. Again using these limits the discrepancy between a projected or grid distance and a sea-level distance will not be greater than 1/10,000. The Mercator projection was designed to meet the requirements of states whose largest distances are in the north–south direction.

24-9 ADOPTION OF STATE PLANE COORDINATE SYSTEMS

The need for establishing plane coordinate systems which could make use of the existing geodetic data of the National Geodetic Survey in the various states led to the development of state plane coordinate systems. The systems adopted were based on the Lambert conformal conic projection and the transverse Mercator projection. The Lambert system is used for the states whose greatest dimension is in the east–west direction while the Mercator system is used for those states whose greatest dimension is in the north–south direction. Actually Alaska, Florida, and New York make use of both projections because of their shapes. For some states one zone of projection is sufficient to cover the entire state while for some larger ones several zones will be needed. California has seven Lambert zones while Texas has five. In Table 24-1 the system or systems used for each of the states of the U.S. is presented.

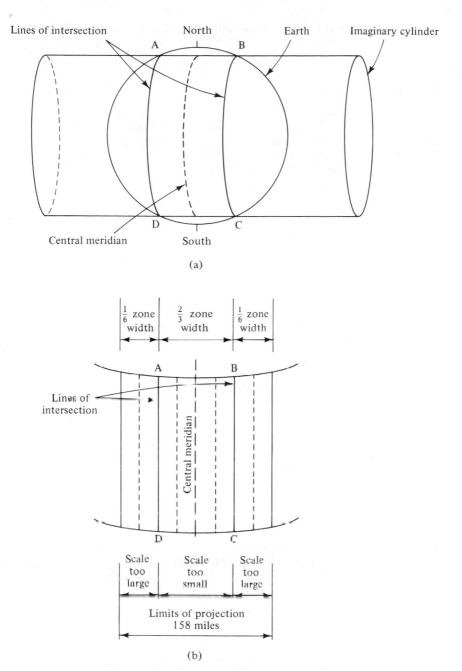

Figure 24-5 Transverse Mercator projection.

TABLE 24-1 States with Their Grid Systems

Lambert System		Transverse Mercator System		Both
Arkansas	North Dakota	Alabama	New Hampshire	Alaska
California	Ohio	Arizona	New Jersey	Florida
Colorado	Oklahoma	Delaware	New Mexico	New York
Connecticut	Oregon	Georgia	Rhode Island	
Iowa	Pennsylvania	Hawaii	Vermont	
Kansas	South Carolina	Idaho	Wyoming	
Kentucky	South Dakota	Illinois		
Louisiana	Tennessee	Indiana		
Maryland	Texas	Maine		
Massachusetts	Utah	Mississippi		
Michigan	Virginia	Missouri		
Minnesota	Washington	Nevada		
Montana	West Virginia			
Nebraska	Wisconsin			
North Carolina				

The National Geodetic Survey has set up state plane coordinate systems for each of the 50 states. In addition they have published computed grid coordinates for various control stations throughout the country. Some states, however, have not passed the necessary legislation to officially recognize the applicable coordinate system. Approximately 36 states have laws permitting the use of state plane coordinates but they do not make their use mandatory. If a point is within 800 m of an NGS control station in North Carolina, the surveyor is supposed to use state plane coordinates to describe its position.

24-10 CALCULATION OF PLANE COORDINATES ON LAMBERT GRID

Before actually considering a traverse with state plane coordinates we shall consider how the geographic coordinates of a point can be transformed into plane coordinates. For a particular Lambert projection zone a central meridian is established near the center of the zone, and is given an x value of 2,000,000 ft. The x axis is placed well below the southern edge of the zone and is given a y value of zero. Thus all x and y positions in the projection area will be positive. This information is shown in Fig. 24-6 together with the other information needed to transform the geographic coordinates of point P into Lambert grid coordinates.

In the figure point O is the apex of the cone from which the area is projected, line OE is the central meridian and C is the x coordinate of that meridian equal to 2,000,000 ft. The meridians are straight lines

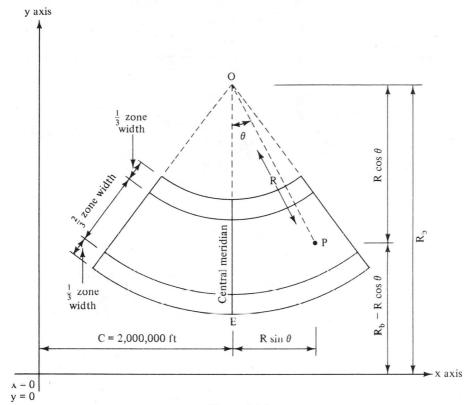

Figure 24-6

which converge at point O. The parallels of latitude are arcs of con-
centric circles which have their centers at point O (see Figs. 24-4 and
24-6).

Line OP represents part of a meridian through point O and has a
length equal to R. The y coordinate of point O is a constant equal to
R_b. The angle θ between lines OE and OP is called the *mapping angle*
or *convergence angle.*

The National Geodetic Survey has available for each state using
the Lambert projection the necessary information needed to use plane
coordinates. Included in this information are values of R for each
whole minute of latitude and values of θ for each whole minute of
longitude as well as the latitudes and longitudes for each of their control stations. Tables N and O in the appendix show this type of information. The calculations necessary to compute state plane coordinates
for these stations are illustrated in this section even though the National
Geodetic Survey provides these coordinates for many of their stations.

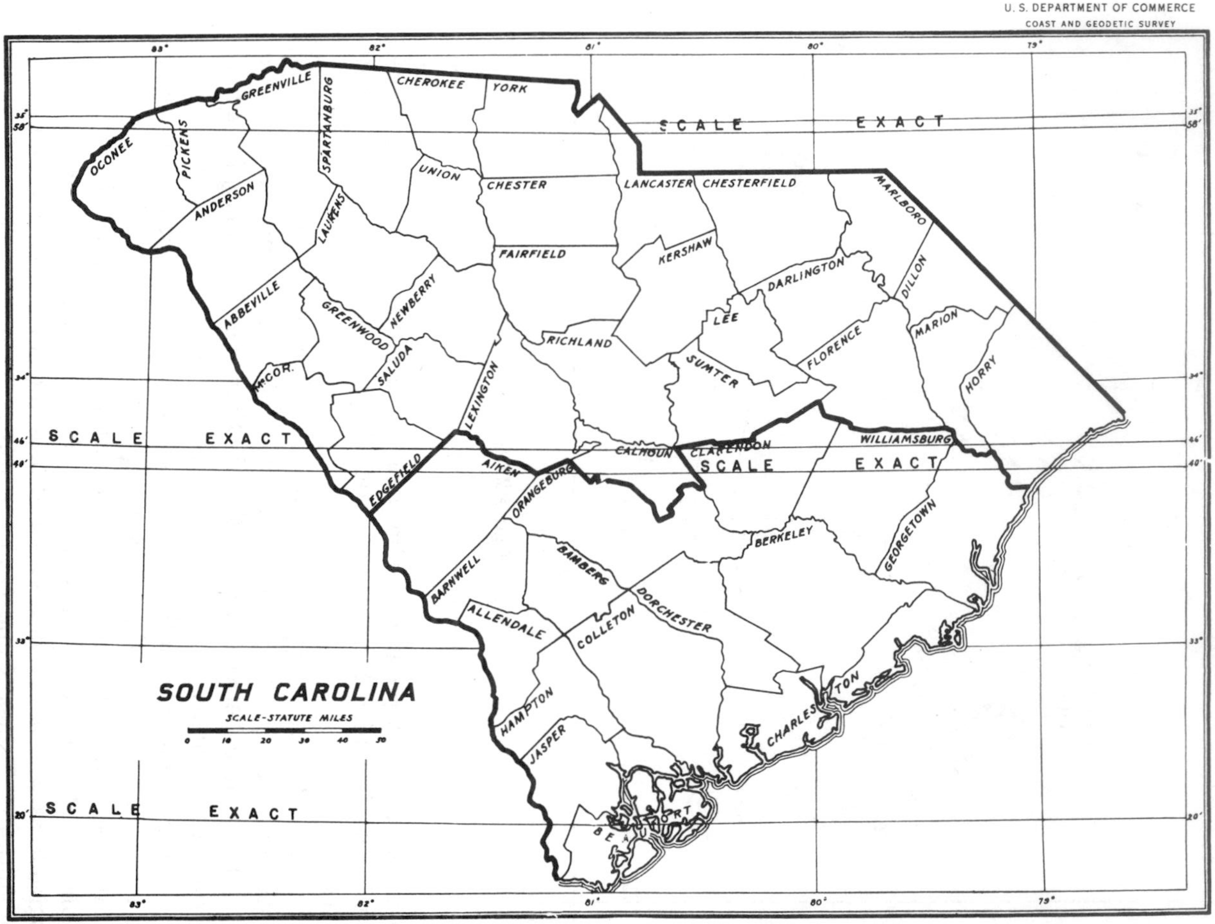

STATE PLANE-COORDINATE ZONES

Figure 24-7

From Fig. 24-6 the x and y plane coordinates of point P can be seen to equal the following:

$$x = R \sin \theta + C$$

$$y = R_b - R \cos \theta$$

If the angle θ is to the left of the central meridian it will have a negative value and thus the value of $R \sin \theta$ will be negative for computing the x coordinate. In computing the y coordinate for a point $R \cos \theta$ will always be subtracted. Example 24-1 which follows illustrates the determination of the Lambert plane coordinates for a particular station in South Carolina.

South Carolina is divided into two Lambert zones, north and south, both having a central meridian of 81°00'00"W. The north zone has standard parallels at 33°46' and 34°58'N latitudes while the south zone has standard parallels at latitudes 32°20' and 33°40'N as shown in Fig. 24-7 where they are labeled *Scale Exact. It should be noted in this figure that the exact division between the north and south zones is not a line of constant latitude. Rather the dividing line follows the somewhat irregular county lines. This means that all counties fall completely in one zone or the other.*

It is considered necessary to make calculations for plane coordinates to at least ten place accuracy and this practice is followed in this chapter. Trigonometric functions can be determined to ten places with some pocket and desk calculators. Special Publication No. 246 of the U.S. Coast and Geodetic Survey entitled "Sines, Cosines, and Tangents Ten Decimal Places With Ten Second Interval 0°-6°," can be helpful in this regard. This publication can be obtained from the U.S. Department of Commerce, National Oceanic and Atmospheric Administration, National Geodetic Survey, Rockville, MD 20852.

Tables N and O of the appendix are taken from Special Publication No. 273 of the U.S. Department of Commerce Coast and Geodetic Survey entitled "Plane Coordinate Projection Tables South Carolina (Lambert)." The reader can obtain the corresponding publication for his state by writing the address given in the preceding paragraph. Table N provides information as to R values for different latitudes while Table O provides values of θ for different longitudes. These values are needed for the solution of Example 24-1.

Example 24-1

Determine the x and y Lambert coordinates for the following NGS monument in the South Carolina North Zone:

Name and location of station: "Blaney," 12 miles southeast of Camden, SC

Geodetic latitude = 34°10'18.873"N

Geodetic longitude = $80°47'30.989''$W

$$C = 2,000,000.00 \text{ ft}$$

R_b for North Zone of SC = 31,127,724.75 ft

Solution

R = 30,703,171.70 at $34°10'$N   (Table N)

minus correction for $18.873''$ further north   (Table N)

= 30,703,171.70 - (18.873)(101.08333)

= 30,701,263.95 ft

θ = $+0°07'20.3080''$ for longitude $80°47'$W   (Table O)

$$\text{minus} \left(\frac{30.989}{60}\right) \text{ (change in } \theta \text{ from } 80°47' \text{ to } 80°48'\text{W longitude)}$$

$$= +0°07'20.3080'' \text{ minus} \left(\frac{30.989}{60}\right) (33.8699)$$

$$= +0°07'2.8148''$$

Then

$x = R \sin \theta + C$

= (30,701,263.95)(0.0020498626) + 2,000,000.00

= **2,062,933.373 ft**

$y = R_b - R \cos \theta$

= 31,127,724.75 - (30,701,263.95)(0.9999978990)

= **426,525.3024 ft**

24-11 GRID AZIMUTHS

At a distance of from about $\frac{1}{4}$ to 2 miles from each triangulation station the National Geodetic Survey has established a monument or azimuth mark which can be sighted on from the station. The grid azimuth from the station to the mark or monument is usually given together with the other data for the station. The reader should be sure that he understands the azimuth definition here. A true or geodetic azimuth is one which takes into account the convergence of the meridians while a plane or grid azimuth is constant for the entire zone (that is, parallel to the central meridian). Thus grid azimuths and geodetic azimuths will be identical only for points located on the central merid-

ian. It will also be remembered that south azimuths are used by the
National Geodetic Survey, although the NGS is planning to change to
north azimuths in 1983 or 1984.

The further a point is located from the central meridian the
greater will be the difference between the grid and geodetic azimuths.
The difference between the two is *substantially* equal to the angular
convergence between the central meridian and a true meridian passing
through the point in question. This difference is denoted as θ in the
Lambert projection and $\Delta\alpha$. in the Mercator projection. In the difference for both systems there is a second term (due to meridian curvature) which may be neglected for all third-order traverses except those
in which the orientation sight is more than 5 miles. For points to the
east of the central meridian the grid azimuth is less than the geodetic
azimuth while for points to the west of the central meridian is greater
than the geodetic azimuth. This situation is illustrated in Fig. 24-8.

To make use of the plane coordinate system in a particular state,
the local surveyor sets up over one of the triangulation stations sights
on the line of known grid azimuth and then runs his survey by traversing (or perhaps by triangulation or trilateration) to the area under
consideration. He then by the usual method of plane coordinates

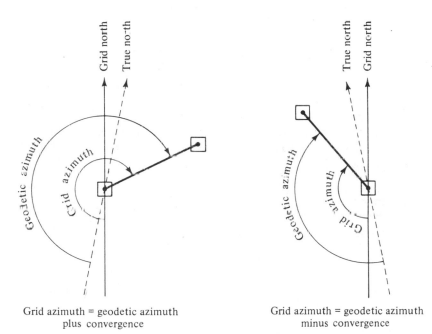

Grid azimuth = geodetic azimuth   Grid azimuth = geodetic azimuth
      plus convergence   minus convergence

Figure 24-8

determines the coordinates of any point in his survey. Of course if the plane coordinates of two stations are known, the grid azimuth between these points can be determined by the following expression:

$$\tan \text{azimuth} = \frac{\Delta x}{\Delta y}$$

where Δx and Δy are the differences between the x coordinates and the y coordinates of the two stations respectively.

If the azimuth mark is not available or if the grid azimuth to that point is not known at a particular station, it will be necessary for the surveyor to establish a true azimuth by means of an astronomic observation. The true azimuth obtained can then be converted to a grid azimuth by one of the expressions:

For Lambert

   Grid azimuth = Geodetic azimuth − θ + Second term

For Mercator

   Grid azimuth = Geodetic azimuth − $\Delta \alpha$ − Second term

24-12 DETERMINATION OF GRID DISTANCES FROM MEASURED GROUND DISTANCES

Before the measured ground distances between traverse points and the grid azimuths or bearings can be used to calculate grid coordinates, it is necessary to reduce the measured distances to sea level or that is to their equivalent geodetic distances. Then these distances must have scale corrections applied for the particular state plane system being used. These two conversions are considered in the paragraphs to follow where the grid distance to be used equals the measured ground distance times the sea-level factor times the scale factor.

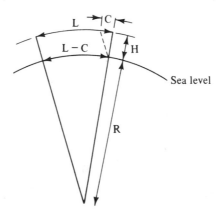

Figure 24-9

Reduction to sea level.   It is necessary to make grid projections on a sea-level basis. For measurements taken at other elevations than sea level the ground distances can be converted to sea-level distances by similar triangles. For this discussion reference is made to Fig. 24-9 in which R, the mean radius of the earth, is assumed to equal 20,906,000 ft (for Alaska and Hawaii different values are used). Also in the figure H is the elevation of the line above sea level and C is the correction to be made to the measured ground distance L to obtain the geodetic distance.

$$\frac{C}{L - C} = \frac{H}{R}$$

$$C = \frac{H}{R}(L - C)$$

$$CR = HL - CH$$

$$CR + CH = HL$$

$$C(R + H) = HL$$

$$C = \frac{HL}{(R + H)}$$

But the value of H is so small as compared to R that it can be neglected and the expression becomes

$$C = \frac{HL}{R} = \frac{HL}{20,906,000}$$

It is common to convert the measured ground distance to sea level by multiplying it by a factor called the *sea-level factor* (SLF). This factor equals the following:

$$SLF = \frac{L - C}{L} = 1 - \underbrace{\frac{HL}{R}}_{L} = 1 - \frac{H}{R}$$

$$SLF = 1 - \frac{H}{20,906,000}$$

Scale correction.   As previously indicated (see Figs. 24-2 and 24-5) there will be errors in scale unless the points in question are exactly at the intersection lines. The scale correction can be taken from prepared tables of the National Geodetic Survey. With the Lambert projection the scale factor tables are entered with the latitude. When working with a particular traverse it is necessary to determine the mean latitude for the line. This can be done by plotting the traverse lines to scale in the appropriate direction and scaling off the distance parallel to grid north from a station of known latitude. The distance can also be com-

puted by multiplying the length of each line by the cosine of its bearing.
The distance from the reference station can be converted to seconds of
latitude with sufficient accuracy by dividing the distance by 100. The
scale factors for each state are published by the National Geodetic
Survey. In Table N of the appendix, scale factors are given for South
Carolina for the Lambert projection. Scale factors for all other states
and the projection used are available from the National Geodetic Survey. A similar discussion can be made for the Mercator scale factor
where the tables are entered with the longitude.

24-13 CALCULATION OF STATE PLANE COORDINATES

Example 24-2 illustrates the calculations of state plane coordinates for
a traverse point in South Carolina. The problem is worked with respect
to a National Geodetic station for which the following detailed description applies.

> Blaney (Kershaw County, C. L. Garner, 1918; 1934).—About 20 miles
> north-east of Columbia, 12 miles southeast of Camden, at village of Blaney, on
> top of high bank, 300 meters (984 feet) east-northeast of railway station, about
> 100 meters (328 feet) north-northwest of center line of U.S. Highway 1,
> directly opposite Blaney High School, and 68.4 feet south-southeast of south-
> southeast rail of main track of Seaboard Air Line Railway. Surface mark,
> standard disk in concrete, note 1a, was reported in 1934 as having been re-
> moved. Underground mark, recovered in 1934, is bottle in concrete, note 7d,
> 2 feet below surface of ground. Reference mark, standard disk in concrete,
> note 11e, is on top of bank. 4.21 meters (13.8 feet) east of east rail of Sea-
> board Air Line Railway, about 4 meters (13 feet) above roadbed, 3 feet east
> of edge of cut and 20.78 meters (68.2 feet) from station in azimuth 80°05′.
> Station *Blaney 2* (see description thereof) is 574.03 feet from station in
> azimuth 304°03′53″.
> Plane coordinates:** (N), x = 2,062,933.37 feet; y = 426,525.30 feet.

For this example the reader should carefully note the manner in
which the grid bearings were obtained. The true bearings of the sides of
the traverse were measured in the field. Each of these bearings was
converted to grid bearings by adding or subtracting the angle θ. To
decide whether to add or subtract θ, a sketch was made for each side
showing true north, grid north, and the true bearing. From these
sketches the correct grid bearings are obvious. One of the sketches is
shown in Fig. 24-10 where θ is + or east 0°07′03″ and the true bearing
of the line is S85°22′36″W. (This is line BC in Example 24-2.) The
grid bearing of the line equals its true bearing minus θ in this case.

Using the grid bearings and distances the east and west projections
of each side of the traverse are computed and the state plane or grid
coordinates of any point of the traverse can be computed. For the

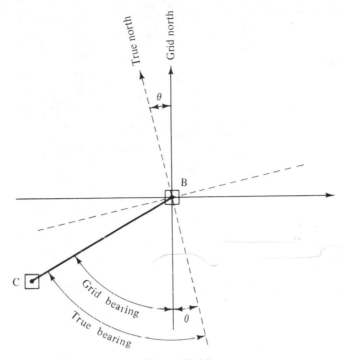

Figure 24-10

solution of the problem use is made of the information contained in Tables N and O of the appendix.

Example 24-2

A traverse is run from NGS monument "Blaney" (a description of which was presented earlier in this section) to a property corner D with the following results:

Line	True Bearing	Distance Measured	Average Elevation
AB	N71°16′20″W	3622.20 ft	400 ft
BC	S85°22′30″W	4736.91 ft	360 ft
CD	S19°46′11″W	2942.41 ft	320 ft

"Blaneys" x and y grid coordinates are 2,062,933.37 ft and 426,525.30 ft respectively as given in the description of the station and as previously calculated in Example 24-1. Its geodetic latitude is 34°10′18.873″N and its geodetic longitude is 80°47′30.989″W. The value of θ is 0°07′03″ as determined from Table O for Example 24-1.

Compute the grid bearing of the traverse lines and the grid coordinates of property corner D.

Solution

Side	Sea-Level Factor = 1 − $\dfrac{\text{Elevation}}{20{,}906{,}000}$	Scale Correction Factor from Table N in Appendix (computed at mean latitude of each line)	Grid Distance = Distance Measured × Sea-Level Factor × Scale-Correction Factor
AB	0.9999808667	0.9999511	3621.953573 ft
BC	0.9999827801	0.9999511	4735.926844 ft
CD	0.9999846934	0.9999513	2942.221668 ft

Grid Bearing = True Bearing ± θ as illustrated in Fig. 24-10 for Line BC	Cosine	Sine	Latitude		Departure	
			N	S	E	W
N71°23′23″W	0.3191293159	0.9477111795	1155.872	X	X	3432.566
S85°15′33″W	0.0826487674	0.9965787381	X	391.419	X	4719.724
S19°39′08″W	0.9417513143	0.3363100683	X	2770.841	X	989.499
			Total = 2006.388S		Total = 9141.789W	

Coordinates for point D:

$x = 2{,}062{,}933.37 - 9141.79 = 2{,}053{,}791.58$ ft.

$y = 426{,}525.30 - 2006.39 = 424{,}518.91$ ft

24-14 DETERMINATION OF ANGLE BETWEEN GEODETIC
AND GRID AZIMUTHS

Example 24-3 shows the calculation of the angle between true north and grid north at point D of the traverse considered in Example 24-2. In Fig. 24-11 point O is the apex of the cone for the Lambert system for the north zone of South Carolina. It has an x coordinate of 2,000,000.00 ft and a y coordinate of $R_b = 31,127,724.75$ ft. The grid or state plane coordinates of point D were determined in Example 24-2 and from them it is possible to determine the x distance (x') and the y distance $(R_b - y)$ from point O to point D as shown in the figure.

With the x' and $R_b - y$ values known it is possible to determine the angle θ which is the angle between true north and grid north at point D:

$$\theta = \tan^{-1} \frac{x'}{R_b - y}$$

Example 24-3

Determine the angle between geodetic north and grid north at point D of the traverse of Example 24-2.

Solution

$$x' = 2,053,791.58 - 2,000,000.00$$

$$= 53,791.58 \text{ ft.}$$

$$R_b - y = 31,127,724.75 - 424,518.91$$

$$= 30,703,205.84 \text{ ft}$$

$$\theta = \tan^{-1} \frac{x'}{R_b - y} = \tan^{-1} \frac{53,791.58}{30,703,205.84}$$

$$= 0.0017519857$$

$$\theta = \tan^{-1} (0.1003812882)$$

$$\theta = 0°06'1.37''$$

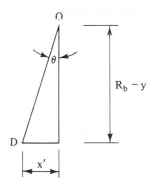

Figure 24-11

24-15 CALCULATION OF GEODETIC POSITION
FROM STATE PLANE COORDINATES

The geodetic position or that is the geodetic latitude and longitude of a point can be determined from the grid coordinates in approximately the reverse order to that used to determine the grid coordinates from the geodetic position illustrated in Example 24-1. Such a procedure is described here.

Geodetic latitude. The grid coordinate y of a particular point has previously been written as follows for the Lambert system:

$$y = R_b - R \cos \theta$$

Thus R is:

$$R = \frac{R_b - y}{\cos \theta}$$

With the value of R known, the latitude ϕ can be determined from Table O in the Appendix.

Geodetic longitude. The convergence angle θ can be determined as illustrated in Example 24-3. Then $\Delta y = \theta / l$ where l is a constant for the particular Lambert zone. The longitude λ is determined as follows:

$$\lambda = \text{Longitude} = \text{Central meridian} - \Delta y$$

Example 24-4 which follows presents the calculations necessary to determine the geodetic position of point D of Example 24-2.

Example 24-4

Determine the latitude and longitude of point D of Example 24-2.

Given values

$$C = 2,000,000.00 \text{ ft}$$

$$R_b = 31,127,724.75 \text{ ft for north zone}$$

$$\left. \begin{array}{l} x = 2,053,791.58 \text{ ft} \\ y = 424,518.91 \text{ ft} \end{array} \right\} \text{from Example 24-2}$$

$$\theta = 0.1003812882°\} \text{ from Example 24-3}$$

Solution

$$R = \frac{R_b - y}{\cos \theta} = \frac{31,127,724.75 - 424,518.91}{0.99999847}$$

$$= 30,703,252.82 \text{ ft}$$

$$\text{Latitude} = 34°09'59.198'' \text{ from Table N in Appendix}$$

$$\Delta y = \frac{\theta}{l} \text{ where } l \text{ is a constant for the north zone}$$

$$= 0.56449738$$

$$\Delta y = \frac{0.1003812882}{0.56449738} = 0.1778234188$$

$$= 0°10'40.167''$$

$$\lambda = \text{Longitude} = 81°00'00.000'' - 0°10'40.167''$$

$$= 80°49'19.833''$$

24-16 CALCULATION OF PLANE COORDINATES FOR MERCATOR GRID

For the Mercator projection the central meridian is set at an x distance usually equal to 500,000 ft for most states while the x axis as in the Lambert projection is placed well below the southern edge of the zone. Again the x and y coordinates of all points in the projected area will be positive.

The x and y Mercator grid coordinates of some point P can be determined with the equation given at the end of this paragraph in which $\Delta\lambda''$ is the difference in seconds between the longitude of the central meridian and point P. The value of $\Delta\lambda''$ will be positive if P is to the east of the central meridian and negative if it's to the west. In the expressions to follow x' is the distance to point P either east or west of the central meridian. In the expression for x and y the terms y_0, H, V, and a are values based on the geographic latitude while b and c are based on $\Delta\lambda''$. The magnitudes of these values are given in the tables available from the NGS for individual states. If the sign of ab is positive it increases the value of x'; a negative value decreases it.

$$x' = H \cdot \Delta\lambda'' \pm ab$$

$$x = x' + 500,000$$

$$y = y_0 + V \left(\frac{\Delta\lambda''}{100}\right)^2 \pm c$$

Space is not taken here to present a transverse Mercator projection example as it is felt that if the surveyor can apply one system he will easily be able to use the other one.

24-17 CONCLUDING REMARKS ON STATE PLANE
COORDINATE SYSTEMS

Sometimes surveys are made in border areas in between different zones or may even extend over into an adjoining state. Such a situation really does not cause a problem as the surveyor can easily change his state plane coordinates into those of the other zone or state. This is done for a point by converting the state plane coordinates of the point to its geodetic latitude and longitude. Then with these values known the state plane coordinates of the other zone or state can be computed.

The Lambert and Mercator projections which have been briefly described in this chapter are not the only coordinate systems available. These two systems were developed to be used for areas primarily running long distances in the east–west or north–south directions respectively. They are not conveniently applied to long areas which run in other directions or for circular areas. Other systems have been proposed for such areas such as the *oblique Mercator projection* (now used for parts of Alaska and by the U.S. Lake Survey for the area of the Great Lakes Erie and Ontario and the St. Lawrence River) and the *horizon stereographic projection* (used in Canada and some other countries).[2] Among other proposals being considered is one that suggests that a national Transverse Mercator system be used using $2°$ bands of longitude.

The use of state plane coordinates is gradually increasing not only by governmental agencies but by private organizations as well. State plane coordinate systems can be used to show the location of important government or private structures such as dams, pipelines, bridges, oil wells, industrial plants, and many more. For such projects good vertical control is needed as well as the horizontal control provided by the coordinates.

[2] R. C. Brinker and P. R. Wolf, *Elementary Surveying*, 6th ed. (New York: Intext Educational Publisher, 1977), pp. 360–361.

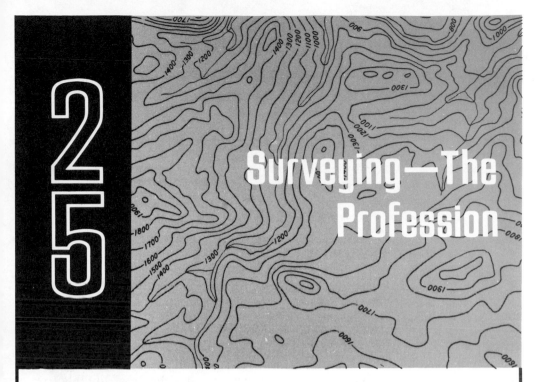

25-1 SURVEYING LICENSES

The purpose of building codes, medical codes, surveying registration requirements, and so on, is to protect the public from unqualified and/or unscrupulous people in these fields. Perhaps the first regulations of this type were contained in the code of laws of Hammurabi, who was king of Babylon in the eighteenth century B.C. His code covered many different subjects (including prohibition), but only his building code is mentioned here. It was famous for its "eye for an eye and tooth for a tooth" section. In general, the code said that if a builder constructed a house that collapsed and killed the owner, the builder would be killed. If the collapse caused the death of the owner's son, the builder's son would be killed and so on. These laws were presumably very effective in making the builder put forth his best efforts.

Through the centuries since the time of Hammurabi many codes and laws have been established to guide the various professions, but it was not until 1883 that registration requirements reached the United States. At that time registration of dentists began and as the years went by doctors, lawyers, pharmacists, and others were required to obtain licenses before they could practice their professions.

Licensing of surveyors is not new. George Washington and Abraham Lincoln held surveyor's licenses, but licensing as it is known today for engineers and surveyors was begun in Wyoming in 1907. Today, the law in all 50 states, as well as Canada, Puerto Rico, and Guam, require that a person must meet certain licensing requirements before he may practice land surveying. For the purposes of such licensing, land surveying is generally said to include: the determination of areas of tracts of land, the surveying needed for preparing descriptions of land for deed conveyance, the surveying necessary for establishing or reestablishing land boundaries, and the preparation of plats for land tracts and subdivisions. A license is usually not required for construction surveys and for route surveys (roads, railroads, pipelines, etc.) unless property corners are set.

25-2 REGISTRATION REQUIREMENTS

In order to obtain a land surveyor's license, it is necessary for a person to meet the requirements of his state board of engineering examiners. (A few states have a separate land surveying board.) Although these requirements vary considerably from state to state, they in general require: (1) graduation from a school or college approved by the state board including the completion of an approved course in surveying, (2) two or more years of surveying experience of a character suitable to the board, and (3) successful completion of a written examination under the supervision of the state board. If the applicant is unable to meet the formal education requirements listed in (1), he will probably be required to obtain additional experience (perhaps as much as four or more years) before he will be permitted to take the written exam.

In 1973 the National Council of Engineering Examiners (NCEE) began offering a semiannual national surveying exam. This fundamental exam, which is of a basic and general type, consists of a half-day of multiple choice questions and a half-day of more detailed problems. Since the laws pertaining to land description (metes and bounds, co-ordinates, public land surveys, and so on) are different from state to state, the individual states using the exam will probably require the taking of an additional exam that relates more specifically to work in their states.

As the years go by, more and more states will probably use the national exam with the result that it will become easier for a person who is registered in one state to become registered in another. Eventually, it is hoped that a surveyor who passes the national exam in one state will be permitted to obtain with little or no delay and few additional requirements a license in another state. Such reciprocity is rapidly becoming a reality for professional engineering registration as a result of the national exams given by the NCEE in which nearly all states participate.

25-3 PENALTIES FOR PRACTICING SURVEYING
WITHOUT A LICENSE

The various states have laws prohibiting a person from practicing land surveying without a license, from using a false or forged license or registration number for same, or from using an expired license. These laws usually call for fines or imprisonment, or both, for such violations. Of course, the ordinary person can measure land as long as he neither represents himself as a registered surveyor nor uses or permits the measurements he makes to be used for deed conveyance or other formal legal purposes.

A surveyor is expected to be competent in his work. He can't be perfect, of course, but if he is grossly negligent, incompetent, or guilty of misconduct, he may lose his license. Furthermore, the state board may revoke a land surveyor's license if it discovers that there was any fraud or deceit involved in obtaining the license.

25-4 REASONS FOR BECOMING REGISTERED

There are several reasons why a person interested in surveying should work toward becoming licensed as soon as possible. These include the following:

1. The desire for obtaining a license encourages a person to study and improve his technical ability and aids in his professional development.
2. A surveyor may be offered a job that requires registration.
3. Registration raises the status of the profession as a whole.
4. If a surveyor is registered, he may be able to perform part-time professional surveying or even enter into full-time practice in his own business.
5. Registration gives a person status as a professional in his community.

25-5 A PROFESSION

The term *profession* has many definitions. In a narrow sense the professions are limited to doctors, lawyers, and clergymen, but in a broad sense they include almost any occupation. In general, a professional is a person who has acquired some special knowledge that he uses to instruct, help, advise, or guide others.

The primary objective of a profession is service to humanity without regard to financial reward. Not only does this mean that a professional has a responsibility to place his job above the amount of money he is to receive, but it also means that he has a responsibility to give voluntary service to his community when needed without any financial compensation. The surveyor should clearly understand that a professional who performs free services (or acts as a gratuitous agent) is legally required to exercise the same high level of care, skill, and judgment that he would if he were being paid.

A true profession is said to have four basic elements: *organization*, *education*, *experience*, and *exclusion*. Organization means membership and participation in a professional organization. For surveyors, this organization might be the American Congress on Surveying and Mapping (ACSM) or perhaps his own state's surveying society. The young surveyor might feel that the dues for these organizations are too high, but his membership in them is a first step toward obtaining the recognition and status of a true professional man. As the old saying goes "You get out of an organization what you put into it," and such participation will lead to that most important feeling of belonging.

Generally, education means the completion of as many surveying courses as possible and obtaining formal school degrees, but it can be, and often is, *self-education*. In addition, education must be continuing through self-study, short courses, and perhaps more formal study.

Experience is obtained over the years and is a gradual transformation obtained by undertaking specific tasks.

A real profession should require the strict exclusion of those who are unfit or unworthy. Exclusion may be accomplished by the state surveying licensing requirements and by a code of ethics or code of professional conduct. Undoubtedly, this has been a weak point in the surveying profession throughout most of the country—the machinery for expelling registered land surveyors for incompetence or unethical behavior is not often applied.

25-6 CODE OF ETHICS

The most important possession a man owes himself is a spotless reputation. No amount of money, fame, or knowledge is an adequate substitute. A man who can be depended on and who will work hard and learn as much as he can about his profession is going to be successful.

The reader is quite well aware that a person may live strictly within the confines of the law and yet be an undesirable citizen. Sadly, there are always people who will operate up to the very limits of and through any loopholes they can find in the law. But to a true professional man a legal right is not a right unless it is also morally right.

Ethics may be defined as the duties that a professional man owes to the public and to his fellow professionals. Such a code is a helpful guide to the members of a profession to help them know what standards they should live up to and what they can expect from their fellows.

The earliest known code of ethics is the Hippocratic oath of the medical profession, which is attributed to Hippocrates, a Greek physician (460–377 B.C.). The present form of the Hippocratic oath dates from approximately 300 A.D., but a detailed code of ethics for the medical profession did not appear in the U.S. until 1912. The code of ethics for the American Society of Civil Engineers was adopted in 1914.

A code of ethics is not intended to be a lengthy detailed statement of "thou shall nots," but rather a few general statements of noble motives expressing concern for the welfare of others and the standing of the profession as a whole. The classic comparison of law and ethics is made by relating them to medicine and hygiene. The object of medicine is to cure diseases and the object of law is to cure or repress evil. The object of hygiene is to prevent illness and the object of ethics is to prevent evil by raising the moral plane on which men deal with each other.

Another classic comparison of law and ethics is frequently made with relation to the Old and New Testaments. Essentially, the Old Testament is a list of "thou shall nots," such as "thou shall not kill, thou shall not steal," and so on; the New Testament is based on a concern for the welfare of our fellow man and a desire to do good because it is a privilege and not a necessity.

Many groups of land surveyors, primarily state societies, have published codes of ethics. In general, these codes are similar to each other and to the codes of the various engineering societies.

Perhaps the purpose and heart of a code of ethics may be summarized in a few sentences: The surveyor must faithfully and impartially perform his work with fidelity to his clients, his employer, and the public. (For instance, if he is marking a property corner, its location will be the same regardless of which property owner is paying him.) He will be seriously concerned with the standing of his profession in the public's eye and will not only strive to live and work according to a high standard of behavior, but, in addition, will avoid association with persons or enterprises of questionable character. In other words, he will not only be concerned with evil but with the very appearance of evil. He will further be actively interested in the welfare of the public and will always be ready to apply his knowledge for the benefit of mankind.

When a Roman soldier encountered a Jew in biblical times, he could require the Jew to carry his pack for a mile. Jesus said (Matthew 5:41) that he should carry it for two miles (thus the origin of the expression "go the extra mile"). This idea is the theme of a code of ethics.

Instead of reproducing one of the codes of ethics, the author is devoting the remainder of this section to the following general statements that summarize what a code of ethics means to him:

1. The surveyor must not place monetary values above other values. Although it may seem difficult to apply this to specific cases, it simply means that the surveyor should never recommend to his client a course of action that is based on the amount of money that the surveyor will thereby receive.

2. In the course of his work he may very well acquire knowledge that could be detrimental to his client if it were revealed to others. His responsibility to his client goes beyond the immediate job and he must not reveal private information concerning his client's business without the client's permission.

3. In his concern for the reputation of his profession he must refrain from speaking badly of other surveyors or he will be lowering the profession in the eyes of the public. This does not mean that there is not a time and place for the honest appraisal of other surveyors. A surveyor is far better able to judge the work of a fellow surveyor than is anyone else, whether he be lawyer, judge, or layman.

4. In his further concern for the standing of his profession he must not become professionally associated with surveyors who do not conform to the standards of ethical practice discussed in this section. Furthermore, he will not become involved in any partnership, corporation, or other business group that is a cloak for unethical behavior. The surveyor must accept full responsibility for his work.

5. He will not be too proud to admit that he needs outside advice in order to solve a particular problem.

6. He will admit and accept his own mistakes.

7. When on a salaried job he will not do outside work to the detriment of his regular job. Further, he will not use such a job to compete unfairly with surveyors in private practice.

8. He will not agree to perform free surveying (except for community service) because he will be taking work away from his profession.

9. He will not advertise in a self-laudatory or blatant manner or in any other way that might be detrimental to the dignity of the profession.

10. The surveyor has a duty to increase the effectiveness of his profession by cooperating in the exchange of information and experience with other surveyors and students, by contributing to the work of surveying societies and by doing all he can to further the public's knowledge of surveying.

11. He will encourage his employees to further their education, to attend and participate in professional meetings, and to become registered. He will do all he can to provide opportunities for the professional development and advancement of surveyors under his supervision.

12. He will not review the work of another surveyor without the knowledge or consent of that surveyor or unless the work has been terminated and the other surveyor has been paid for same.

13. He will always be greatly concerned with the safety and welfare of the public and his employees.

14. He will conform with the registration laws of his state.

Many people say that codes of ethics have not accomplished their purpose. The author likes to think that they have helped because they give surveyors a goal to strive for and because they make them look toward a higher moral plane. The author attended a speech by the late Winston Churchill in Boston in 1949 in which he said that the flame of religious ethics was still our highest guide and that to guard it and cherish it was our first interest in the world.

25-7 TO BE CLASSED AS A PROFESSIONAL

For surveying to be truly classed as a profession, the average standing of the group as a whole must be raised in the eyes of the public. This can only be done by having the title bestowed upon the group by the public because the public recognizes the group's high level of technical and ethical performance.

Although surveyors and engineers are highly respected by the community, they have not fully arrived as a profession in everyone's eyes. As a matter of fact, only doctors, lawyers, and clergymen have achieved this status. These three groups don't have to call themselves professional doctors or professional lawyers or professional clergymen because everyone recognizes them as professionals. But surveyors and engineers have persuaded government bodies to legislate titles for them as "professional surveyors" and "professional engineers."

No amount of self-proclamation or legislative action can achieve a true title as a profession. It can only be obtained by the actions of the group over a long period of time.

25-8 CONCLUSION

This book is an introduction to surveying and anyone planning to follow the subject as a career must continue his studies. Only a small percentage (if any at all) of those entering the surveying field have a completely adequate background for the work that they will face. The answer for most lies in many hours of *self-study*.

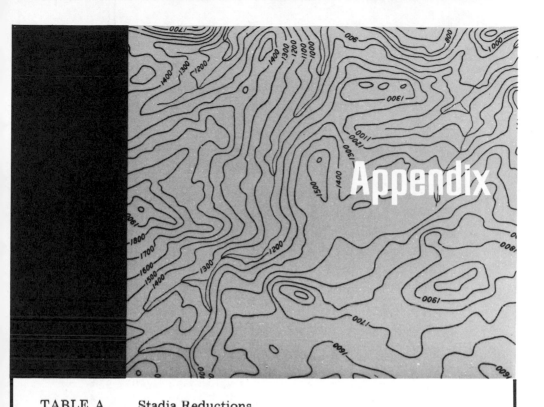

Appendix

TABLE A Stadia Reductions

Minutes	0° Hor. Dist.	0° Diff. Elev.	1° Hor. Dist.	1° Diff. Elev.	2° Hor. Dist.	2° Diff. Elev.	3° Hor. Dist.	3° Diff. Elev.
0	100.00	.00	99.97	1.74	99.88	3.49	99.73	5.23
2	100.00	.06	99.97	1.80	99.87	3.55	99.72	5.28
4	100.00	.12	99.97	1.86	99.87	3.60	99.71	5.34
6	100.00	.17	99.96	1.92	99.87	3.66	99.71	5.40
8	100.00	.23	99.96	1.98	99.86	3.72	99.70	5.46
10	100.00	.29	99.96	2.04	99.86	3.78	99.69	5.52
12	100.00	.35	99.96	2.09	99.85	3.84	99.69	5.57
14	100.00	.41	99.95	2.15	99.85	3.89	99.68	5.63
16	100.00	.47	99.95	2.21	99.84	3.95	99.68	5.69
18	100.00	.52	99.95	2.27	99.84	4.01	99.67	5.75
20	100.00	.58	99.95	2.33	99.83	4.07	99.66	5.80
22	100.00	.64	99.94	2.38	99.83	4.13	99.66	5.86
24	100.00	.70	99.94	2.44	99.82	4.18	99.65	5.92
26	99.99	.76	99.94	2.50	99.82	4.24	99.64	5.98
28	99.99	.81	99.93	2.56	99.81	4.30	99.63	6.04
30	99.99	.87	99.93	2.62	99.81	4.36	99.63	6.09
32	99.99	.93	99.93	2.67	99.80	4.42	99.62	6.15
34	99.99	.99	99.93	2.73	99.80	4.47	99.61	6.21
36	99.99	1.05	99.92	2.79	99.79	4.53	99.61	6.27
38	99.99	1.11	99.92	2.85	99.79	4.59	99.60	6.32
40	99.99	1.16	99.92	2.91	99.78	4.65	99.59	6.38
42	99.99	1.22	99.91	2.97	99.78	4.71	99.58	6.44
44	99.98	1.28	99.91	3.02	99.77	4.76	99.58	6.50
46	99.98	1.34	99.90	3.08	99.77	4.82	99.57	6.56
48	99.98	1.40	99.90	3.14	99.76	4.88	99.56	6.61
50	99.98	1.45	99.90	3.20	99.76	4.94	99.55	6.67
52	99.98	1.51	99.89	3.26	99.75	4.99	99.55	6.73
54	99.98	1.57	99.89	3.31	99.74	5.05	99.54	6.79
56	99.97	1.63	99.89	3.37	99.74	5.11	99.53	6.84
58	99.97	1.69	99.88	3.43	99.73	5.17	99.52	6.90
60	99.97	1.74	99.88	3.49	99.73	5.23	99.51	6.96
*C = .75	.75	.01	.75	.02	.75	.03	.75	.05
C = 1.00	1.00	.01	1.00	.03	1.00	.04	1.00	.06
C = 1.25	1.25	.02	1.25	.03	1.25	.05	1.25	.08

*These C values are discussed in Section 16-9.

TABLE A Stadia Reductions (*Cont.*)

Minutes	4° Hor. Dist.	4° Diff. Elev.	5° Hor. Dist.	5° Diff. Elev.	6° Hor. Dist.	6° Diff. Elev.	7° Hor. Dist.	7° Diff. Elev.
0	99.51	6.96	99.24	8.68	98.91	10.40	98.51	12.10
2	99.51	7.02	99.23	8.74	98.90	10.45	98.50	12.15
4	99.50	7.07	99.22	8.80	98.88	10.51	98.49	12.21
6	99.49	7.13	99.21	8.85	98.87	10.57	98.47	12.27
8	99.48	7.19	99.20	8.91	98.86	10.62	98.46	12.32
10	99.47	7.25	99.19	8.97	98.85	10.68	98.44	12.38
12	99.46	7.30	99.18	9.03	98.83	10.74	98.43	12.43
14	99.46	7.36	99.17	9.08	98.82	10.79	98.41	12.49
16	99.45	7.42	99.16	9.14	98.81	10.85	98.40	12.55
18	99.44	7.48	99.15	9.20	98.80	10.91	98.39	12.60
20	99.43	7.53	99.14	9.25	98.78	10.96	98.37	12.66
22	99.42	7.59	99.13	9.31	98.77	11.02	98.36	12.72
24	99.41	7.65	99.11	9.37	98.76	11.08	98.34	12.77
26	99.40	7.71	99.10	9.43	98.74	11.13	98.33	12.83
28	99.39	7.76	99.09	9.48	98.73	11.19	98.31	12.88
30	99.38	7.82	99.08	9.54	98.72	11.25	98.30	12.94
32	99.38	7.88	99.07	9.60	98.71	11.30	98.28	13.00
34	99.37	7.94	99.06	9.65	98.69	11.36	98.27	13.05
36	99.36	7.99	99.05	9.71	98.68	11.42	98.25	13.11
38	99.35	8.05	99.04	9.77	98.67	11.47	98.24	13.17
40	99.34	8.11	99.03	9.83	98.65	11.53	98.22	13.22
42	99.33	8.17	99.01	9.88	98.64	11.59	98.20	13.28
44	99.32	8.22	99.00	9.94	98.63	11.64	98.19	13.33
46	99.31	8.28	98.99	10.00	98.61	11.70	98.17	13.39
48	99.30	8.34	98.98	10.05	98.60	11.76	98.16	13.45
50	99.29	8.40	98.97	10.11	98.58	11.81	98.14	13.50
52	99.28	8.45	98.96	10.17	98.57	11.87	98.13	13.56
54	99.27	8.51	98.94	10.22	98.56	11.93	98.11	13.61
56	99.26	8.57	98.93	10.28	98.54	11.98	98.10	13.67
58	99.25	8.63	98.92	10.34	98.53	12.04	98.08	13.73
60	99.24	8.68	98.91	10.40	98.51	12.10	98.06	13.78
$C = .75$	.75	.06	.75	.07	.75	.08	.74	.10
$C = 1.00$	1.00	.08	1.00	.10	.99	.11	.99	.13
$C = 1.25$	1.25	.10	1.24	.12	1.24	.14	1.24	.16

Minutes	8°		9°		10°		11°	
	Hor. Dist.	Diff. Elev.	Hor. Dist.	Diff. Elev.	Hor. Dist.	Diff. Elev.	Hor. Dist.	Diff. Elev.
0	98.06	13.78	97.55	15.45	96.98	17.10	96.36	18.73
2	98.05	13.84	97.53	15.51	96.96	17.16	96.34	18.78
4	98.03	13.89	97.52	15.56	96.94	17.21	96.32	18.84
6	98.01	13.95	97.50	15.62	96.92	17.26	96.29	18.89
8	98.00	14.01	97.48	15.67	96.90	17.32	96.27	18.95
10	97.98	14.06	97.46	15.73	96.88	17.37	96.25	19.00
12	97.97	14.12	97.44	15.78	96.86	17.43	96.23	19.05
14	97.95	14.17	97.43	15.84	96.84	17.48	96.21	19.11
16	97.93	14.23	97.41	15.89	96.82	17.54	96.18	19.16
18	97.92	14.28	97.39	15.95	96.80	17.59	96.16	19.21
20	97.90	14.34	97.37	16.00	96.78	17.65	96.14	19.27
22	97.88	14.40	97.35	16.06	96.76	17.70	96.12	19.32
24	97.87	14.45	97.33	16.11	96.74	17.76	96.09	19.38
26	97.85	14.51	97.31	16.17	96.72	17.81	96.07	19.43
28	97.83	14.56	97.29	16.22	96.70	17.86	96.05	19.48
30	97.82	14.62	97.28	16.28	96.68	17.92	96.03	19.54
32	97.80	14.67	97.26	16.33	96.66	17.97	96.00	19.59
34	97.78	14.73	97.24	16.39	96.64	18.03	95.98	19.64
36	97.76	14.79	97.22	16.44	96.62	18.08	95.96	19.70
38	97.75	14.84	97.20	16.50	96.60	18.14	95.93	19.75
40	97.73	14.90	97.18	16.55	96.57	18.19	95.91	19.80
42	97.71	14.95	97.16	16.61	96.55	18.24	95.89	19.86
44	97.69	15.01	97.14	16.66	96.53	18.30	95.86	19.91
46	97.68	15.06	97.12	16.72	96.51	18.35	95.84	19.96
48	97.66	15.12	97.10	16.77	96.49	18.41	95.82	20.02
50	97.64	15.17	97.08	16.83	96.47	18.46	95.79	20.07
52	97.62	15.23	97.06	16.88	96.45	18.51	95.77	20.12
54	97.61	15.28	97.04	16.94	96.42	18.57	95.75	20.18
56	97.59	15.34	97.02	16.99	96.40	18.62	95.72	20.23
58	97.57	15.40	97.00	17.05	96.38	18.68	95.70	20.28
60	97.55	15.45	96.98	17.10	96.36	18.73	95.68	20.34
$C = .75$	.74	.11	.74	.12	.74	.14	.73	.15
$C = 1.00$	.99	.15	.99	.17	.98	.18	.98	.20
$C = 1.25$	1.24	.18	1.23	.21	1.23	.23	1.22	.25

Minutes	12°		13°		14°		15°	
	Hor. Dist.	Diff. Elev.	Hor. Dist.	Diff. Elev.	Hor. Dist.	Diff. Elev.	Hor. Dist.	Diff. Elev.
0	95.68	20.34	94.94	21.92	94.15	23.47	93.30	25.00
2	95.65	20.39	94.91	21.97	94.12	23.52	93.27	25.05
4	95.63	20.44	94.89	22.02	94.09	23.58	93.24	25.10
6	95.61	20.50	94.86	22.08	94.07	23.63	93.21	25.15
8	95.58	20.55	94.84	22.13	94.04	23.68	93.18	25.20
10	95.56	20.60	94.81	22.18	94.01	23.73	93.16	25.25
12	95.53	20.66	94.79	22.23	93.98	23.78	93.13	25.30
14	95.51	20.71	94.76	22.28	93.95	23.83	93.10	25.35
16	95.49	20.76	94.73	22.34	93.93	23.88	93.07	25.40
18	95.46	20.81	94.71	22.39	93.90	23.93	93.04	25.45
20	95.44	20.87	94.68	22.44	93.87	23.99	93.01	25.50
22	95.41	20.92	94.66	22.49	93.84	24.04	92.98	25.55
24	95.39	20.97	94.63	22.54	93.82	24.09	92.95	25.60
26	95.36	21.03	94.60	22.60	93.79	24.14	92.92	25.65
28	95.34	21.08	94.58	22.65	93.76	24.19	92.89	25.70
30	95.32	21.13	94.55	22.70	93.73	24.24	92.86	25.75
32	95.29	21.18	94.52	22.75	93.70	24.29	92.83	25.80
34	95.27	21.24	94.50	22.80	93.67	24.34	92.80	25.85
36	95.24	21.29	94.47	22.85	93.65	24.39	92.77	25.90
38	95.22	21.34	94.44	22.91	93.62	24.44	92.74	25.95
40	95.19	21.39	94.42	22.96	93.59	24.49	92.71	26.00
42	95.17	21.45	94.39	23.01	93.56	24.55	92.68	26.05
44	95.14	21.50	94.36	23.06	93.53	24.60	92.65	26.10
46	95.12	21.55	94.34	23.11	93.50	24.65	92.62	26.15
48	95.09	21.60	94.31	23.16	93.47	24.70	92.59	26.20
50	95.07	21.66	94.28	23.22	93.45	24.75	92.56	26.25
52	95.04	21.71	94.26	23.27	93.42	24.80	92.53	26.30
54	95.02	21.76	94.23	23.32	93.39	24.85	92.49	26.35
56	94.99	21.81	94.20	23.37	93.36	24.90	92.46	26.40
58	94.97	21.87	94.17	23.42	93.33	24.95	92.43	26.45
60	94.94	21.92	94.15	23.47	93.30	25.00	92.40	26.50
$C = .75$	.73	.16	.73	.18	.73	.19	.72	.20
$C = 1.00$	.98	.22	.97	.23	.97	.25	.96	.27
$C = 1.25$	1.22	.27	1.22	.29	1.21	.31	1.20	.33

TABLE A Stadia Reductions (*Cont.*)

Minutes	16° Hor. Dist.	16° Diff. Elev.	17° Hor. Dist.	17° Diff. Elev.	18° Hor. Dist.	18° Diff. Elev.	19° Hor. Dist.	19° Diff. Elev.
0	92.40	26.50	91.45	27.96	90.45	29.39	89.40	30.78
2	92.37	26.55	91.42	28.01	90.42	29.44	89.36	30.83
4	92.34	26.59	91.39	28.06	90.38	29.48	89.33	30.87
6	92.31	26.64	91.35	28.10	90.35	29.53	89.29	30.92
8	92.28	26.69	91.32	28.15	90.31	29.58	89.26	30.97
10	92.25	26.74	91.29	28.20	90.28	29.62	89.22	31.01
12	92.22	26.79	91.26	28.25	90.24	29.67	89.18	31.06
14	92.19	26.84	91.22	28.30	90.21	29.72	89.15	31.10
16	92.15	26.89	91.19	28.34	90.18	29.76	89.11	31.15
18	92.12	26.94	91.16	28.39	90.14	29.81	89.08	31.19
20	92.09	26.99	91.12	28.44	90.11	29.86	89.04	31.24
22	92.06	27.04	91.09	28.49	90.07	29.90	89.00	31.28
24	92.03	27.09	91.06	28.54	90.04	29.95	88.97	31.33
26	92.00	27.13	91.02	28.58	90.00	30.00	88.93	31.38
28	91.97	27.18	90.99	28.63	89.97	30.04	88.89	31.42
30	91.93	27.23	90.96	28.68	89.93	30.09	88.86	31.47
32	91.90	27.28	90.92	28.73	89.90	30.14	88.82	31.51
34	91.87	27.33	90.89	28.77	89.86	30.18	88.78	31.56
36	91.84	27.38	90.86	28.82	89.83	30.23	88.75	31.60
38	91.81	27.43	90.82	28.87	89.79	30.28	88.71	31.65
40	91.77	27.48	90.79	28.92	89.76	30.32	88.67	31.69
42	91.74	27.52	90.76	28.96	89.72	30.37	88.64	31.74
44	91.71	27.57	90.72	29.01	89.69	30.41	88.60	31.78
46	91.68	27.62	90.69	29.06	89.65	30.46	88.56	31.83
48	91.65	27.67	90.66	29.11	89.61	30.51	88.53	31.87
50	91.61	27.72	90.62	29.15	89.58	30.55	88.49	31.92
52	91.58	27.77	90.59	29.20	89.54	30.60	88.45	31.96
54	91.55	27.81	90.55	29.25	89.51	30.65	88.41	32.01
56	91.52	27.86	90.52	29.30	89.47	30.69	88.38	32.05
58	91.48	27.91	90.49	29.34	89.44	30.74	88.34	32.09
60	91.45	27.96	90.45	29.39	89.40	30.78	88.30	32.14
$C = .75$	.72	.21	.72	.23	.71	.24	.71	.25
$C = 1.00$	.96	.28	.95	.30	.95	.32	.94	.33
$C = 1.25$	1.20	.36	1.19	.38	1.19	.40	1.18	.42

Minutes	20° Hor. Dist.	20° Diff. Elev.	21° Hor. Dist.	21° Diff. Elev.	22° Hor. Dist.	22° Diff. Elev.	23° Hor. Dist.	23° Diff. Elev.
0	88.30	32.14	87.16	33.46	85.97	34.73	84.73	35.97
2	88.26	32.18	87.12	33.50	85.93	34.77	84.69	36.01
4	88.23	32.23	87.08	33.54	85.89	34.82	84.65	36.05
6	88.19	32.27	87.04	33.59	85.85	34.86	84.61	36.09
8	88.15	32.32	87.00	33.63	85.80	34.90	84.57	36.13
10	88.11	32.36	86.96	33.67	85.76	34.94	84.52	36.17
12	88.08	32.41	86.92	33.72	85.72	34.98	84.48	36.21
14	88.04	32.45	86.88	33.76	85.68	35.02	84.44	36.25
16	88.00	32.49	86.84	33.80	85.64	35.07	84.40	36.29
18	87.96	32.54	86.80	33.84	85.60	35.11	84.35	36.33
20	87.93	32.58	86.77	33.89	85.56	35.15	84.31	36.37
22	87.89	32.63	86.73	33.93	85.52	35.19	84.27	36.41
24	87.85	32.67	86.69	33.97	85.48	35.23	84.23	36.45
26	87.81	32.72	86.65	34.01	85.44	35.27	84.18	36.49
28	87.77	32.76	86.61	34.06	85.40	35.31	84.14	36.53
30	87.74	32.80	86.57	34.10	85.36	35.36	84.10	36.57
32	87.70	32.85	86.53	34.14	85.31	35.40	84.06	36.61
34	87.66	32.89	86.49	34.18	85.27	35.44	84.01	36.65
36	87.62	32.93	86.45	34.23	85.23	35.48	83.97	36.69
38	87.58	32.98	86.41	34.27	85.19	35.52	83.93	36.73
40	87.54	33.02	86.37	34.31	85.15	35.56	83.89	36.77
42	87.51	33.07	86.33	34.35	85.11	35.60	83.84	36.80
44	87.47	33.11	86.29	34.40	85.07	35.64	83.80	36.84
46	87.43	33.15	86.25	34.44	85.02	35.68	83.76	36.88
48	87.39	33.20	86.21	34.48	84.98	35.72	83.72	36.92
50	87.35	33.24	86.17	34.52	84.94	35.76	83.67	36.96
52	87.31	33.28	86.13	34.57	84.90	35.80	83.63	37.00
54	87.27	33.33	86.09	34.61	84.86	35.85	83.59	37.04
56	87.24	33.37	86.05	34.65	84.82	35.89	83.54	37.08
58	87.20	33.41	86.01	34.69	84.77	35.93	83.50	37.12
60	87.16	33.46	85.97	34.73	84.73	35.97	83.46	37.16
$C = .75$	.70	.26	.70	.27	.69	.29	.69	.30
$C = 1.00$	.94	.35	.93	.37	.92	.38	.92	.40
$C = 1.25$	1.17	.44	1.16	.46	1.15	.48	1.15	.50

Minutes	24°		25°		26°		27°	
	Hor. Dist.	Diff. Elev.	Hor. Dist.	Diff. Elev.	Hor. Dist.	Diff. Elev.	Hor. Dist.	Diff. Elev.
0	83.46	37.16	82.14	38.30	80.78	39.40	79.39	40.45
2	83.41	37.20	82.09	38.34	80.74	39.44	79.34	40.49
4	83.37	37.23	82.05	38.38	80.69	39.47	79.30	40.52
6	83.33	37.27	82.01	38.41	80.65	39.51	79.25	40.55
8	83.28	37.31	81.96	38.45	80.60	39.54	79.20	40.59
10	83.24	37.35	81.92	38.49	80.55	39.58	79.15	40.62
12	83.20	37.39	81.87	38.53	80.51	39.61	79.11	40.66
14	83.15	37.43	81.83	38.56	80.46	39.65	79.06	40.69
16	83.11	37.47	81.78	38.60	80.41	39.69	79.01	40.72
18	83.07	37.51	81.74	38.64	80.37	39.72	78.96	40.76
20	83.02	37.54	81.69	38.67	80.32	39.76	78.92	40.79
22	82.98	37.58	81.65	38.71	80.28	39.79	78.87	40.82
24	82.93	37.62	81.60	38.75	80.23	39.83	78.82	40.86
26	82.89	37.66	81.56	38.78	80.18	39.86	78.77	40.89
28	82.85	37.70	81.51	38.82	80.14	39.90	78.73	40.92
30	82.80	37.74	81.47	38.86	80.09	39.93	78.68	40.96
32	82.76	37.77	81.42	38.89	80.04	39.97	78.63	40.99
34	82.72	37.81	81.38	38.93	80.00	40.00	78.58	41.02
36	82.67	37.85	81.33	38.97	79.95	40.04	78.54	41.06
38	82.63	37.89	81.28	39.00	79.90	40.07	78.49	41.09
40	82.58	37.93	81.24	39.04	79.86	40.11	78.44	41.12
42	82.54	37.96	81.19	39.08	79.81	40.14	78.39	41.16
44	82.49	38.00	81.15	39.11	79.76	40.18	78.34	41.19
46	82.45	38.04	81.10	39.15	79.72	40.21	78.30	41.22
48	82.41	38.08	81.06	39.18	79.67	40.24	78.25	41.26
50	82.36	38.11	81.01	39.22	79.62	40.28	78.20	41.29
52	82.32	38.15	80.97	39.26	79.58	40.31	78.15	41.32
54	82.27	38.19	80.92	39.29	79.53	40.35	78.10	41.35
56	82.23	38.23	80.87	39.33	79.48	40.38	78.06	41.39
58	82.18	38.26	80.83	39.36	79.44	40.42	78.01	41.42
60	82.14	38.30	80.78	39.40	79.39	40.45	77.96	41.45
$C = .75$	.68	.31	.68	.32	.67	.33	.67	.35
$C = 1.00$	.91	.41	.90	.43	.89	.45	.89	.46
$C = 1.25$	1.14	.52	1.13	.54	1.12	.56	1.11	.58

TABLE A Stadia Reductions (*Cont.*)

Minutes	28° Hor. Dist.	28° Diff. Elev.	29° Hor. Dist.	29° Diff. Elev.	30° Hor. Dist.	30° Diff. Elev.
0	77.96	41.45	76.50	42.40	75.00	43.30
2	77.91	41.48	76.45	42.43	74.95	43.33
4	77.86	41.52	76.40	42.46	74.90	43.36
6	77.81	41.55	76 35	42.49	74.85	43.39
8	77.77	41.58	76.30	42.53	74.80	43.42
10	77.72	41.61	76.25	42.56	74.75	43.45
12	77.67	41.65	76.20	42.59	74.70	43.47
14	77.62	41.68	76.15	42.62	74.65	43.50
16	77.57	41.71	76.10	42.65	74.60	43.53
18	77.52	41.74	76.05	42.68	74.55	43.56
20	77.48	41.77	76.00	42.71	74.49	43.59
22	77.42	41.81	75.95	42.74	74.44	43.62
24	77.38	41.84	75.90	42.77	74.39	43.65
26	77.33	41.87	75.85	42.80	74.34	43.67
28	77.28	41.90	75.80	42.83	74.29	43.70
30	77.23	41.93	75.75	42.86	74.24	43.73
32	77.18	41.97	75.70	42.89	74.19	43.76
34	77.13	42.00	75.65	42.92	74.14	43.79
36	77.09	42.03	75.60	42.95	74.09	43.82
38	77.04	42.06	75.55	42.98	74.04	43.84
40	76.99	42.09	75.50	43.01	73.99	43.87
42	76.94	42.12	75.45	43.04	73.93	43.90
44	76.89	42.15	75.40	43.07	73.88	43.93
46	76.84	42.19	75.35	43.10	73.83	43.95
48	76.79	42.22	75.30	43.13	73.78	43.98
50	76.74	42.25	75.25	43.16	73.73	44.01
52	76.69	42.28	75.20	43.18	73.68	44.04
54	76.64	42.31	75.15	43.21	73.63	44.07
56	76.59	42.34	75.10	43.24	73.58	44.09
58	76.55	42.37	75.05	43.27	73.52	44.12
60	76.50	42.40	75.00	43.30	73.47	44.15
$C = .75$	.66	.36	.65	.37	.65	.38
$C = 1.00$	.88	.48	.87	.49	.86	.51
$C = 1.25$	1.10	.60	1.09	.62	1.08	.63

TABLE B Trigonometric Formulas for the Solution of Right Triangles

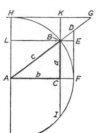

Let $A =$ angle $BAC =$ arc BF, and let radius $AF = AB = AH = 1$. Then,

$\sin A = BC$	$\csc A = AG$
$\cos A = AC$	$\sec A = AD$
$\tan A = DF$	$\cot A = HG$
vers $A = CF = BE$	covers $A = BK = LH$
exsec $A = BD$	coexsec $A = BG$
chord $A = BF$	chord $2\,A = BI = 2\,BC$

In the right-angled triangle ABC, let $AB = c$, $BC = a$, $CA = b$. Then,

1. $\sin A = \dfrac{a}{c}$

2. $\cos A = \dfrac{b}{c}$

3. $\tan A = \dfrac{a}{b}$

4. $\cot A = \dfrac{b}{a}$

5. $\sec A = \dfrac{c}{b}$

6. $\csc A = \dfrac{c}{a}$

7. vers $A = 1 - \cos A = \dfrac{c-b}{c} =$ covers B

8. exsec $A = \sec A - 1 = \dfrac{c-b}{b} =$ coexsec B

9. covers $A = \dfrac{c-a}{c} =$ vers B

10. coexsec $A = \dfrac{c-a}{a} =$ exsec B

11. $a = c \sin A = b \tan A$

12. $b = c \cos A = a \cot A$

13. $c = \dfrac{a}{\sin A} = \dfrac{b}{\cos A}$

14. $a = c \cos B = b \cot B$

15. $b = c \sin B = a \tan B$

16. $c = \dfrac{a}{\cos B} = \dfrac{b}{\sin B}$

17. $a = \sqrt{c^2 - b^2} = \sqrt{(c-b)(c+b)}$

18. $b = \sqrt{c^2 - a^2} = \sqrt{(c-a)(c+a)}$

19. $c = \sqrt{a^2 + b^2}$

20. $C = 90° = A + B$

21. Area $= \frac{1}{2} ab$

TABLE C Trigonometric Formulas for the Solution of Oblique Triangles

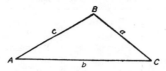

No.	Given	Sought	Formula
22	A, B, a	C, b, c	$C = 180° - (A + B)$
			$b = \dfrac{a}{\sin A} \times \sin B$
			$c = \dfrac{a}{\sin A} \times \sin(A+B) = \dfrac{a}{\sin A} \times \sin C$
		Area	$\text{Area} = \tfrac{1}{2}ab \sin C = \dfrac{a^2 \sin B \sin C}{2 \sin A}$
23	A, a, b	B, C, c	$\sin B = \dfrac{\sin A}{a} \times b$
			$C = 180° - (A + B)$
			$c = \dfrac{a}{\sin A} \times \sin C$
		Area	$\text{Area} = \tfrac{1}{2}ab \sin C$
24	$C, a, b,$	c	$c = \sqrt{a^2 + b^2 - 2ab \cos C}$
25		$\tfrac{1}{2}(A+B)$	$\tfrac{1}{2}(A+B) = 90° - \tfrac{1}{2}C$
26		$\tfrac{1}{2}(A-B)$	$\tan \tfrac{1}{2}(A-B) = \dfrac{a-b}{a+b} \times \tan \tfrac{1}{2}(A+B)$
27		A, B	$A = \tfrac{1}{2}(A+B) + \tfrac{1}{2}(A-B)$
			$B = \tfrac{1}{2}(A+B) - \tfrac{1}{2}(A-B)$
28		c	$c = (a+b) \times \dfrac{\cos \tfrac{1}{2}(A+B)}{\cos \tfrac{1}{2}(A-B)} = (a-b) \times \dfrac{\sin \tfrac{1}{2}(A+B)}{\sin \tfrac{1}{2}(A-B)}$
29		Area	$\text{Area} = \tfrac{1}{2}ab \sin C$
30	a, b, c	A	$\text{Let } s = \dfrac{a+b+c}{2}$
31			$\sin \tfrac{1}{2}A = \sqrt{\dfrac{(s-b)(s-c)}{bc}}$
			$\cos \tfrac{1}{2}A = \sqrt{\dfrac{s(s-a)}{bc}}$
			$\tan \tfrac{1}{2}A = \sqrt{\dfrac{(s-b)(s-c)}{s(s-a)}}$
32			$\sin A = \dfrac{2\sqrt{s(s-a)(s-b)(s-c)}}{bc}$
			$\cos A = \dfrac{b^2 + c^2 - a^2}{2bc}$
33		Area	$\text{Area} = \sqrt{s(s-a)(s-b)(s-c)}$

TABLE D Polaris for the Meridian of Greenwich Latitude 40°N, 1981

Universal Time or Greenwich Civil Time

Date	Upper Culmination	Previous East Elongation	Next West Elongation	Next Lower Culmination	Var. Per Day
1981	h m	h m	h m	h m	m
Jan. 1	19 28.1	13 31.8	1 24.4	6 29.1	
11	18 48.6	12 52.3	0 44.8	5 49.6	3.96
21	18 09.0	12 12.7	0 05.3	5 10.0	3.96
31	17 29.4	11 33.2	23 25.7	4 30.5	3.96
Feb 10	16 49.8	10 53.6	22 46.1	3 50.9	3.96
20	16 10.3	10 14.0	22 06.6	3 11.3	3.96
Mar. 2	15 30.8	9 34.5	21 27.0	2 31.8	3.95
12	14 51.2	8 55.0	20 47.5	1 52.3	3.95
22	14 11.8	8 15.5	20 08.0	1 12.8	3.95
Apr. 1	13 32.4	7 36.1	19 28.6	0 33.4	3.94
11	12 53.0	6 56.7	18 49.2	23 54.0	3.94
21	12 13.6	6 17.4	18 09.9	23 14.7	3.93
May 1	11 34.3	5 38.1	17 30.6	22 35.4	3.93
11	10 55.1	4 58.8	16 51.4	21 56.1	3.92
21	10 15.9	4 19.6	16 12.2	21 16.9	3.92
31	9 36.7	3 40.5	15 33.0	20 37.8	3.92
June 10	8 57.6	3 01.3	14 53.9	19 58.6	3.91
20	8 18.5	2 22.2	14 14.8	19 19.5	3.91
30	7 39.4	1 43.2	13 35.7	18 40.5	3.91
July 10	7 00.4	1 04.1	12 56.6	18 01.4	3.91
20	6 21.3	0 25.0	12 17.6	17 22.3	3.91
30	5 42.2	23 46.0	11 38.5	16 43.3	3.91
Aug. 9	5 03.2	23 06.9	10 59.4	16 04.2	3.91
19	4 24.1	22 27.8	10 20.4	15 25.1	3.91
29	3 45.0	21 48.7	9 41.3	14 46.0	3.91
Sep. 8	3 05.9	21 09.6	9 02.2	14 06.9	3.91
18	2 26.8	20 30.5	8 23.0	13 27.8	3.92
28	1 47.6	19 51.3	7 43.9	12 48.6	3.92
Oct. 8	1 08.4	19 12.2	7 04.7	12 09.5	3.92
18	0 29.2	18 32.9	6 25.5	11 30.2	3.92
27	23 49.9	17 53.6	5 46.2	10 50.9	3.93
Nov. 6	23 10.6	17 14.3	5 06.9	10 11.6	3.93
16	22 31.3	16 35.0	4 27.5	9 32.3	3.93
26	21 51.9	15 55.6	3 48.1	8 52.9	3.94
Dec. 6	21 12.4	15 16.2	3 08.7	8 13.5	3.94
16	20 33.0	14.36.7	2 29.2	7 34.0	3.95
26	19 53.5	13 57.2	1 49.7	6 54.5	3.95

The next lower culmination is on the next day during the intervals: Jan. 01 to April 23, inclusive and Oct. 26 to Jan. 05, inclusive.

The previous East Elongation is on the previous day during the interval: July 27 to Oct. 25, inclusive.

The next West Elongation is on the next day during the intervals: Jan. 01 to Jan. 22 inclusive and Oct. 26 to Jan. 05, inclusive.

TABLE E Polar Distance of Polaris, 1981

(For 0^h Universal Time or Greenwich Civil Time)

1981		Polar Distance Angle ° ′	Cotan	1981		Polar Distance Angle ° ′	Cotan
Jan.	1	0 49.13	69.97	July	10	0 49.62	69.28
	11	0 49.10	70.01		20	0 49.61	69.29
	21	0 49.09	70.02		30	0 49.60	69.30
	31	0 49.09	70.02				
Feb	10	0 49.09	70.02	Aug.	9	0 49.59	69.32
	20	0 49.11	70.00		19	0 49.56	69.36
					29	0 49.52	69.42
Mar.	2	0 49.14	69.95	Sep.	8	0 49.48	69.47
	12	0 49.17	69.91		18	0 49.43	69.54
	22	0 49.21	69.85		28	0 49.37	69.63
Apr.	1	0 49.26	69.78	Oct.	8	0 49.32	69.70
	11	0 49.31	69.71		18	0 49.26	69.78
	21	0 49.36	69.64		28	0 49.19	69.88
May	1	0 49.41	69.57	Nov.	7	0 49.13	69.97
	11	0 49.46	69.50		17	0 49.07	70.05
	21	0 49.51	69.43		27	0 49.01	70.14
	31	0 49.54	69.39				
June	10	0 49.57	69.35	Dec.	7	0 48.95	70.23
	20	0 49.60	69.30		17	0 48.90	70.30
	30	0 49.61	69.29		27	0 48.80	70.35

Declination = $90°$ − Polar Distance.

TABLE F Bearing of Polaris at Elongation, 1981

Polar Dist.	0° 49.00	0° 49.20	0° 49.40	0° 49.60	Polar Dist.	0° 49.00	0° 49.20	0° 49.40	0° 49.60
Lat.	Bearing at Elongation				Lat.	Bearing at Elongation			
°	° ′	° ′	° ′	° ′	°	° ′	° ′	° ′	° ′
10	0 49.8	0 50.0	0 50.2	0 50.4	40	1 04.0	0 04.2	1 04.5	1 04.7
11	0 49.9	0 50.1	0 50.3	0 50.5	41	1 04.9	1 05.2	1 05.5	1 05.7
12	0 50.1	0 50.3	0 50.5	0 50.7	42	1 05.9	1 06.2	1 06.5	1 06.7
13	0 50.3	0 50.5	0 50.7	0 50.9	43	1 07.0	1 07.3	1 07.5	1 07.8
14	0 50.5	0 50.7	0 50.9	0 51.1	44	1 08.1	1 08.4	1 08.7	1 09.0
15	0 50.7	0 50.9	0 51.1	0 51.3	45	1 09.3	1 09.6	1 09.9	1 10.1
16	0 51.0	0 51.2	0 51.4	0 51.6	46	1 10.5	1 10.8	1 11.1	1 11.4
17	0 51.2	0 51.4	0 51.7	0 51.9	47	1 11.9	1 12.1	1 12.4	1 12.7
18	0 51.5	0 51.7	0 51.9	0 52.2	48	1 13.2	1 13.5	1 13.8	1 14.1
19	0 51.8	0 52.0	0 52.2	0 52.5	49	1 14.7	1 15.0	1 15.3	1 15.6
20	0 52.1	0 52.4	0 52.6	0 52.8	50	1 16.2	1 16.5	1 16.9	1 17.2
21	0 52.5	0 52.7	0 52.9	0 53.1	51	1 17.9	1 18.2	1 18.5	1 18.8
22	0 52.8	0 53.1	0 53.3	0 53.5	52	1 19.6	1 19.9	1 20.2	1 20.6
23	0 53.2	0 53.4	0 53.7	0 53.9	53	1 21.4	1 21.8	1 22.1	1 22.4
24	0 53.6	0 53.9	0 54.1	0 54.3	54	1 23.4	1 23.7	1 24.0	1 24.4
25	0 54.1	0 54.3	0 54.5	0 54.7	55	1 25.4	1 25.8	1 26.1	1 26.5
26	0 54.5	0 54.7	0 55.0	0 55.2	56	1 27.6	1 28.0	1 28.3	1 28.7
27	0 55.0	0 55.2	0 55.4	0 55.7	57	1 30.0	1 30.3	1 30.7	1 31.1
28	0 55.5	0 55.7	0 55.9	0 56.2	58	1 32.5	1 32.9	1 33.2	1 33.6
29	0 56.0	0 56.3	0 56.5	0 56.7	59	1 35.1	1 35.5	1 35.9	1 36.3
30	0 56.6	0 56.8	0 57.0	0 57.3	60	1 38.0	1 38.4	1 38.8	1 39.2
31	0 57.2	0 57.4	0 57.6	0 57.9	61	1 41.1	1 41.5	1 41.9	1 42.3
32	0 57.8	0 58.0	0 58.3	0 58.5	62	1 44.4	1 44.8	1 45.2	1 45.7
33	0 58.4	0 58.7	0 58.9	0 59.1	63	1 47.9	1 48.4	1 48.8	1 49.3
34	0 59.1	0 59.3	0 59.6	0 59.8	64	1 51.8	1 52.2	1 52.7	1 53.2
35	0 59.8	1 00.1	1 00.3	1 00.6	65	1 56.0	1 56.4	1 56.9	1 57.4
36	1 00.6	1 00.8	1 01.1	1 01.3	66	2 00.5	2 01.0	2 01.5	2 02.0
37	1 01.4	1 01.6	1 01.9	1 02.1	67	2 05.4	2 05.9	2 06.5	2 07.0
38	1 02.2	1 02.4	1 02.7	1 02.9	68	2 10.8	2 11.4	2 11.9	2 12.4
39	1 03.1	1 03.3	1 03.6	1 03.8	69	2 16.8	2 17.3	2 17.9	2 18.4
40	1 04.0	1 04.2	1 04.5	1 04.7	70	2 23.3	2 23.9	2 24.5	2 25.1

To obtain the Bearing at any other declination compute:

$$\text{Bear. Polaris (in minutes)} = \frac{\text{Polar Dist. (in minutes)}}{\cos \text{Lat.}}$$

TABLE G Corrections to Times of Elongation for Different Latitudes, 1981

Latitude	10°	15°	20°	25°	30°	35°	40°	45°	50°
	m	m	m	m	m	m	m	m	m
West Elongation	+2.2	+1.9	+1.6	+1.2	+0.9	+0.5	0.0	-0.5	-1.2
East Elongation	-2.2	-1.9	-1.6	-1.2	-0.9	-0.5	0.0	+0.5	+1.2

TABLE H Bearing of Polaris at All Local Hour Angles

1981 Computed for a Polar Distance of 0°49.20'
For Local Hour Angles 0° to 180° the Star is West of North
and from 180° to 360° it is East of North

Lat.	10°	20°	26°	30°	32°	34°	36°	38°	Lat.
LHA	° ′	° ′	° ′	° ′	° ′	° ′	° ′	° ′	LHA
0	0 00.0	0 00.0	0 00.0	0 00.0	0 00.0	0 00.0	0 00.0	0 00.0	360
5	0 04.4	0 04.6	0 04.8	0 05.0	0 05.1	0 05.2	0 05.4	0 05.5	355
10	0 08.7	0 09.1	0 09.6	0 09.9	0 10.2	0 10.4	0 10.7	0 11.0	350
15	0 13.0	0 13.6	0 14.3	0 14.8	0 15.1	0 15.5	0 15.9	0 16.3	345
20	0 17.1	0 18.0	0 18.8	0 19.6	0 20.0	0 20.5	0 21.0	0 21.6	340
25	0 21.2	0 22.2	0 23.3	0 24.2	0 24.7	0 25.3	0 25.9	0 26.7	335
30	0 25.0	0 26.3	0 27.5	0 28.6	0 29.2	0 29.9	0 30.7	0 31.5	330
35	0 28.7	0 30.2	0 31.6	0 32.8	0 33.5	0 34.3	0 35.2	0 36.1	325
40	0 32.2	0 33.8	0 35.4	0 36.8	0 37.6	0 38.4	0 39.4	0 40.5	320
45	0 35.4	0 37.2	0 38.9	0 40.4	0 41.3	0 42.3	0 43.3	0 44.5	315
50	0 38.3	0 40.2	0.42.1	0 43.8	0 44.7	0 45.7	0 46.9	0 48.2	310
55	0 41.0	0 43.0	0 45.0	0 46.8	0 47.8	0 48.9	0 50.1	0 51.5	305
60	0 43.3	0 45.5	0 47.6	0 49.4	0 50.5	0 51.6	0 52.9	0 54.4	300
65	0 45.3	0 47.6	0 49.8	0 51.7	0 52.8	0 54.0	0 55.4	0 56.9	295
70	0 47.0	0 49.3	0 51.6	0 53.5	0 54.7	0 56.0	0 57.3	0 58.9	290
75	0 48.3	0 50.6	0 53.0	0 55.0	0 56.2	0 57.5	0 58.9	1 00.5	285
80	0 49.2	0 51.6	0 54.0	0 56.0	0 57.2	0 58.5	1 00.0	1 01.6	280
85	0 49.8	0 52.2	0 54.6	0 56.6	0 57.8	0 59.2	1 00.6	1 02.3	275
90	0 50.0	0 52.4	0 54.7	0 56.8	0 58.0	0 59.3	1 00.8	1 02.4	270
95	0 49.8	0 52.1	0 54.5	0 56.6	0 57.7	0 59.1	1 00.5	1 02.1	265
100	0 49.2	0 51.5	0 53.8	0 55.9	0 57.0	0 58.3	0 59.8	1 01.4	260
105	0 48.2	0 50.5	0 52.8	0 54.8	0 55.9	0 57.2	0 58.6	1 00.1	255·
110	0 46.9	0 49.1	0 51.3	0 53.2	0 54.3	0 55.6	0 56.9	0 58.4	250
115	0 45.2	0 47.3	0 49.5	0 51.3	0 52.4	0 53.6	0 54.9	0 56.3	245
120	0 43.2	0 45.2	0 47.2	0 49.0	0 50.0	0 51.1	0 52.4	0 53.8	240
125	0 40.9	0 42.8	0 44.7	0 46.3	0 47.3	0 48.3	0 49.5	0 50.8	235
130	0 38.2	0 40.0	0 41.7	0 43.3	0 44.2	0 45.2	0 46.3	0 47.5	230
135	0 35.3	0 36.9	0 38.5	0 39.9	0 40.8	0 41.7	0 42.7	0 43.8	225
140	0 32.1	0 33.5	0 35.0	0 36.3	0 37.0	0 37.9	0 38.8	0 39.8	220
145	0 28.0	0 29.9	0 31.2	0 32.4	0 33.0	0 33.8	0 34.6	0 35.5	215
150	0 24.9	0 26.1	0 27.2	0 28.2	0 28.8	0 29.4	0 30.1	0 30.9	210
155	0 21.1	0 22.0	0 23.0	0 23.8	0 24.3	0 24.9	0 25.5	0 26.1	205
160	0 17.0	0 17.8	0 18.6	0 19.3	0 19.7	0 20.1	0 20.6	0 21.1	200
165	0 12.9	0 13.5	0 14.1	0 14.6	0 14.9	0 15.2	0 15.6	0 16.0	195
170	0 08.7	0 09.0	0 09.4	0 09.8	0 10.0	0 10.2	0 10.5	0 10.7	190
175	0 04.3	0 04.5	0 04.7	0 04.9	0 05.0	0 05.1	0 05.2	0 05.4	185
180	0 00.0	0 00.0	0 00.0	0 00.0	0 00.0	0 00.0	0 00.0	0 00.0	180

Lat.	40°	42°	44°	46°	48°	50°	60°	70°	Lat.
LHA	° ′	° ′	° ′	° ′	° ′	° ′	° ′	° ′	LHA
0	0 00.0	0 00.0	0 00.0	0 00.0	0 00.0	0 00.0	0 00.0	0 00.0	360
5	0 05.7	0 05.8	0 06.0	0 06.3	0 06.5	0 06.8	0 08.8	0 13.0	355
10	0 11.3	0 11.6	0 12.0	0 12.5	0 13.0	0 13.5	0 17.5	0 26.0	350
15	0 16.8	0 17.4	0 17.9	0 18.6	0 19.3	0 20.1	0 26.1	0 38.7	345
20	0 22.2	0 22.9	0 23.7	0 24.6	0 25.5	0 26.6	0 34.5	0 51.1	340
25	0 27.4	0 28.3	0 29.3	0 30.3	0 31.5	0 32.9	0 42.5	1 03.0	335
30	0 32.5	0 33.5	0 34.6	0 35.9	0 37.3	0 38.8	0 50.3	1 14.5	330
35	0 37.2	0 38.4	0 39.7	0 41.1	0 42.7	0 44.5	0 57.6	1 25.2	325
40	0 41.7	0 43.0	0 44.4	0 46.0	0 47.8	0 49.9	1 04.5	1 35.3	320
45	0 45.8	0 47.2	0 48.8	0 50.6	0 52.6	0 54.8	1 10.8	1 44.6	315
50	0 49.6	0 51.1	0 52.9	0 54.8	0 56.9	0 59.3	1 16.6	1 53.0	310
55	0 53.0	0 54.6	0 56.5	0 58.5	1 00.8	1 03.3	1 21.8	2 00.5	305
60	0 56.0	0 57.7	0 59.6	1 01.8	1 04.2	1 06.9	1 26.3	2 07.0	300
65	0 58.5	1 00.3	1 02.3	1 04.6	1 07.1	1 09.9	1 30.1	2 12.5	295
70	1 00.6	1 02.5	1 04.6	1 06.9	1 09.5	1 12.3	1 33.2	2 17.0	290
75	1 02.2	1 04.2	1 06.3	1 08.7	1 11.3	1 14.3	1 35.6	2 20.3	285
80	1 03.4	1 05.3	1 07.5	1 09.9	1 12.6	1 15.6	1 37.3	2 22.6	280
85	1 04.0	1 06.0	1 08.2	1 10.6	1 13.3	1 16.4	1 38.2	2 23.7	275
90	1 04.2	1 06.2	1 08.4	1 10.8	1 13.5	1 16.5	1 38.4	2 23.8	270
95	1 03.9	1 05.9	1 08.0	1 10.5	1 13.1	1 16.1	1 37.8	2 22.7	265
100	1 03.1	1 05.1	1 07.2	1 09.6	1 12.2	1 15.1	1 36.5	2 20.6	260
105	1 01.8	1 03.7	1 05.8	1 08.1	1 10.7	1 13.6	1 34.4	2 17.5	255
110	1 00.1	1 01.9	1 04.0	1 06.2	1 08.7	1 11.5	1 31.7	2 13.3	250
115	0 57.9	0 59.7	1 01.6	1 03.8	1 06.2	1 08.9	1 28.2	2 08.2	245
120	0 55.3	0 57.0	0 58.8	1 00.9	1 03.2	1 05.7	1 24.2	2 02.1	240
125	0 52.3	0 53.8	0 55.6	0 57.5	0 59.7	1 02.1	1 19.5	1 55.2	235
130	0 48.8	0 50.3	0 51.9	0 53.7	0 55.8	0 58.0	1 14.2	1 47.5	230
135	0 45.0	0 46.4	0 47.9	0 49.6	0 51.4	0 53.5	1 08.4	1 38.9	225
140	0 40.9	0 42.1	0 43.5	0 45.0	0 46.7	0 48.6	1 02.1	1 29.7	220
145	0 36.5	0 37.6	0 38.8	0 40.1	0 41.6	0 43.3	0 55.3	1 19.9	215
150	0 31.8	0 32.7	0 33.8	0 35.0	0 36.3	0 37.7	0 48.2	1 09.6	210
155	0 26.9	0 27.7	0 28.5	0 29.5	0 30.6	0 31.9	0 40.7	0 58.7	205
160	0 21.7	0 22.4	0 23.1	0 23.9	0 24.8	0 25.8	0 32.9	0 47.4	200
165	0 16.4	0 16.9	0 17.5	0 18.1	0 18.7	0 19.5	0 24.9	0 35.9	195
170	0 11.0	0 11.4	0 11.7	0 12.1	0 12.6	0 13.1	0 16.7	0 24.0	190
175	0 05.5	0 05.7	0 05.9	0 06.1	0 06.3	0 06.6	0 08.4	0 12.1	185
180	0 00.0	0 00.0	0 00.0	0 00.0	0 00.0	0 00.0	0 00.0	0 00.0	180

TABLE I Correction to Bearing of Polaris for Other Polar Distances, 1981

Bearing	0°	0°20′	0°40′	1°00′	1°20′	1°40′	2°00′	2°20′	2°40′
Polar Dist.									
° ′	′	′	′	′	′	′	′	′	′
0 49.70	0.0	+0.2	+0.4	+0.6	+0.8	+1.0	+1.2	+1.4	+1.6
0 49.60	0.0	+0.2	+0.3	+0.5	+0.7	+0.8	+1.0	+1.1	+1.3
0 49.50	0.0	+0.1	+0.2	+0.4	+0.5	+0.6	+0.7	+0.9	+1.0
0 49.40	0.0	+0.1	+0.2	+0.2	+0.3	+0.4	+0.5	+0.6	+0.7
0 49.30	0.0	+0.0	+0.1	+0.1	+0.2	+0.2	+0.2	+0.3	+0.3
0 49.20	0.0	0.0	0.0	0.0	0.0	0.0	0.0	0.0	0.0
0 49.10	0.0	− 0.0	− 0.1	− 0.1	− 0.2	− 0.2	− 0.2	− 0.3	− 0.3
0 49.00	0.0	− 0.1	− 0.2	− 0.2	− 0.3	− 0.4	− 0.5	− 0.6	− 0.7
0 48.90	0.0	− 0.1	− 0.2	0.4	0.5	0.6	− 0.7	− 0.9	− 1.0
0 48.80	0.0	− 0.2	− 0.3	− 0.5	− 0.7	− 0.8	− 1.0	− 1.1	− 1.3
0 48.70	0.0	− 0.2	− 0.4	− 0.6	− 0.8	− 1.0	− 1.2	− 1.4	− 1.6

TABLE J Solar Ephemeris, October 1981

(For 0^h Universal Time or Greenwich Civil Time)

Day of Month & Week		The Sun's Apparent Declination	Diff. in Declin. for 1 hour	Equation of Time			GHA of Polaris	
				True Sol. Time = LCT + Eq. of Time	Differ. for 1 hour			
		° ′	′	m s	s	°	′	
1	TH	S03 03.9	0.97	+10 10.5	0.80	335	58.6	
2	FR	S03 27.2	0.97	+10 29.7	0.79	336	57.5	
3	SA	S03 50.4	0.97	+10 48.6	0.77	337	56.4	
4	SU	S04 13.6	0.97	+11 07.2	0.76	338	55.3	
5	M	S04 36.8	0.96	+11 25.5	0.75	339	54.2	
6	TU	S04 59.8	0.96	+11 43.4	0.73	340	53.2	
7	W	S05 22.9	0.95	+12 00.9	0.72	341	52.1	
8	TH	S05 45.8	0.95	+12 18.1	0.70	342	51.1	
9	FR	S06 08.7	0.95	+12 34.9	0.68	343	50.1	
10	SA	S06 31.5	0.95	+12 51.2	0.66	344	49.1	
11	SU	S06 54.2	0.94	+13 07.1	0.64	345	48.1	
12	M	S07 16.8	0.94	+13 22.6	0.62	346	47.2	
13	TU	S07 39.3	0.93	+13 37.5	0.60	347	46.2	
14	W	S08 01.7	0.93	+13 51.9	0.58	348	45.2	
15	TH	S08 24.0	0.92	+14 05.8	0.56	349	44.2	
16	FR	S08 46.2	0.92	+14 19.1	0.53	350	43.1	
17	SA	S09 08.2	0.92	+14 31.9	0.51	351	42.1	
18	SU	S09 30.2	0.91	+14 44.0	0.48	352	41.1	
19	M	S09 52.0	0.90	+14 55.5	0.45	353	40.1	
20	TU	S10 13.6	0.90	+15 06.4	0.43	354	39.1	
21	W	S10 35.1	0.89	+15 16.6	0.40	355	38.1	
22	TH	S10 56.4	0.88	+15 26.2	0.37	356	37.2	
23	FR	S11 17.6	0.88	+15 35.0	0.34	357	36.3	
24	SA	S11 38.6	0.87	+15 43.2	0.31	358	35.4	
25	SU	S11 59.5	0.86	+15 50.6	0.28	359	34.5	
26	M	S12 20.1	0.85	+15 57.3	0.25	000	33.6	
27	TU	S12 40.6	0.85	+16 03.3	0.22	001	32.7	
28	W	S13 00.9	0.83	+16 08.6	0.19	002	31.8	
29	TH	S13 20.9	0.83	+16 13.1	0.16	003	30.8	
30	FR	S13 40.8	0.82	+16 16.8	0.12	004	29.9	
31	SA	S14 00.4	0.81	+16 19.8	0.09	005	29.0	
32	SU	S14 19.8		+16 22.0		006	28.0	

Hourly differences in declination and equation of time are for the 24-hours following 0-hours of date in left column.

TABLE J (*Cont.*) Solar Ephemeris, November 1981

(For 0^h Universal Time or Greenwich Civil Time)

Day of Month & Week	The Sun's Apparent Declination	Diff. in Declin. for 1 hour	Equation of Time True Sol. Time = LCT + Eq. of Time	Differ. for 1 hour	GHA of Polaris
	° ′	′	m s	s °	′
1 SU	S14 19.8	0.80	+16 22.0	0.06 006	28.0
2 M	S14 39.0	0.79	+16 23.4	0.03 007	27.1
3 TU	S14 58.0	0.78	+16 24.0	0.01 008	26.2
4 W	S15 16.7	0.77	+16 23.8	0.04 009	25.4
5 TH	S15 35.1	0.76	+16 22.8	0.07 010	24.5
6 FR	S15 53.3	0.75	+16 21.1	0.11 011	23.7
7 SA	S16 11.2	0.74	+16 18.5	0.14 012	22.9
8 SU	S16 28.9	0.72	+16 15.0	0.18 013	22.1
9 M	S16 46.3	0.71	+16 10.8	0.21 014	21.3
10 TU	S17 03.4	0.70	+16 05.7	0.25 015	20.5
11 W	S17 20.2	0.69	+15 59.8	0.28 016	19.6
12 TH	S17 36.7	0.67	+15 53.0	0.32 017	18.8
13 FR	S17 52.9	0.66	+15 45.4	0.35 018	17.9
14 SA	S18 08.7	0.65	+15 37.0	0.39 019	17.0
15 SU	S18 24.3	0.64	+15 27.6	0.42 020	16.2
16 M	S18 39.6	0.62	+15 17.5	0.46 021	15.4
17 TU	S18 54.5	0.61	+15 06.4	0.50 022	14.6
18 W	S19 09.1	0.59	+14 54.5	0.53 023	13.8
19 TH	S19 23.3	0.58	+14 41.8	0.57 024	13.1
20 FR	S19 37.2	0.56	+14 28.2	0.60 025	12.4
21 SA	S19 50.7	0.55	+14 13.8	0.64 026	11.6
22 SU	S20 03.9	0.53	+13 58.5	0.67 027	10.9
23 M	S20 16.7	0.52	+13 42.5	0.70 028	10.2
24 TU	S20 29.1	0.50	+13 25.6	0.73 029	09.4
25 W	S20 41.2	0.49	+13 08.0	0.77 030	08.7
26 TH	S20 52.9	0.47	+12 49.6	0.80 031	07.9
27 FR	S21 04.2	0.45	+12 30.4	0.83 032	07.1
28 SA	S21 15.1	0.44	+12 10.6	0.86 033	06.4
29 SU	S21 25.6	0.42	+11 50.0	0.89 034	05.6
30 M	S21 35.6	0.40	+11 28.7	0.91 035	04.9
31 TU	S21 45.3		+11 06.8	036	04.2

Hourly differences in declination and equation of time are for the 24-hours following 0-hours of date in left column.

TABLE J (*Cont.*) Solar Ephemeris, December 1981

(For O^h Universal Time or Greenwich Civil Time)

Day of Month & Week		The Sun's Apparent Declination	Diff. in Declin. for 1 hour	Equation of Time True Sol. Time = LCT + Eq. of Time	Differ. for 1 hour		GHA of Polaris
		° ′	′	m s	s	°	′
1	TU	S21 45.3	0.39	+11 06.8	0.94	036	04.2
2	W	S21 54.6	0.37	+10 44.2	0.97	037	03.6
3	TH	S22 03.4	0.35	+10 21.1	0.99	038	02.9
4	FR	S22 11.8	0.33	+09 57.3	1.01	039	02.3
5	SA	S22 19.8	0.31	+09 33.0	1.04	040	01.6
6	SU	S22 27.3	0.30	+09 08.1	1.06	041	01.0
7	M	S22 34.4	0.28	+08 42.7	1.08	042	00.4
8	TU	S22 41.1	0.26	+08 16.9	1.10	042	59.7
9	W	S22 47.3	0.24	+07 50.5	1.12	043	59.0
10	TH	S22 53.1	0.22	+07 23.8	1.13	044	58.3
11	FR	S22 58.4	0.20	+06 56.6	1.15	045	57.6
12	SA	S23 03.2	0.18	+06 29.0	1.16	046	57.0
13	SU	S23 07.6	0.17	+06 01.1	1.18	047	56.3
14	M	S23 11.6	0.15	+05 32.8	1.19	048	55.7
15	TU	S23 15.1	0.13	+05 04.2	1.20	049	55.1
16	W	S23 18.1	0.11	+04 35.3	1.21	050	54.5
17	TH	S23 20.7	0.08	+04 06.2	1.22	051	53.9
18	FR	S23 22.7	0.07	+03 36.9	1.23	052	53.4
19	SA	S23 24.4	0.05	+03 07.3	1.24	053	52.8
20	SU	S23 25.5	0.03	+02 37.6	1.24	054	52.3
21	M	S23 26.2	0.01	+02 07.8	1.25	055	51.7
22	TU	S23 26.4	0.01	+01 37.9	1.25	056	51.1
23	W	S23 26.2	0.03	+01 07.9	1.25	057	50.5
24	TH	S23 25.4	0.05	+00 38.0	1.25	058	49.9
25	FR	S23 24.2	0.07	+00 08.1	1.24	059	49.3
26	SA	S23 22.6	0.09	− 00 21.8	1.24	060	48.7
27	SU	S23 20.4	0.11	− 00 51.6	1.23	061	48.1
28	M	S23 17.8	0.13	− 01 21.2	1.23	062	47.6
29	TU	S23 14.7	0.15	− 01 50.6	1.22	063	47.0
30	W	S23 11.2	0.17	− 02 19.8	1.21	064	46.5
31	TH	S23 07.2	0.19	− 02 48.8	1.20	065	46.0
32	FR	S23 02.7		− 03 17.5		066	45.5

Hourly differences in declination and equation of time are for the 24-hours following 0-hours of date in left column.

TABLE K Refraction and Sun's Parallax

(To be applied to observed altitudes.)
Bar. = 29.6 in.   Temp. = 50°F

Measured Altitude	Refraction	Sun's Par.	Measured Altitude	Refraction	Sun's Par.
° ′	′	′	° ′	′	′
7 30	6.88	0.15	17 30	3.02	0.14
7 40	6.75	0.15	18 00	2.93	0.14
7 50	6.62	0.15	18 30	2.85	0.14
8 00	6.50	0.15	19 00	2.77	0.14
8 10	6.37	0.15	19 30	2.70	0.14
8 20	6.25	0.15	20 00	2.62	0.14
8 30	6.13	0.15	21 00	2.48	0.14
8 40	6.02	0.15	22 00	2.36	0.14
8 50	5.92	0.15	23 00	2.25	0.14
9 00	5.82	0.15	24 00	2.15	0.14
9 10	5.72	0.15	25 00	2.05	0.14
9 20	5.63	0.15	26 00	1.96	0.13
9 30	5.53	0.15	27 00	1.88	0.13
9 40	5.43	0.15	28 00	1.80	0.13
9 50	5.34	0.15	29 00	1.73	0.13
10 00	5.26	0.15	30 00	1.66	0.13
10 20	5.10	0.15	32 00	1.53	0.13
10 40	4.95	0.14	34 00	1.42	0.12
11 00	4.81	0.14	36 00	1.32	0.12
11 20	4.67	0.14	38 00	1.23	0.12
11 40	4.54	0.14	40 00	1.15	0.11
12 00	4.42	0.14	42 00	1.07	0.11
12 30	4.25	0.14	44 00	1.00	0.11
13 00	4.09	0.14	46 00	0.93	0.10
13 30	3.93	0.14	48 00	0.86	0.10
14 00	3.78	0.14	50 00	0.80	0.09
14 30	3.65	0.14	55 00	0.67	0.08
15 00	3.53	0.14	60 00	0.55	0.07
15 30	3.42	0.14	65 00	0.45	0.06
16 00	3.32	0.14	70 00	0.35	0.05
16 30	3.22	0.14	80 00	0.17	0.03
17 00	3.12	0.14	90 00	0.00	0.00

The refraction values in Table K are corrected by multiplying them by the multipliers in Table L when the barometric pressure and the temperature differ from those on which Table K is based, i.e., 29.6 inches and 50°F.

If the barometric pressure is not known, it may be estimated from the elevation of the locality in accordance with the values given in Table L. Otherwise the elevations are disregarded.

TABLE L Multipliers for Observed Barometric Pressure or Elevation

(To correct Table K, see Examples below.)

Bar. (in.)	Elev. (ft)	Multi- plier	Bar. (in.)	Elev. (ft)	Multi- plier
30.5	– 451	1.03	23.9	+ 6194	0.81
30.2	– 181	1.02	23.6	6538	0.80
30.0	00	1.01	23.3	6887	0.79
29.9	+ 91	1.01	23.0	7239	0.78
29.6	366	1.00	22.7	7597	0.77
29.3	643	0.99	22.4	7960	0.76
29.0	924	0.98	22.1	8327	0.75
28.7	1207	0.97	21.8	8700	0.74
28.4	1493	0.96	21.5	9077	0.73
28.1	1783	0.95	21.2	9460	0.72
27.8	2075	0.94	20.9	9848	0.71
27.5	2371	0.93	20.6	10242	0.70
27.2	2670	0.92	20.3	10642	0.69
26.9	2972	0.91	20.0	11047	0.68
26.6	3277	0.90	19.7	11458	0.67
26.3	3586	0.89	19.4	11875	0.66
26.0	3899	0.88	19.1	12299	0.65
25.7	4215	0.87	18.8	12729	0.64
25.4	4535	0.86	18.5	13165	0.63
25.1	4859	0.85	18.2	13608	0.62
24.8	5186	0.84	17.9	14058	0.61
24.5	5518	0.83			
24.2	5854	0.82			

Multipliers for Temperature

Temp. (deg. F)	Multi- plier	Temp. (deg. F)	Multi- plier	Temp. (deg. F)	Multi- plier
– 20	1.16	+ 30	1.04	+ 80	0.94
– 10	1.13	+ 40	1.02	+ 90	0.93
0	1.11	+ 50	1.00	+100	0.91
+ 10	1.08	+ 60	0.98	+110	0.90
+ 20	1.06	+ 70	0.96	+120	0.88

Example. Sun: Meas. Alt. = 30°; Bar. = 26 in. or Elev. 3900 ft; Temp. 70°F.
Refraction = 1.66' (0.88) (0.96) = 1.40'. Parallax = 0.13'.
True Alt. = 30°00.00' – 1.40' + 0.13' = 29°58.73'.

Example. Star: Meas. Alt. = 25°; Bar. = 24.5 or Elev. 5518 ft; Temp. 10°F.
Refraction = 2.05' (0.83) (1.08) = 1.84'.
True Alt. = 25° 00.00' – 1.84' = 24°58.16'.

TABLE M The Sun's Semi-Diameter, 1981

(For O^h Universal Time or Greenwich Civil Time)

Date	Semi-Diam.	Date	Semi-Diam.	Date	Semi-Diam.
1981		1981		1981	
Jan. 1	16.29	May 1	15.90	Sep. 8	15.90
11	16.29	11	15.86	18	15.94
21	16.28	21	15.83	28	15.99
31	16.26	31	15.80		
Feb. 10	16.23	June 10	15.78	Oct. 8	16.03
20	16.20	20	15.76	18	16.08
		30	15.76	28	16.12
Mar. 2	16.16	July 10	15.76	Nov. 7	16.17
12	16.12	20	15.77	17	16.20
22	16.08	30	15.78	27	16.24
Apr. 1	16.03	Aug. 9	15.80	Dec. 7	16.26
11	15.98	19	15.83	17	16.28
21	15.94	29	15.86	27	16.29

TABLE N Lambert Projection for South Carolina—North

Part I

Lat.	R (ft)	Y' Y Value on Central Meridian (ft)	Tabular Difference for 1 sec of Lat. (ft)	Scale in Units of 7th Place of Logs	Scale Expressed as a Ratio
33°00′	31,127,724.75	0	101.09167	+987.1	1.0002273
01	31,121,659.25	6,065.50	101.09117	+957.5	1.0002205
02	31,115,593.78	12,130.97	101.09083	+928.3	1.0002137
03	31,109,528.33	18,196.42	101.09033	+899.5	1.0002071
04	31,103,462.91	24,261.84	101.09000	+871.0	1.0002006
05	31,097,397.51	30,327.24	101.08967	+842.9	1.0001941
33°06′	31,091,332.13	36,392.62	101.08933	+815.1	1.0001877
07	31,085,266.77	42,457.98	101.08883	+787.7	1.0001814
08	31,079,201.44	48,523.31	101.08850	+760.6	1.0001751
09	31,073,136.13	54,588.62	101.08833	+733.9	1.0001690
10	31,067,070.83	60,653.92	101.08783	+707.6	1.0001629
33°11′	31,061,005.56	66,719.19	101.08750	+681.6	1.0001569
12	31,054,940.31	72,784.44	101.08733	+656.0	1.0001510
13	31,048,875.07	78,849.68	101.08683	+630.7	1.0001452
14	31,042,809.86	84,914.89	101.08667	+605.8	1.0001395
15	31,036,744.66	90,980.09	101.08633	+581.3	1.0001338
33°16′	31,030,679.48	97,045.27	101.08600	+557.1	1.0001283
17	31,024,614.32	103,110.43	101.08583	+533.3	1.0001228
18	31,018,549.17	109,175.58	101.08550	+509.8	1.0001174
19	31,012,484.04	115,240.71	101.08517	+486.7	1.0001121
20	31,006,418.93	121,305.82	101.08500	+464.0	1.0001068
33°21′	31,000,353.83	127,370.92	101.08483	+441.6	1.0001017
22	30,994,288.74	133,436.01	101.08450	+419.6	1.0000966
23	30,988,223.67	139,501.08	101.08433	+398.0	1.0000916
24	30,982,158.61	145,566.14	101.08400	+376.7	1.0000867
25	30,976,093.57	151,631.18	101.08400	+355.8	1.0000819
33°26′	30,970,028.53	157,692.22	101.08367	+335.2	1.0000772
27	30,963,963.51	163,761.24	101.08350	+315.0	1.0000725
28	30,957,898.50	169,826.25	101.08317	+295.2	1.0000680
29	30,951,833.51	175,891.24	101.08317	+275.7	1.0000635
30	30,945,768.52	181,956.23	101.08300	+256.6	1.0000591
33°31′	30,939,703.54	188,021.21	101.08283	+237.9	1.0000548
32	30,933,638.57	194,086.18	101.08267	+219.5	1.0000505
33	30,927,573.61	200,151.14	101.08250	+201.4	1.0000464
34	30,921,508.66	206,216.09	101.08233	+183.8	1.0000423
35	30,915,443.72	212,281.03	101.08233	+166.5	1.0000383
33°36′	30,909,378.78	218,345.97	101.08217	+149.6	1.0000344
37	30,903,313.85	224,410.90	101.08200	+133.0	1.0000306
38	30,897,248.93	230,475.82	101.08200	+116.8	1.0000269
39	30,891,184.01	236,540.74	101.08183	+100.9	1.0000232
40	30,885,119.10	242,605.65	101.08183	+ 85.4	1.0000197

Part I (*Cont.*)

Lat.	R (ft)	Y' Y Value on Central Meridian (ft)	Tabular Difference for 1 sec of Lat. (ft)	Scale in Units of 7th Place of Logs	Scale Expressed as a Ratio
33°41'	30,879,054.19	248,670.56	101.08167	+ 70.3	1.0000162
42	30,872,989.29	254,735.46	101.08167	+ 55.5	1.0000128
43	30,866,924.39	260,800.36	101.08150	+ 41.1	1.0000095
44	30,860,859.50	266,865.25	101.08150	+ 27.1	1.0000062
45	30,854,794.61	272,930.14	101.08150	+ 13.4	1.0000031
33°46'	30,848,729.72	278,995.03	101.08150	0.0	1.0000000
47	30,842,664.83	285,059.92	101.08150	- 12.9	0.9999970
48	30,836,599.94	291,124.81	101.08150	- 25.5	0.9999941
49	30,830,535.05	297,189.70	101.08133	- 37.7	0.9999913
50	30,824,470.17	303,254.58	101.08150	- 49.6	0.9999886
33°51'	30,818,405.28	309,319.47	101.08150	- 61.1	0.9999859
52	30,812,340.39	315,384.36	101.08133	- 72.3	0.9999834
53	30,806,275.51	321,449.24	101.08150	- 83.1	0.9999809
54	30,800,210.62	327,514.13	101.08167	- 93.5	0.9999785
55	30,794,145.72	333,579.03	101.08150	-103.5	0.9999762
33°56'	30,788,080.83	339,643.92	101.08167	-113.2	0.9999739
57	30,782,015.93	345,708.82	101.08167	-122.5	0.9999718
58	30,775,951.03	351,773.72	101.08183	-131.5	0.9999697
59	30,769,886.12	357,838.63	101.08183	-140.1	0.9999677
34°00'	30,763,821.21	363,903.54	101.08200	-148.3	0.9999659
34°01'	30,757,756.29	369,968.46	101.08217	-156.2	0.9999640
02	30,751,691.36	376,033.39	101.08217	-163.7	0.9999623
03	30,745,626.43	382,098.32	101.08217	-170.8	0.9999607
04	30,739,561.50	388,163.25	101.08250	-177.6	0.9999591
05	30,733,496.55	394,228.20	101.08250	-184.0	0.9999576
34°06'	30,727,431.60	400,293.15	101.08267	-190.0	0.9999563
07	30,721,366.64	406,358.11	101.08283	-195.7	0.9999549
08	30,715,301.67	412,423.08	101.08300	-201.0	0.9999537
09	30,709,236.69	418,488.06	101.08317	-205.9	0.9999526
10	30,703,171.70	424,553.05	101.08333	-210.5	0.9999515
34°11'	30,697,106.70	430,618.05	101.08350	-214.7	0.9999506
12	30,691,041.69	436,683.06	101.08367	-218.6	0.9999497
13	30,684,976.67	442,748.08	101.08400	-222.1	0.9999489
14	30,678,911.63	448,813.12	101.08417	-225.2	0.9999481
15	30,672,846.58	454,878.17	101.08433	-227.9	0.9999475
34°16'	30,666,781.52	460,943.23	101.08450	-230.3	0.9999470
17	30,660,716.45	467,008.30	101.08483	-232.3	0.9999465
18	30,654,651.36	473,073.39	101.08500	-234.0	0.9999461
19	30,648,586.26	479,138.49	101.08533	-235.3	0.9999458
20	30,642,521.14	485,203.61	101.08550	-236.2	0.9999456

Part I (*Cont.*)

Lat.	R (ft)	Y' Y Value on Central Meridian (ft)	Tabular Difference for 1 sec of Lat. (ft)	Scale in Units of 7th Place of Logs	Scale Expressed as a Ratio
34°21'	30,636,456.01	491,268.74	101.08583	−236.8	0.9999455
22	30,630,390.86	497,333.89	101.08617	−237.0	0.9999454
23	30,624,325.69	503,399.06	101.08633	−236.8	0.9999455
24	30,618,260.51	509,464.24	101.08667	−236.3	0.9999456
25	30,612,195.31	515,529.44	101.08700	−235.4	0.9999458
34°26'	30,606,130.09	521,594.66	101.08733	−234.1	0.9999461
27	30,600,064.85	527,659.90	101.08767	−232.5	0.9999465
28	30,593,999.59	533,725.16	101.08800	−230.5	0.9999469
29	30,587,934.31	539,790.44	101.08817	−228.1	0.9999475
30	30,581,869.02	545,855.73	101.08867	−225.4	0.9999481
34°31'	30,575,803.70	551,921.05	101.08900	−222.3	0.9999488
32	30,569,738.36	557,986.39	101.08933	−218.9	0.9999496
33	30,563,673.00	564,051.75	101.08983	−215.1	0.9999505
34	30,557,607.61	570,117.14	101.09000	−210.9	0.9999514
35	30,551,542.21	576,182.54	101.09050	−206.3	0.9999525
34°36'	30,545,476.78	582,247.97	101.09100	−201.4	0.9999536
37	30,539,411.32	588,313.43	101.09133	−196.1	0.9999548
38	30,533,345.84	594,378.91	101.09167	−190.4	0.9999562
39	30,527,280.34	600,444.41	101.09217	−184.4	0.9999575
40	30,521,214.81	606,509.94	101.09267	−178.0	0.9999590
34°41'	30,515,149.25	612,575.50	101.09300	−171.2	0.9999606
42	30,509,083.67	618,641.08	101.09350	−164.1	0.9999622
43	30,503,018.06	624,706.69	101.09383	−156.6	0.9999639
44	30,496,952.43	630,772.32	101.09450	−148.7	0.9999658
45	30,490,886.76	636,837.99	101.09483	−140.5	0.9999676
34°46'	30,484,821.07	642,903.68	101.09533	−131.9	0.9999696
47	30,478,755.35	648,969.40	101.09600	−122.9	0.9999717
48	30,472,689.59	655,035.16	101.09633	−113.6	0.9999738
49	30,466,623.81	661,100.94	101.09683	−103.9	0.9999761
50	30,460,558.00	667,166.75	101.09733	− 93.8	0.9999784
34°51'	30,454,492.16	673,232.59	101.09800	− 83.3	0.9999808
52	30,448,426.28	679,298.47	101.09850	− 72.5	0.9999833
53	30,442,360.37	685,364.38	101.09900	− 61.3	0.9999859
54	30,436,294.43	691,430.32	101.09950	− 49.8	0.9999885
55	30,430,228.46	697,496.29	101.10017	− 37.9	0.9999913
34°56'	30,424,162.45	703,562.30	101.10067	− 25.6	0.9999941
57	30,418,096.41	709,628.34	101.10117	− 13.0	0.9999970
58	30,412,030.34	715,694.41	101.10183	0.0	1.0000000
59	30,405,964.23	721,760.52	101.10250	+ 13.4	1.0000031
35°00	30,399,898.08	727,826.67	101.10317	+ 27.2	1.0000063

TABLE N Lambert Projection for South Carolina—North

Part I (*Cont.*)

Lat.	R (ft)	Y' Y Value on Central Meridian (ft)	Tabular Difference for 1 sec of Lat. (ft)	Scale in Units of 7th Place of Logs	Scale Expressed as a Ratio
35°01′	30,393,831.89	733,892.86	101.10367	+ 41.3	1.0000095
02	30,387,765.67	739,959.08	101.10417	+ 55.8	1.0000128
03	30,381,699.42	746,025.33	101.10500	+ 70.7	1.0000163
04	30,375,633.12	752,091.63	101.10567	+ 85.9	1.0000198
05	30,369,566.78	758,157.97	101.10617	+101.5	1.0000234
35°06′	30,363,500.41	764,224.34	101.10683	+117.4	1.0000270
07	30,357,434.00	770,290.75	101.10750	+133.8	1.0000308
08	30,351,367.55	776,357.20	101.10833	+150.5	1.0000347
09	30,345,301.05	782,423.70	101.10883	+167.6	1.0000386
10	30,339,234.52	788,490.23	101.10950	+185.0	1.0000426
35°11′	30,333,167.95	794,556.80	101.11033	+202.8	1.0000467
12	30,327,101.33	800,623.42	101.11100	+221.0	1.0000509
13	30,321,034.67	806,690.08	101.11183	+239.6	1.0000552
14	30,314,967.96	812,756.79	101.11233	+258.5	1.0000595
15	30,308,901.22	818,823.53	101.11317	+277.8	1.0000640
35°16′	30,302,834.43	824,890.32	101.11400	+297.4	1.0000685
17	30,296,767.59	830,957.16	101.11467	+317.5	1.0000731
18	30,290,700.71	837,024.04	101.11550	+337.9	1.0000778
19	30,284,633.78	843,090.97	101.11617	+358.7	1.0000826
20	30,278,566.81	849,157.94	101.11700	+379.8	1.0000875
35°21′	30,272,499.79	855,224.96	101.11783	+401.3	1.0000924
22	30,266,432.72	861,292.03	101.11850	+423.2	1.0000974
23	30,260,365.61	867,359.14	101.11950	+445.5	1.0001026
24	30,254,298.44	873,426.31	101.12017	+468.1	1.0001078
25	30,248,231.23	879,493.52	101.12100	+491.1	1.0001131
35°26′	30,242,163.97	885,560.78	101.12183	+514.4	1.0001184
27	30,236,096.66	891,628.09	101.12267	+538.2	1.0001239
28	30,230,029.30	897,695.45	101.12350	+562.3	1.0001295
29	30,223,961.89	903,762.86	101.12450	+586.8	1.0001351
30	30,217,894.42	909,830.33		+611.6	1.0001408

TABLE O Lambert Projection for South Carolina—North

Part II

$1''$ of long. $= 0''.56449738$ of θ

Long.	θ	Long.	θ	Long.	θ
78°20′	+1°30′19″.1748	79°01′	+1°07′10″.5113	79°41′	+0°44′35″.7176
21	+1 29 45.3050	02	+1 06 36.6415	42	+0 44 01.8477
22	+1 29 11.4352	03	+1 06 02.7716	43	+0 43 27.9779
23	+1 28 37.5653	04	+1 05 28.9018	44	+0 42 54.1081
24	+1 28 03.6955	05	+1 04 55.0319	45	+0 42 20.2382
25	+1 27 29.8256	79°06′	+1 04 21.1621	79°46′	+0 41 46.3684
78°26′	+1 26 55.9558	07	+1 03 47.2922	47	+0 41 12.4985
27	+1 26 22.0859	08	+1 03 13.4224	48	+0 40 38.6287
28	+1 25 48.2161	09	+1 02 39.5526	49	+0 40 04.7588
29	+1 25 14.3463	10	+1 02 05.6827	50	+0 39 30.8890
30	+1 24 40.4764	79°11′	+1 01 31.8129	79°51′	+0 38 57.0192
78°31′	+1 24 06.6066	12	+1 00 57.9430	52	+0 38 23.1493
32	+1 23 32.7367	13	+1 00 24.0732	53	+0 37 49.2795
33	+1 22 58.8669	14	+0 59 50.2033	54	+0 37 15.4096
34	+1 22 24.9970	15	+0 59 16.3335	55	+0 36 41.5398
35	+1 21 51.1272	79°16′	+0 58 42.4637	79°56′	+0 36 07.6699
78°36′	+1 21 17.2574	17	+0 58 08.5938	57	+0 35 33.8001
37	+1 20 43.3875	18	+0 57 34.7240	58	+0 34 59.9303
38	+1 20 09.5177	19	+0 57 00.8541	59	+0 34 26.0604
39	+1 19 35.6478	20	+0 56 26.9843	80°00′	+0 33 52.1906
40	+1 19 01.7780	79°21′	+0°55′53″.1144	80°01′	+0 33 18.3207
78°41′	+1 18 27.9081	22	+0 55 19.2446	02	+0 32 44.4509
42	+1 17 54.0383	23	+0 54 45.3748	03	+0 32 10.5810
43	+1 17 20.1685	24	+0 54 11.5049	04	+0 31 36.7112
44	+1 16 46.2986	25	+0 53 37.6351	05	+0 31 02.8414
45	+1 16 12.4288	79°26′	+0 53 03.7652	80°06′	+0 30 28.9715
78°46′	+1 15 38.5589	27	+0 52 29.8954	07	+0 29 55.1017
47	+1 15 04.6891	28	+0 51 56.0255	08	+0 29 21.2318
48	+1 14 30.8192	29	+0 51 22.1557	09	+0 28 47.3620
49	+1 13 56.9494	30	+0 50 48.2859	10	+0 28 13.4921
50	+1 13 23.0796	79°31′	+0 50 14.4160	80°11′	+0 27 39.6223
78°51′	+1 12 49.2097	32	+0 49 40.5462	12	+0 27 05.7525
52	+1 12 15.3399	33	+0 49 06.6763	13	+0 26 31.8826
53	+1 11 41.4700	34	+0 48 32.8065	14	+0 25 58.0128
54	+1 11 07.6002	35	+0 47 58.9366	15	+0 25 24.1429
55	+1 10 33.7304	79°36′	+0 47 25.0668	80°16′	+0 24 50.2731
78°56′	+1 09 59.8605	37	+0 46 51.1970	17	+0 24 16.4032
57	+1 09 25.9907	38	+0 46 17.3271	18	+0 23 42.5334
58	+1 08 52.1208	39	+0 45 43.4573	19	+0 23 08.6636
59	+1 08 18.2510	40	+0 45 09.5874	20	+0 22 34.7937
79°00′	+1 07 44.3811				

Part II (*Cont.*)

$1''$ of long. $= 0''56449738$ of θ

Long.	θ	Long.	θ	Long.	θ
80°21′	+0°22′00″9239	81°01′	−0°00′33″8698	81°41′	−0°23′08″6636
22	+0 22 27.0540	02	−0 01 07.7397	42	−0 23 42.5334
23	+0 20 53.1842	03	−0 01 41.6095	43	−0 24 16.4032
24	+0 20 19.3143	04	−0 02 15.4794	44	−0 24 50.2731
25	+0 19 45.4445	05	−0 02 49.3492	45	−0 25 24.1429
80°26′	+0 19 11.5747	81°06′	−0 03 23.2191	81°46′	−0 25 58.0128
27	+0 18 37.7048	07	−0 03 57.0889	47	−0 26 31.8826
28	+0 18 03.8350	08	−0 04 30.9587	48	−0 27 05.7525
29	+0 17 29.9651	09	−0 05 04.8286	49	−0 27 39.6223
30	+0 16 56.0953	10	−0 05 38.6984	50	−0 28 13.4921
80°31′	+0°16′22″2254	81°11′	−0 06 12.5683	81°51′	−0 28 47.3620
32	+0 15 48.3556	12	−0 06 46.4381	52	−0 29 21.2318
33	+0 15 14.4858	13	−0 07 20.3080	53	−0 29 55.1017
34	+0 14 40.6159	14	−0 07 54.1778	54	−0 30 28.9715
35	+0 14 06.7461	15	−0 08 28.0476	55	−0 31 02.8414
80°36′	+0 13 32.8762	81°16′	−0 09 01.9175	81°56′	−0 31 36.7112
37	+0 12 59.0064	17	−0 09 35.7873	57	−0 32 10.5810
38	+0 12 25.1365	18	−0 10 09.6572	58	−0 32 44.4509
39	+0 11 51.2667	19	−0 10 43.5270	59	−0 33 18.3207
40	+0 11 17.3969	20	−0 11 17.3969	82°00	−0 33 52.1906
80°41′	+0 10 43.5270	81°21′	−0 11 51.2667	82°01′	−0 34 26.0604
42	+0 10 09.6572	22	−0 12 25.1365	02	−0 34 59.9303
43	+0 09 35.7873	23	−0 12 59.0064	03	−0 35 33.8001
44	+0 09 01.9175	24	−0 13 32.8762	04	−0 36 07.6699
45	+0 08 28.0476	25	−0 14 06.7461	05	−0 36 41.5398
80°46′	+0 07 54.1778	81°26′	−0 14 40.6159	82°06′	−0 37 15.4096
47	+0 07 20.3080	27	−0 15 14.4858	07	−0 37 49.2795
48	+0 06 46.4381	28	0 15 48.3556	08	−0 38 23.1493
49	+0 06 12.5683	29	−0 16 22.2254	09	−0 38 57.0192
50	+0 05 38.6984	30	−0 16 56.0953	10	−0 39 30.8890
80°51′	+0 05 04.8286	81°31′	−0 17 29.9651	82°11′	−0 40 04.7588
52	+0 04 30.9587	32	−0 18 03.8350	12	−0 40 38.6287
53	+0 03 57.0889	33	−0 18 37.7048	13	−0 41 12.4985
54	+0 03 23.2191	34	−0 19 11.5747	14	−0 41 46.3684
55	+0 02 49.3492	35	−0 19 45.4445	15	0 42 20.2382
80°56′	+0 02 15.4794	81°36′	−0 20 19.3143	82°16′	−0 42 54.1081
57	+0 01 41.6095	37	−0 20 53.1842	17	−0 43 27.9779
58	+0 01 07.7397	38	−0 21 27.0540	18	−0 44 01.8477
59	+0 00 33.8698	39	−0 22 00.9239	19	−0 44 35.7176
81°00	0 00 00.0000	40	−0 22 34.7937	20	−0 45 09.5874

TABLE O Lambert Projection for South Carolina—North

Part II (*Cont.*)

$1''$ of long. $= 0''.56449738$ of θ

Long.	θ	Long.	θ	Long.	θ
82°21′	−0°45′43″.4573	82°46′	−0°59′50″.2033	83°11′	−1°13′56″.9494
22	−0 46 17.3271	47	−1 00 24.0732	12	−1 14 30.8192
23	−0 46 51.1970	48	−1 00 57.9430	13	−1 15 04.6891
24	−0 47 25.0668	49	−1 01 31.8129	14	−1 15 38.5589
25	−0 47 58.9366	50	−1 02 05.6827	15	−1 16 12.4288
82°26′	−0 48 32.8065	82°51′	−1°02′39″.5526	83°16′	−1 16 46.2986
27	−0 49 06.6763	52	−1 03 13.4224	17	−1 17 20.1685
28	−0 49 40.5462	53	−1 03 47.2922	18	−1 17 54.0383
29	−0 50 14.4160	54	−1 04 21.1621	19	−1 18 27.9081
30	−0 50 48.2859	55	−1 04 55.0319	20	−1 19 01.7780
82°31′	−0 51 22.1557	82°56′	−1 05 28.9018	82°21′	−1 19 35.6478
32	−0 51 56.0255	57	−1 06 02.7716	22	−1 20 09.5177
33	−0 52 29.8954	58	−1 06 36.6415	23	−1 20 43.3875
34	−0 53 03.7652	59	−1 07 10.5113	24	−1 21 17.2574
35	−0 53 37.6351	83°00′	−1 07 44.3811	25	−1 21 51.1272
82°36′	−0 54 11.5049	83°01′	−1 08 18.2510	83°26′	−1 22 24.9970
37	−0 54 45.3748	02	−1 08 52.1208	27	−1 22 58.8669
38	−0 55 19.2446	03	−1 09 25.9907	28	−1 23 32.7367
39	−0 55 53.1144	04	−1 09 59.8605	29	−1 24 06.6066
40	−0 56 26.9843	05	−1 10 33.7304	30	−1 24 40.4764
82°41′	−0 57 00.8541	83°06′	−1 11 07.6002	83°31′	−1 25 14.3463
42	−0 57 34.7240	07	−1 11 41.4700	32	−1 25 48.2161
43	−0 58 08.5938	08	−1 12 15.3399	33	−1 26 22.0859
44	−0 58 42.4637	09	−1 12 49.2097	34	−1 26 55.9558
45	−0 59 16.3335	10	−1 13 23.0796	35	−1 27 29.8256

Index

Index

3-11-91 dc